Problem Plants
and Alien Weeds
of South Africa
Clive Bromilow

BRIZA

Published by
BRIZA PUBLICATIONS
CK/1990/011690/23

www.briza.co.za

PO Box 56569
Arcadia 0007
Pretoria
South Africa

Third edition, first impression 2010

Copyright © in text and maps: Clive Bromilow
Copyright © in photographs: Clive Bromilow and others as credited
Copyright © in published edition: Briza Publications CC

© All rights reserved. No part of this publication may be reproduced or transmitted in any form or by any means without written permission of the copyright holders.

ISBN 978-1-920217-30-3

Project management by Johan Steenkamp
Text edited by Emsie du Plessis
Inside design and typesetting by Karen Pretorius, Nudog Design, Pretoria
Distribution maps by Steyn Pretorius, Nudog Design, Pretoria
Cover design by Steyn Pretorius, Nudog Design, Pretoria
Cover photograph: *Lilium formosanum*, formosa lily * trompetlelie
Printed and bound by Tien Wah Press (Pte.) Ltd., Singapore

> **Disclaimer**
> Although care has been taken to be as accurate as possible, neither the author nor the publisher or sponsors make any express or implied representation as to the accuracy of the information contained in his book and cannot be held legally responsible or accept liability for any errors or omissions.

Foreword

Dr Klaus Eckstein
Country Head South and Southern Africa, Bayer CropScience

The United Nations predicts that by the year 2030 the world population will expand by more than 1,7 billion people, all of whom have to be fed. This means that on a global scale the ratio of arable land to the population is continuously declining. In order to meet the challenges of food production and optimise use of our agricultural resources against the background of a continuously growing world population, we need intensive agricultural research. This means that more and more farmers will need to be provided with the tools to produce more, while preserving natural resources.

In the last 50 years, the application of sound scientific principles in agriculture has already seen the doubling of the production of the world's food calories, tripling the production of resource intensive foods caused by changing dietary habits and increasing the per-capita food supply by more than 25%.

Farmers now grow more food per given area than ever before. In 1980, one hectare of arable land produced 1,8 tons of food per annum on average, while today the production of one hectare exceeds an average of 2,5 tons.

Integrated crop protection contributes substantially to these increased yields and by so doing allow many millions of hectares around the world to be excluded from cultivation, thus allowing natural habitats to remain undisturbed. Although we have seen a doubling of the world's population since the 1970s, the area of land utilised for food production has remained virtually constant. Therefore, Bayer CropScience further intensifies its focus on innovation and will put substantial investments behind the drive to help bring about a second green revolution in agriculture.

Today, about a third of the world's food production potential is still lost annually owing to the effects of plant diseases, pests and weeds.

Weeds can account for crop losses in excess of 75%. The judicious use of herbicide programmes not only affords the farmer a cost-effective means of managing weed populations but, more importantly, plays a vital role in preserving natural resources through the optimal utilisation of land.

Inherent to the success of these programmes, is the correct identification of the problem plant species and subsequent selection of the most appropriate remedy.

This very well illustrated and comprehen-sive publication on the problem plants of South Africa will provide great assistance in identifying weeds to agriculturists, students, farmers and gardeners alike.

Bayer CropScience is again proud to be associated with this weed identification guide.

Endorsement

South Africa has a rich botanical heritage, with one of the world's six floral kingdoms contained within its boundaries. Many of the plant species of this rich flora are under threat of extinction. One of the main threats comes from the spread of introduced species. These plants were either introduced by accident, or were deliberately brought over as crops, shade trees or ornamental plants. Deliberate introductions were usually, but not entirely, carried out long before the potential danger was realised. The risks posed by alien and invasive species are now known to be enormous as they are capable of transforming landscapes, eliminating indigenous species and threatening the entire ecosystem. Whereas the UK has around a dozen alien weeds that could be called threatening, South Africa has several hundred; the effort required to control them differs accordingly.

In South Africa, the task of creating awareness of this problem, the plant culprits and future threats remains an enormous one and awareness at all levels is needed. I am familiar with Clive's previous books and I know they are the ideal way to bring this problem to the fore. BGCI, with over 500 member botanic gardens and headquarters here in Kew, considers it vital to promote an interest in plants, how to grow them and how to protect them. I am sure Clive's new book will further stimulate the interest required in South Africa and beyond. Colour photographs, interesting facts and background information along with easy identification guides will bring this subject not only to the farmers and experts but also to amateur gardeners and the wider public. I am also well aware that one cannot fully appreciate the beauty of indigenous gardens or even ecosystems unless one is aware of which species do not belong there.

BGCI supports Clive's new book, feeling sure that it will be a vital weapon in the battle to protect and preserve plant diversity.

Dr Sara Oldfield
Secretary General
Botanic Gardens Conservation International
Kew
England

RESPONSIBLE USE OF HERBICIDES – A MESSAGE FROM THE AGROCHEMICAL INDUSTRY

Herbicides play a very important role in the effective control of weeds and unwanted plants in agriculture, silviculture and environmental management. The range of herbicides available on the market is vast and poses a daunting challenge to land managers and farmers to select the appropriate products for their plant control requirements. CropLife South Africa's member companies manufacture this vast selection of herbicides while the ACDASA members distribute these essential products to their clients. Both associations uphold the principles of responsible herbicide use and appeal to the end users of herbicides to:
- Always seek expert advice from the manufacturers and distributors when selecting herbicides.
- Always study the label for the correct application information and never divert from the label.
- Never overdose or underdose as it compromises the integrity of the herbicide, leads to weed resistance against herbicides and damages the environment.
- Enforce the use of personal protective equipment by all individuals who work with herbicides despite the benign toxicity of most of the products.
- Recycle empty herbicide containers and never allow workers to use them for any other purposes.
- Keep yourself informed about weed resistance and seek expert advice on the management thereof.

CropLife South Africa's members are committed to manufacturing high-quality products and also offer advice to end users of their products. Members abide by a code of conduct to ensure good service, good product stewardship and environmental responsibility. A list of members is available on the website www.croplife.co.za. A list of herbicides that are registered in South Africa is also available on this website under the CropLife SA Initiatives icon.

ACDASA's members are competent persons who have successfully completed the prescribed AVCASA course. They abide by the ACDASA Code of Conduct to ensure service excellence and commit to distributing registered, top-quality products from CropLife SA members. ACDASA members also provide comprehensive support in the form of on-farm training in the use of agrochemicals, equipment calibrationf, problem assessment and solutions before use and follow-up visits after use. A list of accredited members are available on the website, www.acdasa.co.za.

Acknowledgements

This book is based on Problem Plants of South Africa, which was first published in 1995 and with a revised edition in 2001. Extending the scope of the book to include all plants considered alien and invasive has been the challenge of this new revision, which is now called Problem Plants and Alien Weeds of South Africa.

My gratitude to Briza Publications for having faith in the book and for creating a beautiful, quality publication. I am also very grateful to everyone who helped me find these additional plants and gave me valuable and interesting information on them.

I am also very grateful to Bayer CropScience for their confidence and faith in me and in the new book and for becoming the principal sponsor. Without this help, a book like this would cost several times more than what it actually does in the bookstores and I hope their support is recognised and well rewarded. Also sincere gratitude to CropLife South Africa and ACDASA for being very valuable support sponsors.

Thanks to Emsie du Plessis, who did both the English editing and Afrikaans translation. And finally my gratitude to Johan Steenkamp (Briza) and Karen Pretorius (Nudog Design cc) for putting it all together in such a professional, colourful and artistic way.

The maps in this book are mainly those created for the 2001 edition, that used data from the PRECIS system of the National Herbarium as well as personal experience and other informative data. New maps were created in a similar fashion, but using a variety of published and accredited sources as well as personal experience.

I dedicate this book to my children, Jonathan, Geoffrey, Caitlin and Richard.

TABLE OF CONTENTS

INTRODUCTION .... p. 8
HOW TO USE THIS BOOK .... p. 9
THE ELIMINATION KEY .... p. 11
ALIEN INVASIONS AND GLOBALISATION .... p. 12
CURRENT, NEW AND REVISED LEGISLATION .... p. 13
VOLUNTEER CROPS .... p. 14
BUSH ENCROACHMENT VS ALIEN INVADERS .... p. 15
BIOLOGICAL CONTROL .... p. 24
CHEMICAL WEED CONTROL .... p. 29
HERBICIDE RESISTANCE .... p. 36
INDIGENOUS PLANTS .... p. 38
THE THREAT OF ORNAMENTAL AND CROP PLANTS .... p. 39
APPENDIX: LIST OF DECLARED WEEDS AND INVADER PLANTS .... p. 44
GRASSES AND SEDGES .... p. 60–111
TREES .... p. 112–162
SHRUBS .... p. 163–196
HERBS .... p. 197–307
SPREADING OR FLAT-GROWING HERBS .... p. 308–349
CREEPERS AND CLIMBERS .... p. 350–370
SUCCULENTS .... p. 371–381
WATER WEEDS .... p. 382–393
GLOSSARY .... p. 394–397
BIBLIOGRAPHY .... p. 398
INDEX .... p. 399
BIOCONTROL AGENTS.. p. 423

INTRODUCTION

What is a weed?

A weed is a plant in the wrong place at the wrong time. Any plant can be a weed: pretty ones, ugly ones, rare ones, even crop plants such as maize or wheat. A weed must be a nuisance, just sometimes.

Plants that become weeds are usually vigorous growers that compete for water, light, space and nutrients. They are adaptable, being able to easily invade a wide range of ecological niches. Most of them are of exotic or foreign origin. They are tough and can withstand unfavourable conditions. They are easily spread, producing an abundance of fertile seeds or other propagules, often with efficient methods of dispersal. They are aggressive competitors, being able to eliminate other species. The most frustrating characteristic is that, for a number of reasons, they are often difficult to control.

Any plant with one or more of these characteristics may easily become a nuisance. These plants compete with crops for precious water and nutrients. They also clog machines, poison people and animals, act as hosts for crop diseases and insect pests, upset the natural indigenous ecology, destroy biodiversity, interfere with people's activities and spoil the natural beauty of the environment.

Which plants are in this book?

The plants in this book are as many as possible of those that occur as 'weeds' in South Africa. Many relatively common weeds may not be included and there will be others where inclusion may not seem to be justified. The general criteria used to judge whether a species warrants inclusion are the following:

- **Economic importance.** Some indigenous or even alien plants that are of economic importance have never appeared on a herbicide label. This is sometimes a factor of personal experience.

- **A herbicide registration.** This usually indicates that a plant represents a sufficient market to justify the investment required to obtain the registration. It appears, however, that sometimes a plant can appear on a label because it just happened to have occurred in a herbicide trial, was formally identified and then added to the label simply because it was there. Although it cannot be claimed that these plants are never a problem, they are not covered in detail in this book, but nevertheless deserve a mention (see page 38). Indigenous poisonous plants are also not the principal subject of this book.

- **Exotic origin.** Not all alien plants are invasive, since many introduced species cannot reproduce or spread rapidly without continued human help, for example, many crop plants and ornamentals. Some alien plants can indeed reproduce and survive in the wild and these plants would be called 'naturalised' but only some of these naturalised plants are capable of replacing indigenous species and transforming the environment. The exact dividing line is not clear, so some relatively obscure naturalised plants may be included and some that are starting to be invasive could have been omitted.

- **Alien weeds and invasive species** that have been or are classified as alien weeds or invasive plants by the current CARA or proposed CARA or NEMBA regulations.

The plants in this book therefore comprise mainly exotic (alien) invaders, weeds of cultivated lands, gardens and sports turf, and weeds of waterways. Although dealt with in the introductory chapters, indigenous species involved in bush encroachment and veld degradation are not covered in the main body of this book.

Even plants such as the attractive indigenous Cape marigold (*Arctotheca calendula*) can become a problem under certain circumstances.

HOW TO USE THIS BOOK

How do I find the plant I am looking for?
- **Use the elimination key:** Turn to page 11 for an explanation of how to use this key.
- **Search through the photographs:** The plants in this book are colour-coded and grouped according to the following categories:

  - Grasses and sedges
  - Trees or large, erect, woody shrubs
  - Shrubs (woody or herbaceous), not distinctly tree-like or succulent
  - Herbs (not woody)
  - Spreading or flat-growing herbs
  - Creepers and climbers
  - Succulents
  - Water weeds

  Within each section, the plants are arranged alphabetically according to family and genus.
- **Use the index:** All the plant names mentioned in the book can be found in the index. This includes the botanical or Latin names as well as the common names in English, Afrikaans and, in many cases, names in several indigenous languages.

What do I find in the text?
Detailed botanical descriptions are avoided. Only important or interesting characteristics are mentioned in the text and, where possible, identification is left to the photographs. The text includes such information as origin, distribution, impact or pest status, methods of control, and anything else of specific interest.

What do the photographs show?
The photographs are of typical specimens and are intended to highlight the unique features of the plant. Many plants have distinctive seedlings and, in many cases, early identification is important so that suitable control measures can be initiated before a weed has become established. Where appropriate, photographs of the seedlings are included. Likewise, where a plant has distinctive vegetative growth or characteristic flowers, the main features are shown in the photograph.

What names are used?
Where available, the English and Afrikaans common names as well as the names in several indigenous languages are given. Many farmers refer to weeds by their names in the indigenous languages or other common names not included here. For instance, *Cynodon dactylon* is known to have some 65 English and Afrikaans names as well as 14 recorded names in the indigenous languages – and this just in South Africa! The common names included here are those by which, in my experience and according to most of the publications listed, a plant is most commonly known. 'Other common names' includes various names on record in this country and occasionally a very descriptive name from elsewhere in the world. There will, of course, be other names not mentioned here. The following abbreviations for the various indigenous languages are used:

N Ndebele
P Pedi
Sh Shona
Si Siswati
S Sotho
Ss Southern Sotho
To Tsonga
T Tswana
V Venda
X Xhosa
Z Zulu

The botanical or Latin names follow *Plants of southern Africa: names and distribution* (Arnold & De Wet 1993). Recent name changes were found on the POSA online checklist available on the SANBI website, www.sanbi.org. For the sake of accuracy and precision in communication, it is better to use the Latin names. However, for various reasons these names are also subject to change. The non-botanist may find it difficult to understand why plant names are sometimes changed. Botanists are continuously studying all plant species and sometimes find new information requiring them to make adjustments. For example, large families may be subdivided, smaller families may be combined or a species can be moved to a family to which it is more closely related than to another. A recent series of changes involves the mistaken gender of the Latin names given to plants by Linnaeus over 250 years ago! In this book, previous or outdated botanical names are given in brackets after the current name, preceded by an equal sign. In most cases, name changes older than about 10 years are not mentioned.

The distribution maps

The maps were prepared using information from various sources:
- Various published and unpublished data.
- Personal observations.
- Herbarium records in the National Herbarium in Pretoria, which were downloaded from its database and checked by specialists.

The maps are intended as a guide. They are not a definitive indication of the actual distribution of a plant. Herbaria only keep records of those plants actually sent to them for identification, so large gaps are bound to exist. Ornamental plants, for example, may occur throughout the country in gardens, but are only recorded as being naturalised in very restricted areas. These ornamentals are simply seen as being part of urban vegetation, specimens are not sent for identification, no records exist, so officially the plant does not occur there!

The Southern African Plant Invaders Atlas (SAPIA) is a project of the Plant Protection Research Institute (PPRI). It aims to collect information on the distribution, abundance and habitat types of alien invader plants in South Africa. This atlas contains more reliable distribution maps of the species it covers.

Plant identification

The identity of the plants in the photographs was usually determined by or with the assistance of the staff of herbaria around the country, by referring to literature and comparing collected specimens with material in the herbarium collections. Although every effort has been made to ensure that the identifications are accurate, they are, however, not guaranteed. Remember also, that a plant species may appear considerably different in colour or form under different growing conditions. Conversely, some closely related species can appear superficially identical. If a positive identification of a weed is required, refer to a detailed botanical book or take a specimen to an expert. The background information contained in the text was drawn from published literature, colleagues in the agrochemical industry, farmers, researchers and from personal experience. To the best of my knowledge, this information is true and correct. However, this book is a guide, not a scientific or botanical reference – please treat it as such.

Plant size and scale

To assist with determining scale, the species descriptions include the average height of the plant. This is not the maximum height, but the height a species usually achieves when it is growing unhindered and under favourable conditions. The height given for a flat-growing or spreading species is the usual height of the aerial parts above the ground and, for a climbing species, the usual height to which a specimen could climb.

Key to pictograms

- Exotic origin.
- The subject of a herbicide registration.
- Subject of bio-control – agents already released.

Blue 1 is the existing 'CARA' Category

Red 1b is the proposed 'NEMBA' Category

(See p 13 for an explanation of the CARA Categories, and p 14 for the NEMBA Categories.)

THE ELIMINATION KEY

This key provides a quick and easy way to find a plant in this book. The key is based on a very simple process of elimination. The following are the simple steps in the process:

1. Decide in which broad category the plant falls. Each section is colour-coded. Turn to the key at the front of the section.

   ▪ Grasses and sedges
   ▪ Trees or large, erect, woody shrubs
   ▪ Shrubs (woody or herbaceous), not distinctly tree-like or succulent
   ▪ Herbs (not woody)
   ▪ Spreading or flat-growing herbs
   ▪ Creepers and climbers
   ▪ Succulents
   ▪ Water weeds

Note: Grasses and sedges are not included in the key.

SPECIES	Compound leaves	Phyllodes (stalkless leaves)	Needle-like leaves	Thorny stems	Deciduous or semideciduous	Red flowers	Yellow flowers	Purple flowers	White flowers	Green or indistinct flowers
Ailanthus altissima, p. 158	■				■					■
Albizia spp., p. 127	■							■	■	
Gleditsia triacanthos, p. 129	■			■	■		■			
Jacaranda mimosifolia, p. 118	■				■			■		
Melia azedarach, p. 137	■				■			■		
Paraserianthes lophantha, p. 131	■								■	
Phytolacca dioica, p. 146	■								■	

2. All the species in the section are listed down the left-hand side.
3. Across the top are various simple criteria for elimination.
4. Simply answer 'YES' or 'NO' to each criterion. If your answer is 'NO', move across to the next criterion.
5. When you come to a 'YES' answer, the plant you are searching for is in this section. It is not before it and it is not after it.
6. Within the chosen section, move across the remaining criteria and eliminate as many plants as possible.
7. You should end up with one possible answer with a page number next to it. Turn to that page.
8. If you cannot distinguish between species, you will have to use the photographs.

ALIEN INVASIONS AND GLOBALISATION

The concept of 'globalisation' has been an ongoing process, with increasing momentum, for the past 500 years. The ability of pests, pathogens and weeds to move around the world is truly staggering. There is hardly a corner of the earth that is not suffering from the presence or imminent threat of an invasion of an alien species. Even if we look solely at botanical threats, they are still enormous. The natural flora of South Africa, although unique and one of the most recognisable in the world, is also one of the most polluted and threatened.

Before the phenomenon of globalisation, physical barriers kept species apart and protected ecosystems from disturbance by species that had not evolved with them. Today these barriers are losing their effect and organisms can easily cross them and take up residence in ecosystems in which they did not evolve. The spread of exotics, or biological pollution, is one of the greatest threats to the earth's biological diversity and as a threat of extinction, bio-invasion may rank just behind habitat loss, which covers almost any kind of physical disruption of existing habitats.

Most of the alien invaders in South Africa (and hence in this book) will suppress or replace indigenous species. On a larger scale, of course, the indigenous species can be pushed out of certain areas and are ultimately threatened with total extinction. As the indigenous types decline, so too do the fauna that have evolved to depend on them. The exotic species increase and the absence of cohabiting fauna becomes increasingly noticeable. For example, the enormous diversity of the Cape fynbos is currently threatened by huge monospecific stands of exotic vegetation. As the diversity of the fynbos declines, so too does the diversity of the birds and butterflies. The irony of this situation can be clearly seen: the introduction of new, exotic species actually causes the *reduction* of species diversity.

As an ecosystem weakens, its stability declines. It becomes more susceptible to catastrophic events such as fires, floods and disease epidemics. Exotic vegetation is often blamed for some of the catastrophic fires that periodically rage through the Western Cape. The destruction of ground cover by exotic species is also blamed for increasing the severity of floods and the associated soil erosion.

Even as our knowledge increases and we learn to understand the dynamics of these invasions, they are still difficult or impossible to predict. Carelessness and ignorance are still the cause of introductions and they still occur almost on a daily basis.

Even though modern agriculture is often the victim of exotic invasions, it has also often been the main instigator of many introductions. Many weeds were accidental introductions with crop seeds; this has been going on for hundreds of years. The cultivation of exotic species as crops (e.g. maize, wheat, potatoes) meant that their associated pests, diseases and weeds spread with them. These pests and weeds flourished in their new habitats, and became a universal problem, even in the absence of the crops.

Sometimes the difference between weeds and a crop is small enough for interbreeding to occur. This is a topical problem with the possibility of genetically engineered traits escaping into the wild, but the reverse can also cause problems. If it is possible for certain crops to interbreed with weeds or wild types, then it is quite possible for the genes to go the other way and for the crop to acquire some of the weeds' less desirable characteristics. This is not a new phenomenon and has been going on as long as humans have bred and cultivated improved varieties.

Reduction of ecological diversity is a global problem. It is caused by many things, but it is also a result of the removal or bypassing of ecological barriers in the process we know as globalisation. We cannot stop it but we have to learn to live with it and manage it. It has been said that loss of species diversity could ultimately threaten our very existence on this planet, so it is something that should be considered very seriously. It is important that we recognise the problem and its causes. Only then, and with study and education, can we hope to at least face up to one of the major threats to planet earth.

CURRENT, AND PROPOSED LEGISLATION

The current legislation on weeds and invasive plants is part of the Conservation of Agricultural Resources Act (CARA), 1983 (Act 43 of 1983). This Act is administered by the National Department of Agriculture, Directorate Agricultural Land Resource Management. Regulations 15 and 16 under this Act were revised and amended in March 2001. This Act was updated because:
- In 1996 South Africa became a signatory to the Convention on Biological Diversity and undertook to 'prevent the introduction of, control or eradicate those alien species which threaten ecosystems, habitats and species' (Article 8(h)).
- It was deemed necessary to support the 'Working for Water' campaign, which is South Africa's largest alien plant-clearing programme, started in 1995 and at the time administered by the then Department of Water Affairs and Forestry.
- Since 1983 many more plants have become invasive and problematic.

The main changes involve Regulations 15 and 16, where the old terms 'Declared Weeds' and 'Declared Invader Plants' were replaced with three categories of alien plants:

Category 1: Declared weeds
These are prohibited plants, which must be controlled or eradicated (except in biocontrol reserves that are designated for the breeding of their biocontrol agents). These plants serve no economic purpose and possess characteristics that are harmful to humans, animals or the environment.

Category 2: Declared invader plants with a commercial or utility value
These are 'invaders' with certain useful qualities, such as a commercial use or for woodlots, animal fodder, soil stabilisation, etc. These plants are allowed in demarcated areas under controlled conditions and in biocontrol reserves.

Category 3: Mostly ornamental plants
These are alien plants that are currently growing in or have escaped from areas such as gardens, but that are proven invaders. No further planting is allowed (except with special permission), nor trade in propagative material. Existing plants may remain (except those within the flood line of watercourses or wetlands or as directed by the executive officer) but must be prevented from spreading.

Regulation 16 concerns the naming of indigenous species that are implicated in 'bush encroachment'. Further additions to the regulations include:
- A water use tax on plantations of Category 2 plants.
- Prohibition of Category 2 and 3 plants within 30 m of the 'one-in-fifty-year' flood line of watercourses or wetlands (except where authorised).
- Category 2 and 3 plants may occur in areas set aside as 'biocontrol reserves'.
- Biocontrol reserves must not be disturbed by other control methods.
- Various exemptions may be granted on application.

These regulations under the CARA Act of 1983 were amended in 2001 and are in the process of being revised again. Also, drafting of new legislation under the National Environmental Management: Biodiversity Act, No. 10 of 2004 (otherwise known as 'NEMBA' – Department of Water and Environmental Affairs) is still in progress. These new regulations will have additional species and different categories and were not final at the time of going to print. However, many of the new species that are mentioned in these lists are included in this book. There will be an additional 47 listed, invasive species, bringing the total to 345. In addition, CARA will list 243 prohibited species that may not be brought into South Africa. They do not yet occur and are therefore, likewise, not covered in this book.

The table on pages 44–58 combines the names and categories as they appear in both the CARA and the proposed NEMBA lists. The CARA categories have been in existence for a while and the NEMBA ones are still 'proposed'. It is possible therefore, that there could yet be changes to this list. For up-to-date news on the issue, refer to the website www.dwaf.co.za/wfw/legal.

In order to give as much information as possible on these weeds, both categories are included in the pictograms in this book:
1/1b

Where the blue **1** is the CARA category and the red **1b** is the proposed NEMBA category (where different categories have been proposed for different regions, only the lowest one is included in the pictogram – refer to the table for more detail. The explanation of these numbers can be seen on page 59.

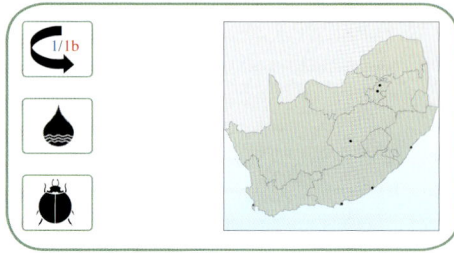

Example

VOLUNTEER CROPS

Many crop plants can be found growing in other crops and can become serious pests if not controlled. Plants that grow in this way are called 'volunteers'. This term is usually limited to plants that regenerate from a crop that was grown in the same land in recent seasons. Many alien problem plants in this book were originally introduced for use as a crop of some sort and have escaped into the wild, establishing themselves as weeds of both cropland and non-cropland. These plants cannot be called volunteers.

There are also crop plants that grow along roads and in waste areas that come from spilled seed or nearby fields. Some of these plants may even have partly established themselves in this environment and can regenerate from one season to the next. These are not considered true 'problem plants'. Such plants include:
- *Medicago sativa* (lucerne)
- *Lupinus* sp. (lupin)
- *Brassica napus* (rape or canola)
- *Zea mays* (maize)
- *Sorghum vulgare* (sorghum)
- *Helianthus annuus* (sunflower)

Medicago sativa

Maize in a potaoto field

Lupinus luteus

ALIEN INVADERS VS BUSH ENCROACHMENT

Alien invaders

Alien invaders are organisms – in this case plants – that are of exotic, non-native or of foreign origin and are invading previously pristine areas or ecological niches. Of course, not all weeds or problem plants are alien or exotic in origin but, as mentioned previously, this is just one criterion used to try and categorise a problem plant. It is usually the more dramatic examples that catch people's attention, such as lantana and triffid weed, but it could also refer to less well-known examples such as chickweed, redstar zinnia, etc.

The threat of alien plants is enormous and awareness of what they are and what they do is just the start, but a vital part, of the battle. *Problem Plants and Alien Weeds of South Africa* now includes as many species as possible of plants of foreign origin that have established themselves in the wild. Some are of minor importance and some are already seriously threatening the indigenous ecology, biodiversity and productivity of the land. However, plant invasion is a dynamic process and the species composition and prominence will continue to change. New species will arrive and some will come under control or even be eliminated.

Invasions of alien terrestrial plants cause:

- **A decline in species diversity (biodiversity).** Many alien plants are capable of creating monospecific stands over large areas.
- **Changes in the fauna.** Many of our indigenous birds, insects and other animals are not adapted to feed on or nest in alien plants and consequently leave the area.
- **Local and even total extinction of indigenous species (loss of genetic pool).** Pines (*Pinus* spp.), wattles (*Acacia* spp.) and *Hakea* spp. have already caused the total extinction of several fynbos species.
- **Ecological imbalance,** and therefore an increased risk of catastrophic events (e.g. fire risk, flooding).
- **Increased fire hazard.** Some aliens are very flammable (e.g. *Acacia* spp., *Chromolaena*), enhancing the chance of runaway fires and increasing the fuel load, thereby creating hotter fires, which can sterilise the soil and kill deep-growing roots.
- **Prevention of access.** Thorny or spiny aliens (e.g. *Opuntia* spp., *Pereskia aculeata*) can form impenetrable barriers, thereby preventing access to streams, pastures, shade trees or plantations, whether by game, stock or workers.
- **Decreased productivity of rangeland.** Unpalatable or poisonous species will be promoted by selective grazing (e.g. *Nassella trichotoma*). These species can then cause suffering and even the death of stock. They can also contaminate and damage the coats, feet and mouths of the animals.
- **Reduction in land value.** Often it would cost more to clear a farm of invasive species than the land is worth, thereby rendering it worthless.
- **Reduction in conservation and tourism value.** Monotonous stands of tall alien trees can obscure views of the scenery and natural species-rich vegetation, thereby detracting from a tourist's experience and limiting the scope of the tourism industry.
- **Soil erosion and the consequent siltation of dams and rivers.** Alien trees (e.g. *Sesbania* spp. and wattles) are easily ripped out during floods, exposing bare soil. The dislodged trees can then block the watercourse, thereby causing even more flood damage.
- **Depletion of water resources.** Invasive alien plants usually use more water than the plants they replace. Gums and wattles in catchment areas are often implicated in the drying up of rivers and the lowering of the water table.
- **Changing the natural soil composition.**

to sandy, nitrogen-poor soils and cannot survive under the changed conditions.
- **Increased agricultural input costs.** Many aliens interfere directly with agricultural activities such as crop production and in pastures. The costs incurred in controlling them increase the overall costs of production.

Invasions of alien aquatic weeds can:
- **Cause an oxygen deficiency.** Submerged weeds will remove oxygen from the water during respiration, thereby causing oxygen deficiencies, which will threaten other life forms. Chemical control of aquatic weeds can cause the mass decay of organic matter in the water, which will also remove oxygen, killing other organisms and affecting the smell and taste of the water.
- **Create dense, floating mats,** which can block pumps, prevent access by boats and fishermen, reduce water flow in canals, increase siltation in rivers and increase water loss from evaporation.
- **Provide breeding sites** for mosquitoes and snails that carry diseases such as malaria and bilharzia.
- **Cause cattle to drown** when they try to walk over a seemingly solid mass of vegetation.
- **Prevent the access of sunlight,** thereby affecting the entire food chain and seriously affecting biodiversity.

Bush encroachment

Bush encroachment is a term that describes a phenomenon that is a serious but not uniquely South African problem. In South Africa there are areas of countryside that are referred to as bushveld (savannah), which is grassland or veld with a variable density of indigenous bush vegetation. Under natural conditions the bush, grass and animals exist in a stable balance, as they have done for all of time, and such areas are a highly valued part of our ecosystem. However, under certain conditions, of which overgrazing is the main culprit, the indigenous bushes and trees increase in density to such an extent that virtually all other vegetation is excluded. Grasses cannot compete with the tall, deep-rooted bushes for light and moisture and they wither and die. The open veld becomes a mass of dense, impenetrable bushes and thorn trees, which can support progressively fewer animals. The numbers and variety of game and wild animals are drastically reduced, visibility is almost zero and the carrying capacity of domesticated stock becomes totally uneconomic.

This phenomenon is a symptom of humans' impact on their environment. If one compares the landscape in photographs taken during the Anglo-Boer War, for example, with the landscape of today, the difference is often dramatic. Many of the hills around Ladysmith in KwaZulu-Natal were once open grassland but are now (and have been for some time) covered in dense bush, a transformation that took a mere 60 years to occur. The Springbok Flats can still be remembered as open grassland, but are now bushveld. Estimates vary, but up to 53 million hectares of the 95 million hectares of bushveld area in South Africa and Namibia are suffering from various degrees of encroachment and as much as 13 million hectares can be described as severely encroached.

There are certain species of indigenous bushes and shrubs that tend to become problems in terms of encroachment. They are the first ones to respond to the changes in ecological balance and to increase in density. There are many species that have the ability to encroach and they cannot all be mentioned here, but perhaps a good indicator of their propensity to do this would be the existence of a herbicide registration, or having been declared an 'Indicator of Bush Encroachment' under Regulation 16.A. **The plants themselves are not the problem, they are not outlawed, but they are symptoms and indicators of poor land management.**

These are the indigenous bush and tree species involved:

Indicators of Bush Encroachment

Botanical name	English name	Afrikaans name	Herbicide Registration	Declared in Regulation 16 (+ Area)
Acacia ataxacantha	Flame thorn	Vlamdoring		(G, KZN, M, EC, NP)
Acacia borleae	Sticky thorn	Kleefdoring		(KZN, M)
Acacia burkei	Black monkey thorn	Swartapiesdoring	✓	
Acacia caffra	Common hook-thorn	Haakdoring	✓	(M, G, NP, NW)
Acacia erioloba	Camel thorn	Kameeldoring	✓	
Acacia erubescens	Blue thorn	Blouhaak	✓	(M, G, NW, NP)
Acacia exuvialis	Flaky thorn	Skilferdoring	✓	(M, NP)
Acacia fleckii	Blade thorn	Bladdoring	✓	(NP, NW)
Acacia galpinii	Monkey thorn	Apiesdoring	✓	
Acacia gerrardii var. gerrardii	Red thorn	Rooidoring	✓	(KZN, M, NP)
Acacia grandicornuta	Horned thorn	Horingdoring		(KZN, M, NP)
Acacia hebeclada	Candle thorn	Trassiedoring	✓	(G, NW, NP, FS)
Acacia hereroensis	Mountain thorn	Bergdoring	✓	
Acacia karroo	Sweet thorn	Soetdoring	✓	RSA
Acacia luederitzii var. luederitzii	False umbrella thorn	Baster haak-en-steek		(NC, NW)
Acacia luederitzii var. retinens	Belly thorn	Baster haak-en-steek	✓	(NC, NW)
Acacia mellifera subsp. detinens	Black thorn	Swarthaak	✓	(G, NC, NW, NP)
Acacia nebrownii	Water thorn	Waterdoring	✓	
Acacia nigrescens	Knob thorn	Knoppiesdoring	✓	(KZN, M, NW, NP)
Acacia nilotica subsp. kraussiana	Scented thorn	Lekkerruikpeul	✓	(KZN, M, G, NW, NP)
Acacia permixta	Slender thorn	Slapdoring		(NP)
Acacia reficiens	Red umbrella thorn	Rooihaak-en-steek	✓	(M, G, EC, NC, WC, NW, NP)
Acacia robusta	Splendid acacia	Enkeldoring	✓	
Acacia senegal var. rostrata	Three-hook thorn	Geelhaak	✓	(KZN, M, NP)
Acacia sieberiana var. woodii	Paperbark thorn	Papierbasdoring		(KZN, M, NP)
Acacia tenuispina	Turf thorn	Fyndoring	✓	(G, NW, NP)
Acacia tortilis subsp. heteracantha	Umbrella thorn	Haak-en-steek	✓	(KZN, M, G, NC, NP, NW, FS)

Botanical name	English name	Afrikaans name	Herbicide Registration	Declared in Regulation 16 (+ Area)
Asparagus spp.	Wild asparagus	Katbos		(NW, FS)
Azima tetracantha	Needle bush	Speldedoring		(KZN, M, EC, NP)
Burkea africana	Wild seringa	Wildesering	✓	
Catophractes alexandri	Trumpet thorn	Trompetdoring	✓	
Colophospermum mopane	Mopane	Mopanie	✓	(NP)
Combretum apiculatum	Red bush willow	Rooibos (wilg)	✓	(KZN, M, G, NW, NP)
Commiphora pyracanthoides	Cork tree	Kanniedood		(NC)
Dichrostachys cinerea	Sickle bush	Sekelbos	✓	(M, G, NP, NW)
Diospyros lycioides	Blue bush	Bloubos		(KZN, M. NP, NW)
Dodonaea angustifolia	Sand olive	Sandolien	✓	(NP, NW)
Dombeya rotundifolia	Wild pear	Blompeer	✓	
Ehretia rigida	Puzzle bush	Deurmekaarbos	✓	
Euclea crispa	Blue guarri	Bloughwarrie		(RSA)
Euclea divinorum	Magic guarri	Towerghwarrie	✓	(KZN, M, NP)
Euclea undulata	Common guarri	Gewone ghwarrie		(RSA)
Grewia bicolor	Bastard raisin	Basterrosyntjie	✓	(M, NC, NW)
Grewia flava	Wild raisin	Wilderosyntjie	✓	(KZN, M, G, NC, NW, NP)
Grewia flavescens	Rough-leaved raisin	Kruisbessie	✓	(M, NP, NW)
Grewia monticola	Grey raisin	Vaalrosyntjie		(NP)
Lippia javanica	Fever tea	Koorsbossie	✓	
Leucosidea sericea	Oldwood	Ouhout	✓	RSA
Lopholaena coriifolia	Fluff bush	Pluisbossie	✓	
Philenoptera violacea	Apple-leaf	Appelblaar	✓	
Lopholaena coriifolia	Small-leaf fluff bush	Pluisbossie	✓	(M, G, NW, NP)
Gymnosporia polyacantha	Hedge spike-thorn	Kraaldoring		(NP)
Gymnosporia senegalensis	Red spike-thorn	Rooipendoring	✓	(KZN)
Ochna pulchra	Peeling plane	Lekkerbreek	✓	
Peltophorum africanum	African weeping wattle	Huilboom	✓	

Botanical name	English name	Afrikaans name	Herbicide Registration	Declared in Regulation 16 (+ Area)
Rhigozum trichotomum	Three thorn rhigozum	Wildegranaat	✓	(NC)
Strychnos madagascariensis	Spineless monkey-orange	Swartklapper		(KZN, M, NP, NW)
Tarchonanthus camphoratus	Camphor bush	Kanferbos	✓	(NC, NW)
Terminalia prunioides	Purple cluster-leaf	Sterkbos	✓	
Terminalia sericea	Silver cluster-leaf	Sandgeelhout	✓	(KZN, M, G, NC, NP, NW)
Ziziphus mucronata	Buffalo thorn	Blinkblaar-wag-'n-bietjie	✓	
Ziziphus zeyheriana	Dwarf buffalo thorn	Dwerg-blinkblaar-wag-'n-bietjie	✓	

There is a fine distinction between a 'bush' and a 'shrub', but it is generally accepted that a shrub is shorter and less woody. In the context of encroachment, they are caused and controlled by the same factors. Shrubs perhaps do not cause the same problems in terms of forming impenetrable stands, but their aggressive nature and inferior grazing value are all causes of veld retrogression.

Again, there are a great number of shrub species that are known to encroach and cause veld deterioration. Here are some of the more serious ones, which appear on herbicide labels or are declared invaders:

Botanical name	English name	Afrikaans name	Herbicide Registration	Declared in Regulation 16
Anthospermum aethiopicum		*Katstert*	✓	
Athanasia crithmifolia	Klaaslouw bush	Klaaslouwbos	✓	
Chrysanthemoides monilifera	Brother berry	Boetebessie	✓	
Chrysocoma tenuifolia	Bitterbush	Bitterbos	✓	
Cliffortia ruscifolia	Prickly bush	Doringbos	✓	
Elytropappus rhinocerotis	Rhinoceros bush	Renosterbos	✓	
Eriocephalus africanus	Rosemary	Kapokbossie	✓	
Helichrysum auriculatum			✓	
Leucadendron rubrum		Tolletjiesbos	✓	
Leucadendron ericifolium	Erica-leaved yellow bush	Heideblaargeelbos	✓	
Lycium ferocissimum	Box thorn	Driedoring	✓	
Metalasia muricata		Blombos	✓	

Botanical name	English name	Afrikaans name	Herbicide Registration	Declared in Regulation 16
Myrsine africana	Wild myrtle	Wildemirt	✓	
Pentzia globosa	Bitter karoo bush	Bitterkaroo	✓	
Seriphium plumosum	Bankrupt bush	Bankrotbossie	✓	RSA
Stoebe spiralis		Slangbos	✓	

Under normal conditions these species are a desirable component of the bushveld and with correct veld management they would never have the chance to become a problem. They are usually not considered a problem in terms of the criteria used to justify inclusion in this book (see page 14).

Veld retrogression

Veld retrogression, which includes the thickening of the woody component, is further displayed by a shift from the palatable, nutritious grass species to those that are less palatable, less vigorous and with a lower nutritional value. Pasture scientists categorise these grasses into 'decreaser' species (those that decrease when the veld is badly managed) and 'increaser' species (those that increase when the veld is badly managed). There are various categories of increasers, depending on the level of their undesirability and the bio-climatic region they are from. The following species are some of those that are considered undesirable when they occur in any quantity and that indicate veld mismanagement and therefore retrogression:

Botanical name	English name	Afrikaans name
Aristida spp.	Three-awn grasses (ngongoni)	Steekgras
Diheteropogon filifolius	Thread-leaved bluestem	Smalblaarblougras
Elionurus muticus	Wire grass	Koperdraad
Enneapogon cenchroides	Nine-awned grass	Negenaaldgras
Eragrostis spp.	Love grasses	
Perotis patens	Cat's tail	Katstertgras
Setaria pallide-fusca	Garden bristle grass	Tuinmannagras
Sporobolus africanus	Rat's tail dropseed	Taaipol
Sporobolus pyramidalis	Cat's tail dropseed	Katstert-fynsaadgras
Tragus berteronianus	Common carrot-seed grass	Gewone wortelsaadgras
Tragus racemosus	Large carrot-seed grass	Grootwortelsaadgras
Tricholaena monachne	Blue seed grass	Blousaadgras
Urochloa mosambicensis	Bushveld signal grass	Bosveldbeesgras

These extreme examples generally indicate that severe veld damage has already occurred. There are many species other than these that can indicate that veld deterioration is in progress. They would be monitored and compared to nearby healthy veld to estimate the degree and speed of the retrogression.

Causes of bush encroachment
The main causes of the thickening of the woody component of bushveld are:

Overstocking
This leads to overgrazing, which depletes the grass cover. This allows bush seedlings

to develop unhindered and eventually outcompete the remaining grass. Poor grass cover also leads to greater run-off of rainwater, which in turn depletes the moisture reserves further and facilitates soil erosion.

Fires
A high fuel load is required to produce a hot fire that can destroy many of the developing scrub seedlings. Overgrazing depletes this fuel load and limits the efficacy of fires. (Burning does not stimulate grass production but simply removes the old material, making it easier for the grass tuft to push up the new tillers during the following growing season.) Furthermore, fire is sometimes deliberately excluded from veld management systems for various, and often misguided, reasons. When used properly, fire is a natural and effective tool for managing veld and should be used as such.

Inadequate grazing management
Central to all issues is the implementation of a proper system for managing the veld so that encroachment does not occur, and if it has already, to reverse it.

Grazing management
The main components of an efficient grazing system are:
- Fencing off 'camps' of suitable size and ecological similarity.
- Rotation of these camps in terms of grazing and resting.
- Correct stocking rate.
- Correct mix of animals (e.g. cattle and sheep) to make optimum use of existing vegetation.
- The incorporation of browsers (e.g. goats) into the system.
- Sufficient and suitably placed drinking points to avoid trampling and localised overgrazing.
- Correct use of fire.

Once an area of open veld has succumbed to the thickening of the scrub, it is very difficult and expensive to redress the balance. In extreme cases the cost of the operation would be more than the land is worth, so a farmer would find it more cost-effective just to buy another farm! Any veld management system is a long-term and continuous operation. A healthy bushveld area has plenty of grass (of the right species) and some well-spaced trees for shade. Efforts in bush (and weed) control must be concentrated on maintaining these areas. When resources become available, then clearing new areas can be considered.

The control or reversal of bush encroachment

Perhaps it stands to reason that the first step would be to correct all those factors that caused the problem in the first place. This would entail the implementation of an efficient grazing management system as outlined above. In many instances, stock is allowed to roam a farm in densities that become more and more damaging as the encroachment intensifies. A decision has to be made to remove stock totally from an area, set this area aside and implement a grazing system that would start with a rest period. From then on grazing is managed, not only to maximise the economic returns of the enterprise, but also as a tool to ensure a healthy and sustainable grassland ecosystem. As the balance returns to this pilot area, then further areas can be included.

Control of already encroached bush, however, will require a degree of extra input. There are various methods available and it is probably best to incorporate elements of them all as the situation arises:
- Mechanical: this includes the use of bulldozers, tractors, etc.
- Physical: this entails the use of axes, etc. for felling and girdling.
- Fire.
- Chemical herbicides.

Mechanical methods
These are expensive and often leave the soil bare and susceptible to erosion, creating an ideal seedbed for further seed germination. For example, the sickle bush (*Dichrostachys cinerea*) relishes this scenario. These methods are possibly only suitable for extreme cases as they are more cost-effective as the density increases.

Physical methods
These are also relatively expensive, especially in severe cases, but become more cost-effective as the bush density decreases. Bushes are cut down and left on the ground or piled and burned. Leaving them on the ground can assist to protect the new grass growth for a while. The stumps of those species that coppice must be removed, treated with a herbicide or killed by some other means.

Girdling entails the removal of bark from a complete band around a stem, thus cutting off the transport of water and nutrients, causing the plant to die. Some species can regrow from the remaining stem below the girdle.

Fire
This method is probably useless in very dense bush as there is a lack of fuel in terms of grass cover. The more fuel there is available, the hotter and more effective the fire. A hot fire can remove much of the top growth of bushes and will destroy the seedlings, but it will not necessarily kill the stumps. The stumps must be removed or killed by some other means. It is not easy to effectively apply a herbicide to a burned stump. It is possible to kill a stump by piling old maize cobs or twigs around it and setting fire to them, but this technique can only be used on a limited scale.

Herbicides
Herbicides can be used alone or in combination with the above-mentioned methods. There are various techniques to choose from:

Aerial application
For large areas there are soil-acting granular herbicides that can be spread by hand or from the air. The granules fall through the canopy onto the soil and the herbicide is washed into the soil by rain and is absorbed by the roots of the bush. Highly susceptible species can be reduced in number before significant damage to the grasses occurs.

Hand application
There are several methods to hand-apply the herbicide:
- Granules as applied from the air can be spread around individual bushes by hand or thrown into the canopy.
- It is possible to dose individual bushes with a soil-acting herbicide using a dosing gun. The herbicide is applied in small, concentrated doses onto the soil at the base of a bush. A young, susceptible bush requires very small amounts of herbicide and grass damage is negligible. Total death of the bush can take months and even years, but grass regeneration commences as soon as the bush defoliates, which can be almost immediate.
- Stem treatment can be done to standing bushes. There are herbicides that are mixed with water or with diesel. (The diesel acts as a penetrant to assist with the penetration of the herbicide through the bark.) The chemical can be sprayed or painted on. There have been various attempts at producing equipment that can cut the bark and inject a dose of herbicide all in one process. Sometimes the stems can be scarified to aid in the penetration of the chemical.
- Stump application is made to bushes that have been cut down. The herbicide is applied to the fresh cut for rapid uptake and translocation. The chemical formulation often contains a dye so that it is easy to see if a stump has been treated or not.

It may be necessary to combine aspects of all these methods, depending on terrain, availability of labour, soil type and species composition. It is also important to weigh up all the aspects of the control operation when trying to decide on the best and most cost-effective approach. It is important to know how each species reacts to the various methods employed, as they will all react differently. Labour costs, chemical costs and secondary costs such as time, grass damage and management costs must all be considered. When implementing a plan, it is usually best to start on the least encroached areas and work inwards to the denser areas, adjusting the techniques accordingly. It must be remembered that the aim is usually not to eliminate all bushes, but to selectively thin them. Once an area has been returned to an acceptable bush density, the balance is then maintained with the veld management system.

Destruction of grass cover by encroaching bush

Rapid renewal of grass cover after destruction of bush

BIOLOGICAL CONTROL

What is biological control?

One reason why exotic plants rapidly establish themselves and spread throughout a suitable new habitat is the absence of natural enemies that would have existed in their native homes and did not transfer with them. For example, triffid weed (*Chromolaena odorata*) is not considered a weed in its place of origin in South America, but has reached virtually uncontrollable and devastating proportions along the eastern seaboard of KwaZulu-Natal. In South America, the plant would have existed in a natural balance with a wide range of herbivorous insects, mites, nematodes and microbial pathogens, each of which would be an integral part of the local ecology. But here these organisms are absent. The plant therefore flourishes and can out-compete indigenous vegetation, which has its own range of indigenous enemies that keep it under 'control'.

The obvious reaction to this is to consider introducing a range of such enemies to help control these plant pests, but unfortunately it is not that easy. Consider the implications if, just for the sake of illustration, an insect was brought from South America to control triffid weed, and after successful introduction it was found to prefer the proteas!

The awareness of this problem has led to the precise science of biological control, not just of weeds but all types of pests and diseases. Indeed, many exotic pests have reached such epidemic proportions that perhaps the only realistic hope of long-term, sustainable control is through the use of biological control agents. Biological control entails the introduction of natural enemies from the country of origin of the problem plant with the aim of reducing the plant population to manageable levels but not necessarily eliminating it. A residual population of plants would have to remain on which the biocontrol agents could survive and thereby achieve sustainable control.

Biological control agents for plants are mostly confined to herbivorous insects, mites and plant pathogens mainly because they often exhibit a very high level of host specificity. Confirmation of this property is a crucial part of the research needed before any agent is introduced. Once identified, the agents can be mass reared and released into the field and from then on their performance is closely monitored.

Once established, these agents can work in one of two ways:
- They can prevent the further spread of the plants by attacking propagative or reproductive material such as seeds and do not actually kill the host plants. This is a popular choice where the weeds are also of commercial or utility value.
- They can reduce the vigour of individual plants by attacking nutritive or structural tissues such as stems, leaves and roots. These plants will be weakened and become less competitive and will die if the damage is dramatic.

Unfortunately, not all biocontrol programmes are effective and the reasons for these failures can include:
- **Climate.** Sometimes, the biocontrol agents do not survive in all climatic conditions. This can happen in their home ranges too, but there is always a large and versatile range there. In a control programme of an alien weed, the plant can still flourish in areas where the single agent cannot survive.
- **Varieties.** Biocontrol agents are very host-specific and sometimes will not survive if the host plant has interbred or altered its genetic makeup in any way from what it was in its home range.
- **Indigenous predators and parasites.** These organisms can keep populations of biocontrol agents low and ineffective.
- **Other methods of control** can interfere with biocontrol. Extensive chemical control can kill populations of plants that are acting as a haven and a source of the biocontrol agent, from where they can infect residual and resurgent populations.

The history of biological weed control in South Africa

The first record in the world of successful biological weed control dates back to 1836, when a cochineal bug (*Dactylopius ceylonicus*) from Brazil was introduced into southern India to combat the smooth prickly pear (*Opuntia vulgaris*). This bug was so successful that within a few years the smooth prickly pear was virtually eliminated from India. After this dramatic success, the cochineal bug was introduced into other countries where these plants from South America were threatening the local ecology, including South Africa in 1913. *O. vulgaris* was first recorded in South Africa in 1772 and by the late 1800s vast areas of the current Eastern Cape and KwaZulu-Natal were covered with impenetrable thickets. Within a few short years of the introduction of the cochineal bug, this weed had been reduced to a few small, isolated pockets.

A great deal was learnt from this dramatic success, including the principle that a highly species-specific insect, such as the cochineal bug, will not necessarily switch to a 'non-target' plant once its original host has disappeared. The development and survival of a host-specific insect herbivore is closely tied to characteristics of their host plants: the plant must feel, smell and taste right before it is recognised as a suitable host; the seasonal cycles of the plant and insect must be synchronised and the insects must be able to cope with the combination of toxins that characterise most plant species.

Nothing else was seriously attempted in South Africa until the 1930s when, in 1933, an Argentinean moth (*Cactoblastis cactorum*) was introduced to combat two other cactus weeds, the prickly pear (*Opuntia ficus-indica*) and the jointed cactus (*O. aurantiaca*). Success was limited, but in 1938 another cochineal species (*Dactylopius opuntiae*) was introduced and this again produced spectacular results on the prickly pear, clearing about 80% of the one million infected hectares in the Eastern Cape and Karoo. In 1935 another species of cochineal (*D. austrinus*) was released on the jointed cactus. Although it initially caused massive reductions in the density of the jointed cactus over most of its range, there have been cyclical resurgences of the weed. Usually these are suppressed again by the insect and never reach the proportions they did before the implementation of biological control, but as a result success was considered by many to have been limited.

Opuntia plants devastated by cochineal.

Cochineal on *Opuntia* sp.

The partial success of *Dactylopius austrinus* appears to have discouraged the science of biological control as little progress was made throughout the 1940s and 1950s It was not until the 1960s that significant advances were made and the momentum gained from them has continued to today (2009). More than 88 biocontrol agents have been released onto 47 weed species.

Biological weed control in South Africa today

The Plant Protection Research Institute (PPRI) is responsible for laying down guidelines for the testing and importation of potential biocontrol agents. Great care is taken to ensure that a particular agent will not pose a threat to any indigenous species. Each candidate is closely scrutinised and must pass a series of stringent tests over a period of several years. Care must also be taken that an agent works in an acceptable manner and that there are no conflicts of interest. For instance, it is important that pines and black wattles, which are both economically important and very invasive, are controlled by agents that prevent their spread but not their growth. This could be achieved by finding an agent that attacks the seeds or prevents seeding, thereby preventing the spread of plants without affecting existing plantations. The black wattle seed weevil, *Melanterius maculatus*, was released in 1995/6 and became established. Further releases are expected.

Eight species or 17% of the weeds subjected to these imported agents are now considered under complete control (virtually no other control measures are needed or recommended). Altogether 14 species or 30% are under substantial control (alternative methods of control needed but at reduced rates) and four species or 8% per cent are under negligible control despite the fact that, in many instances, the biocontrol agents have established themselves on the plants. The remaining 45% or 21 species remain undecided, mainly because the release of the biocontrol agents has been too recent to determine their impact.

Some of the notable successes of biological control of weeds in South Africa are mentioned below.

St John's Wort (*Hypericum perforatum*) was successfully prevented from becoming the serious problem as it had in Australia and California. The few isolated infestations in the southwestern Cape have been kept in check for more than 30 years by the defoliating beetle (*Chrysolina quadrigemina*) and more recently the stem-boring fly (*Zeuxidiplosis giardi*).

Kariba weed (*Salvinia molesta*) is now well controlled in all areas of southern Africa into which it had spread and where it was choking dams and rivers. A leaf-feeding and stem-boring weevil (*Cyrtobagous salviniae*) was released in South Africa in 1985 and dramatic success was seen within a year.

The water hyacinth (*Eichhornia crassipes*) is controlled by four agents, *Neochetina bruchi, N. eichhorniae, Sameodes albiguttalis* and *Eccritotarsus catarinensis,* although supplementary methods of control must be used in most areas. The more recent introduction of the sap sucker, *E. catarinensis*, has yet to prove effective.

Red sesbania (*Sesbania punicea*) is now under effective control from the depredations of three insects: the flower bud weevil (*Trichapion lativentre*) whose larvae develop in and destroy flower buds, a seed-feeding weevil (*Rhyssomatus marginatus*), and a very destructive stem-boring weevil (*Neodiplogrammus quadrivittatus*). These weevils were released in 1984.

Biocontrol agents have been successfully introduced against several of the invasive Australian acacias. Long-leaved wattle (*Acacia longifolia*) is being well controlled by the gall-forming wasp (*Trichoglaster acacialongifoliae*). This wasp, which was released in 1982, lays its eggs on the flowers, which then turn into galls instead of seeds. Under very favourable conditions, such as on a stream bank, or in a high-rainfall area such as KwaZulu-Natal, these galls are not sufficient on their own to deplete the moisture and nutrient reserves of the plant. In such situations it is necessary to introduce a second biocontrol agent, a seed weevil (*Melanterius ventralis*).

Galls on *Acacia longifolia*

The rooikrans seed weevil (*Melanterius servulus*) was released in 1994 and seems to be effective in controlling rooikrans (*Acacia cyclops*). In fact, this same weevil, albeit a different biotype, was released onto stinkbean (*Paraserianthes lophantha*) in 1990. The two biotypes are highly host-specific, so that they will only attack the species onto which they were released. A seed weevil (*M. acaciae*) was released onto blackwood (*A. melanoxylon*) in 1985.

The hakeas of the southern Cape are now well suppressed by moths and weevils: the hakea fruit weevil (*Erytenna consputa*), the hakea seed moth (*Carposina autologa*), and the hakea gummosis fungus (*Colletotrichum gloeosporioides*), which was not intentionally introduced but which is also having an effect.

In 1996 the leaf-mining moth (*Parectopa thalassias*) was released onto Australian myrtle (*Leptospermum laevigatum*).

A notable first in South Africa is the release in March 1999 of the world's first biological control agent against cat's claw (*Macfadyena unguis-cati*). Leaf-feeding beetles from Brazil (*Charidotis auroguttata* – the golden-spotted tortoise beetle) were released near Magoebaskloof. To date success has not been proven.

Also in 1999 a Brazilian lace-bug (*Gargaphia decoris*) was released near Sabie in Mpumalanga to combat bugweed (*Solanum mauritianum*). Further releases followed, but success is proving elusive.

It may be difficult to evaluate the long-term success of more recent efforts, but it seems as if South Africa may have developed the world's first successful biological control of satansbos or the silver leaf bitter apple (*Solanum elaeagnifolium*). It appears to be succumbing to two species of leaf feeding beetles (*Leptinotarsa texana* and *L. defecta*). The beetles defoliate the plants, which weaken them, reduce their seed production and generally reduce their ability to spread. Although not spectacular in the short term, populations are definitely declining.

The use of fungal pathogens has recently gained momentum and several fungal herbicides have been registered. One is an indigenous stump-colonising fungus

(*Cylindrobasidium laeve*), which has been registered to control the stumps of black and golden wattles (*Acacia mearnsii* and *A. pycnantha*). The formulated fungus must be applied to each individual stump. The fungus then colonises the stump, eventually destroying it.

A gall rust fungus (*Uromycladium tepperianum*) was recently introduced onto Port Jackson (*Acacia saligna*) with promising results. The fungus spreads by means of airborne spores that are produced on large, fungus-induced galls. These galls act as 'nutrient sinks' on the plant, which eventually succumbs to their effects. Unfortunately, many local communities have grown to depend on Port Jackson as a source of firewood and fodder and the indiscriminate loss of this plant will cause them hardship if they cannot find a suitable replacement. (Biological control of Port Jackson appears to have caused a conflict of interests.)

Rust gall on Port Jackson (*Acacia saligna*)

Not all weeds are suitable for biocontrol and there are some plants that have thwarted all attempts at finding a damaging agent. Common lantana (*Lantana camara*) has received much attention in this regard, with at least 19 insect agents having been tried without any real success even though many of them managed to establish themselves. It is thought that all the different hybrids formed by crosses with horticultural cultivars have made a species-specific insect almost useless. Over 50 variants of the species *Lantana camara* have been recorded in South Africa, out of a total of 650 worldwide. As a result, the insects have suppressed the wild type of lantana on the KwaZulu-Natal coast, but have had negligible effect on other variants, particularly in colder, inland regions. Many different agents are being assessed for the control of triffid weed (*Chromolaena odorata*). None of them seem to be sufficiently damaging on their own, but a combined effort of a range of these agents could be the answer.

It is clear that the success rate of biological control varies from total to negligible and that the 8% of total success is relatively low. However, biological control is now being regarded more as a means of improving the efficacy of conventional control methods and in fact, those agents that cause even moderate levels of damage to a weed, can contribute substantially to the control achieved by an overall, integrated approach.

Plants that are weakened in such a way are more susceptible to the conventional tactics, such as those used by the WfW Programme. Conversely, great care must be taken that conventional methods of weed control do not interfere with or even destroy the biological ones. This can happen if:
- the entire infestation of an invasive plant is destroyed, then the highly specific bio-control agent will perish as well.
- some herbicides or adjuvants are directly toxic to the biocontrol agent.

It is important then, to preserve the presence of the biocontrol agent throughout the weed's distribution. To this end, 'agent reserves' are demarcated within a patch of weeds from where the agents can survive and recolonise any regrowth of the weed.

Biological control of weeds is usually considered environmentally friendly, selfsus-taining and cost-effective. On the down side, it can be unpredictable, slow and some-times it can be relatively expensive.

There is little doubt that biological control is the only long-term, safe and sustainable method of controlling some of South Africa's most problematic plants, but its long-term nature must be understood and accepted. Every step forward must be scrutinised and double-checked for fear of introducing something that could become yet another alien invader. Meanwhile all control methods currently available must be put to their best and most cost-effective use, while raising awareness of biocontrol, its theory and practicalities, in order to maximise its effect and to attract the resources that it will surely require.

CHEMICAL WEED CONTROL

Chemicals and the law

Today a wide range of sophisticated chemical herbicides is sold under an even wider range of trade names. The use of these products requires specialised knowledge of both the products involved and the weeds to be con-trolled. The label on the container should always be read carefully and understood. Anyone intending to use a herbicide must bear in mind the proclamation made by the Minister of Agriculture in *Government Gazette* No. 13424 dated 26-07-92, which states that it is an offence to:

> ... acquire, dispose, sell or use an agri-cultural or stock remedy for a purpose or in a manner other than that specified on the label on a container thereof or on such a container.

Any claim on a herbicide label must be approved by the Registrar, backed up by conclusive data and registered in accordance with the Act known as Fertilizer, Farm Feeds, Agricultural Remedies and Stock Remedies Act No. 36 of 1947. This Act is intended to co-ver and protect the user, manufacturer, and supplier of a product, the general public and the environment. The implications of the Act are wide-ranging, but as far as product labels are concerned, the label is in effect a legal document. Any deviation from the prescribed label is therefore a contravention of the Act.

Although there is a wide range of products with a wide variety of weeds specified on the labels, there is still a large number of common and often extremely troublesome weeds that are not mentioned on any product label. There are also many minor crops or situations for which products have not been specifically registered.

Most of the weeds in this book have one or more herbicides registered for their control and in a variety of crops. If possible, this is mentioned in the text, along with a mention of techniques needed to apply the herbicide. Where no herbicides are actually registered for a weed and it is known that one of the other methods is effective, this is also mentioned. This book must also conform to Act 36 of 1947, therefore no claims can be made about chemicals or chemical control that do not appear on a current herbicide label.

Historical overview

Chemicals have been used for weed control for over 120 years. The earliest were used in massive doses and were aimed at total weed control, mainly on industrial sites, railway tracks, etc. Such chemicals included crushed arsenical ores, oil wastes and creosote. They were toxic, messy and wasteful.

The first selective herbicides, in the early 1900s, were used in cereals and were based on soluble copper salts and sulphuric acid. Copper salts are toxic to all forms of plant life although at the trace level, copper is an essential micronutrient. Sulphuric acid was used until very recently for haulm killing of potatoes and for weed control in onions. The selective action was physical and was dependent on the different wetting properties of onion leaves for example, as opposed to those of the weeds.

The first widely used synthetic herbicide was 2,4-D, which was first commercialised in the late 1940s and it remains one of the most commonly used herbicides in the world today.

The 1950s saw the introduction of the triazines, a group of residual herbicides that greatly improved weed control options at that time. In recent years, however, there is growing concern over groundwater contamination and they are losing favour.

Glyphosate was introduced in 1974 and remains the world's largest selling herbicide in terms of volume.

Since then, a huge range of sophisticated, broad-spectrum and selective chemical products has been developed and commercialised.

Modern herbicides depend for their selectivity on biochemical factors. For instance, many plants possess enzyme systems that can metabolise certain chemicals. This enables the plant to tolerate the chemical's normally toxic effects. Some plants have unique enzyme systems that are only affected by very specific chemicals. These systems can be interrupted or stimulated, which makes the plant more or less susceptible to the herbicidal effects.

Today, herbicide rates are measured in grams per hectare and are used to control perhaps just one species of weed in a very closely related crop.

Chemicals for the control of weeds

Below is an outline of the types of chemical products currently available and some of the common terms used to describe them:

Herbicides

A herbicide is a substance, either naturally occurring or man-made, that alters the metabolic processes of a plant so that the plant is suppressed, killed or its growth altered in such a way that it becomes less of a problem. Herbicides may be divided into groups according to their general mode of action:

Non-selective vs selective herbicides
- **Non-selective:** These herbicides will affect any plant with which they come into contact. For instance, some of these chemicals can act by destroying chlorophyll.
- **Selective:** These chemicals can kill a weed without harming a crop, even if it is sprayed over the crop. (One can use a non-selective chemical in a selective manner such as for inter-row treatment, but this does not alter these basic divisions.) Most selective herbicides will lose their selectivity at a high enough dose.

Contact vs systemic herbicides
- **Contact:** These are products that act by affecting only the plant tissue with which they come into direct contact. Thorough coverage and wetting of the weed is necessary for effective control. Many plants become more tolerant of such chemicals as they mature. Seedlings are usually very sensitive, whereas perennial plants, that can regrow from the remaining roots, are able to recover.
- **Systemic:** These chemicals are translocated throughout the plant from the initial site of application. For instance, a chemical applied to the foliage can be translocated to the roots and if sufficient quantities have been applied, the whole plant is destroyed, including tough perennial root systems.

Pre-emergence vs post-emergence herbicides
- **Pre-emergence:** A pre-emergence herbicide is applied to the soil before the weeds emerge. Uptake is usually by the growing coleoptile (shoot) or by the developing roots.
- **Post-emergence:** These herbicides are

applied after emergence of the weeds and usually have a high degree of leaf uptake. To a certain extent this term is also used to describe products applied to mature plants.

Long vs short residual action herbicides
Some products can remain active in the soil for many months or even years, whereas others have a relatively short life. Some chemicals are deactivated almost immediately on contact with the soil.

Modern chemical groups
There are many different chemical groupings at present and no doubt new ones will continue to be discovered. The following are some of the more important chemical groups:

Triazines and ureas	The members of these groups are usually soil-applied, residual herbicides, acting by the inhibition of photosynthesis. They do not affect germination but are taken up by the developing roots, with a plant often only showing symptoms of poisoning after emergence and actually dying of starvation. These chemicals normally have a fairly wide range of activity, but are usually better at controlling broadleaf weeds than grasses.
Acetanelides (amines)	This group acts by inhibiting protein synthesis, disturbing cell division and affecting the cell membrane. They can therefore prevent germination but will also work by inhibiting root and overall growth. The members of this group are primarily grass killers with a variable effect on broadleaf weeds and sedges (e.g. nut-grass).
Phenoxy compounds (synthetic auxins)	This group is also sometimes referred to as the 'hormone' group, acting primarily as artificial plant hormones and upsetting the hormone balance within the weed. This imbalance causes uncontrolled cell division and enlargement in the growth point of the plant. It consequently induces a large number of biochemical and metabolic changes that lead to abnormal plant development. The phenoxy compounds are principally killers of broadleaf weeds.
Thiocarbamates	The chemicals in this group are volatile and must be incorporated into the soil to prevent loss by evaporation. They show a high degree of selectivity and have a relatively long residual action. They are absorbed by the growing coleoptile, so they do not prevent germination but should prevent emergence. They are active mainly on grasses and sedges and are often used to control red and yellow nut-grass.
Dinitroanalines	The dinitroanalines are somewhat less volatile than the thiocarbamates. Some, however, still require incorporation into the soil and have a long residual action. These chemicals are mainly grass killers.
Sulphonyl ureas	The first products in this group were only developed during the 1980s and started a whole new era in chemical weed control. They are highly active at very low dosage rates, often being effective at rates as low as 10 gram per hectare. Although the first sulphonyl ureas were mainly post-emergence broadleaf weed killers, this group now contains products with a wide range of herbicidal actions.
Phosphorous herbicides	Typified by glyphosate and its salts. Discovered in the early 1970s, this group currently enjoys the largest volume of usage of any group of herbicides. They are post-emergence, non-selective and systemic products and act by inhibiting the formation of amino acids. On contact with the soil these products lose their herbicidal qualities and are broken down by soil microorganisms into carbon dioxide, nitrogen, water and phosphate.
Bipyridylium compounds	A small, but very widely used group of herbicides typified by the chemical paraquat. They are non-selective, post-emergence and contact herbicides. They are rapidly absorbed by green plant tissue, with only partial translocation. On contact with the soil they are immediately deactivated by being adsorbed onto clay particles.

In light of the occurrence of resistance to certain herbicides or herbicide groups, emphasis is sometimes placed on grouping products according to their biochemical mode of action. Two groups that are identified this way are the following:
- **ALS inhibitors:** These herbicides work by inhibiting acetolactate synthase.
- **ACCase inhibitors:** These herbicides inhibit acetyl-coenzyme A carboxylase.

Herbicide formulations

Herbicides are presented or formulated in many ways. The formulation will depend largely on the properties of the chemical, but also where and how it is to be applied. Some versatile herbicides will have a choice of formulations. These are today's usual herbicide formulations:
- **Liquid (AL):** To be applied undiluted.
- **Capsule suspension (CS):** Microcapsules of a chemical suspended in a liquid.
- **Soluble salt (SS):** Compounds that simply dissolve in water forming a solution.
- **Emulsifiable concentrate (EC):** Liquid formulations that have an emulsifying agent that breaks up into small droplets when added to water and remains suspended. They become milky when added to water and will normally remain in suspension for several hours.
- **An emulsion, oil in water (EW):** A milky coloured formulation already emulsified, but can be diluted in water.
- **Fumigant (GE):** Usually liquids or gas-generating products that are injected into the soil in horticultural nurseries or seedbeds.
- **Gel (GL):** Used to apply directly to a stem or stump.
- **Macro-granules (GG):** The active herbicide is attached to a carrier such as clay or even fertiliser. These are larger granules for manual application.
- **Granules (GR):** These granules are applied directly to the soil usually using mechanical equipment. They usually require rainfall for activation and do not have drift problems.
- **Oil-miscible liquid (OL):** A formulation suitable for diluting in diesel.
- **Suspension concentrate (SC):** A wettable powder already suspended in water so it can be poured.
- **Water-soluble granules (SG).** Granules that dissolve.
- **Soluble liquid concentrate (SL):** Simply added to water and will dissolve instantly.
- **Water-soluble powder (SP):** Powders that form true solutions and require no agitation.
- **Water-dispersible granules (WG):** Granulated product for easy handling. It is added to water, the product disperses and is suspended in the water in the same way as a wettable powder.
- **Wettable powder (WP):** Finely ground, insoluble materials that are suspended in water. They contain a wetting agent to assist in the mixing of the powder with the water, but will need constant agitation to stay in suspension.

'Organic' and 'biological' herbicides

'Organic' in this sense refers to herbicides that are 'non-chemical' since many conventional herbicides are actually organic molecules as they contain carbon as the primary molecular component. Vinegar (acetic acid) and spices such as oil of cloves can kill plants and so can ammoniated soap of fatty acids and corn gluten meal. Citrus oil is commercially available as a herbicide and works by stripping away the waxy cuticle of the leaf, causing dehydration of the plant and ultimately death. Sometimes such products are effective enough to achieve registration but rarely exhibit any selective action. Organic growers can make effective use of such products, but they are usually too expensive or cumbersome for large-scale use.

Mycoherbicides

A mycoherbicide is a formulation of infective propagules such as fungal spores, which can be applied to weeds in the same way as a chemical herbicide. It is promoted wherever possible as an eco-friendly approach. The spores germinate on the plant, penetrate the plant tissues and cause a disease, which will weaken and even kill the plant. Since the

mycoherbicide is indigenous to the country of use, it is already naturally present in the environment but usually in sub-lethal amounts. This is different from 'biological control' agents, which are introduced from the country of the plant's origin. Applying large numbers of the spores of the mycoherbicide at once causes a local epidemic, killing susceptible plants over a short period of time, but they cannot maintain the epidemic under natural conditions, and have to be re-applied at regular intervals. A few mycoherbicides are or have been available in South Africa and include:
- *Cylindrobasidium leave* is an indigenous wood-rotting fungus used for the control of alien black and golden wattle (*Acacia* spp.). It is applied to cut stumps, causes rotting of the wood and prevents regrowth of the stumps.
- *Colletotrichum acutatum* is an indigenous fungus that causes cankers and gummosis (the oozing of gum) in the silky hakea (*Hakea sericea*). The shrubs are inoculated by applying the spores to cuts in the bark or by firing coated shotgun pellets into inaccessible stands. The infected plants eventually die.

Out of hundreds of potential bioherbicides that are researched worldwide, very few have become commercially viable. There are several reasons for this:
- High cost of mass production.
- Limited markets (only one or two target species, environmental constraints, etc.).
- Incompatibility with conventional pesticides.
- The appearance of resistant bio-types.
- The introduction of new herbicide chemistries.
- Little, if any, patent protection.
- Inadequate, expensive and confusing regulatory conditions.

On the positive side, there are several strong reasons for research to continue:
- The growing threat of resistance to certain groups of chemical herbicides.
- The banning or loss of registration of some chemicals.
- The lack of registrations on minor crops.
- The costs of controlling non-agricultural and alien weeds, especially in low-maintenance areas.
- The need to control weeds in organic production systems.

Adjuvants or surfactants

This refers to surface-active chemicals such as wetters, stickers, spreaders, etc. This is a group of chemicals that are added to spray mixtures to enhance the effect or properties of the chemicals to which they are added. Sometimes the use of one of these products is specifically recommended, and sometimes it is important that they are not used. (Most herbicides have various types of surfactants already in the formulation, therefore it is important to read the labels carefully.) Sometimes a suitable adjuvant or penetrant can reduce the amount of active herbicide required, which could thereby potentially benefit soil health and the environment.

Wetters
In general terms, a wetter is a substance that reduces the surface tension of a spray droplet. This, for example, allows the droplet to spread over and adhere to a waxy or hairy leaf surface. This has implications for the water volume applied and for the possibility of runoff. Many of these products are based on soaps and detergents.

Stickers
A sticker is a surfactant that improves the retention of spray droplets once good wetting and coverage has been achieved. Stickers tend to dry and should not dissolve again too quickly in water. They can improve the rainfastness of applied products.

Penetrants
A penetrant increases the penetration of the active ingredient of an applied product into the target. This requires absorption into the cuticle, movement across the cuticular membrane and absorption by the underlying cells.

Stabilisers
A stabiliser promotes and maintains uniform

distribution of the active ingredient throughout the spray tank. Most herbicide formulations already contain a range of chemicals such as emulsifiers, dispersants, solubilisers, etc. There are, however, products available that can be added to a spray mixture to enhance the effect described above.

Compatibility aids
There are one or two products that can be added to a spray mixture to prevent a chemical reaction or physical changes from occurring. They may influence the efficacy of the other products in the mixture.

Buffers
Buffers maintain the desired pH of spray mixtures in the tank. Some pesticides, particularly some types of insecticides and fungicides, degrade rapidly under alkaline conditions (high pH). Most modern products are relatively stable in a pH of 6 but the optimum pH would appear to be about 4,5.

Herbicides are generally less sensitive to a low pH and do not normally require the addition of a buffer. Note that a buffer is not an acidifier.

Drift control agent
Such an agent controls the size of spray droplets by various means, one of which is by the reduction of evaporation. These products are particularly valuable for aerial application where low spray volumes with small droplets are used and droplet size must be maintained in order to reduce wastage and drift.

Factors that influence effective weed control

Successful weed control depends on the correct choice of method or product and the proper application thereof. There are many factors that influence the efficacy of chemical herbicides.

Factors that affect many soil-applied herbicides

- **Fineness of the seedbed:** A fine seedbed encourages even germination and even distribution of the herbicide.
- **Soil moisture or rainfall:** Soil-applied herbicides usually need a certain level of moisture or even need to be washed into the soil by rain or irrigation before they can start to work.
- **Clay percentage:** Most soil-applied herbicides are chemically adsorbed by clay particles, thereby rendering them inactive. In general, the higher the clay content, the more herbicide must be applied to compensate for this loss.
- **Humus or organic matter content of the soil:** The organic fraction of soil acts in the same way as clay but is an even stronger adsorber of chemical ions.
- **Soil pH:** This can affect the rate of breakdown and thus the residual effect of some chemicals. Chemicals hydrolise more readily at a higher pH. A pH of 4 to 5 is considered optimum.
- **Timing in relation to weed germination:** Weeds that have already germinated may not be controlled, especially by pre-emergence herbicides.
- **Depth of germination:** A weed germinating very close to the surface may escape the herbicide if the chemical has been washed into the soil. Conversely, a weed germinating at depth may escape if uptake of that particular product is mainly by the roots and the chemical is not washed in.
- **Application method:** The type and accuracy of the equipment used is of the utmost importance.

Factors that affect foliar-applied herbicides

- **Development stage of the weed:** Seedlings are very sensitive to these chemicals, especially the contact type herbicides. Systemic herbicides require a large leaf area and active plant growth for efficient translocation.
- **Vitality of the plant:** Plants under stress cannot efficiently absorb or translocate a herbicide.
- **Climate:** This affects plant vitality. Rain, for example, can wash a chemical off before it has been taken up.
- **Canopy:** Shorter plants are shaded and protected by taller plants.
- **Product mixtures:** Sometimes wetting

agents must be added to foliar-applied herbicides, thereby enhancing the adherence of spray droplets to the leaves, especially waxy ones. Sometimes, if a mixture of products is used, one of the products may interfere with the action of another. Conversely, some chemicals have a synergistic effect on others, with the result that a mixture can be more active than the sum of the individual components.

- **Application method and degree of wetting:** Suitable equipment and sufficient water if necessary, should always be used for required coverage.

Pre-emergent application. Note the poor spray distribution – this could lead to poor control or even crop damage.

Aerial application – accuracy is essential.

HERBICIDE RESISTANCE

Herbicide-resistant weeds

Whereas resistance of pests and diseases to agricultural chemicals has been known for nearly 100 years, resistance of weeds to herbicides was only documented for the first time in the late 1960s. In 1968 *Senecio vulgaris* in parts of North America was found to be able to tolerate atrazine. Although the list of resistant weed species has since grown to well over 200, it is still very short compared to that of the resistant insects and fungi. The 1998 International Survey of Herbicide Resistant Weeds recorded 216 herbicide-resistant weed biotypes in 45 countries. By 2009 this number has grown to 327.

The wider a particular chemical is used, the greater the selection pressure that is applied. It is no surprise therefore that the first confirmed case of herbicide resistance in 1968 involved atrazine, which was, and still is, one of the most widely used herbicides in the world. For many years thereafter most of the reported cases of resistance involved atrazine and other triazines. By 1983, 68% of all cases worldwide involved triazines, 13% involved bipyridiliums, 12% synthetic auxins and 8% were others. In recent times, the number of different herbicides involved has grown and resistance to the triazines, as a group, now only represents about 28% of all cases worldwide to date. It is estimated, however, that there are over 3 million hectares infested with triazine-resistant weeds in the world. It is also clear that weeds evolve resistance to herbicides at different rates, and that certain chemical modes of action seem easier for weeds to bypass than others. By 2000, ALS inhibitors had become the herbicide group with the largest number of resistant biotypes. ALS inhibitors include the sulphonyl ureas, the imidazolines, the triazolopyrimidines, pyrimidinyl oxybenzolates, and the sulphonylamino carbonyl triazolines, a pathway found only in plants, making them some of the safest herbicides available. Their popularity, however, will also expose them to a higher risk of the development of resistance and this is clearly the case.

In South Africa, the first case of a resistant weed was documented in 1985. Wild oats were found to have become tolerant to the 'dims' and 'fops' in the Western Cape. It was followed by *Lolium* sp. in 1996. Since then, in 1996, *Amaranthus* in the Free State was found to be resistant to atrazine and simazine and, in 1998, *Phalaris* was shown to be resistant to the ACCases. Also in 1998, *Raphanus raphanistrum* had become resistant to the sulphonyl ureas. All these cases except one were from the wheat lands of the Western Cape. More recently suspected glyphosate resistance has appeared in vineyards and orchards and populations of *Lolium* sp., *Plantago lanceolata*, *Stellaria media* and *Conyza bonariensis* are suspected of being resistant to this nonselective product. Also, suspected resistant *Raphanus raphanistrum*, *Stellaria media* and *Bromus diandrus* have been found in annual croplands. It was in 2001 that glyphosate-tolerant *Lolium* sp. was first discovered in a vineyard at Tulbagh, a phenomenon that appears to be increasing dramatically since then. What is most perturbing, is that nowhere else in the world has any weed biotype developed resistance to glyphosate, paraquat and the ACCase inhibitors simultaneously.

The widespread use of 'dims', 'fops', 'ACCase' and ALS inhibitors in the Western and southern Cape is the reason for this concentration of cases. Nearly 80% of all herbicides used in this area belong to these groups, which have been shown to be very susceptible to the build up of resistance.

Many weeds that become resistant to one group of chemicals can usually still be controlled by products from another group. Triazine resistance is therefore not too much of a problem as atrazine is losing favour to alternative, more environmentally favourable products anyway. These other products have different modes of action. Resistance to the newer herbicides, especially grass killers, presents more of a problem, such as in the Western Cape. All the systemic products used to control grasses in cereal crops are ACCase and ALS inhibitors. This means that when resistance eventually occurs, there

are no alternative products easily available. If the occurrence of such resistance should become severe and widespread, then it could become almost impossible to grow the crops involved. Resistance cannot be reversed, only its spread can be contained. Where these herbicides are extensively used, as in the Western Cape, farmers must make every effort to reduce the risks of resistance. Where possible, they should:
- Rotate the herbicides used, ensuring that they alternate products with different modes of action.
- Introduce crop rotation to increase the herbicide options that are available.
- Employ other methods of weed control such as hand weeding, burning, tillage, fallow or competitive crops/cultivars, etc.
- Modify the harvesting operations to divert and isolate the weed seeds.
- Isolate and destroy populations of weeds that are known or suspected to have developed resistance.
- Scout lands regularly, identify and monitor those weeds that are present.
- Ensure that suspected resistance is not due to inefficient weed control.
- Ensure that tillage and harvest equipment is cleaned before moving from one land to another in order to prevent the transfer of resistant populations to new areas.
- Any suspected resistant population should be reported immediately to the relevant authorities.

The natural shift in susceptibility of a weed species to a group of herbicides is very different to the latest gene splicing technology that gives us herbicide-resistant crops. Although the gene might originally have been found in a wild plant, it is deliberately inserted into the genetic material of a crop. Resistance is then bestowed onto the crop plant, which can then be sprayed with a chemical that would normally have killed it. This means that a nonselective herbicide, such as glyphosate, can be sprayed safely over the top of a crop such as maize, killing the weeds but not the crop. In theory this will reduce the cost and increase the efficiency of the grower's weed control programme.

There are several chemicals and many crops that are being developed in this way. However, the whole issue of genetically manipulated organisms (GMOs) is a complicated and controversial one that lies beyond the scope of this book. As far as weed control is concerned, two issues that are immediately raised, are:
- Will the gene find its way into normally susceptible weeds, thereby making their control difficult if not impossible?
- Will these resistant crops become serious problem plants themselves?

These are just two examples of the issues involved in the science of genetic modification, all of which must be thoroughly researched before the technology can continue and the world can eventually benefit on the scale that is envisaged.

Amaranthus sp., that was found to be resistant to atrazine and simazine in the Free State.

INDIGENOUS PLANTS

Although the existence of a herbicide registration is one of the criteria used to determine if a species warrants inclusion in this book as a 'problem plant', there are quite a few plants that do have registrations but are still not included. This could be because:
- Although some indigenous poisonous plants are included, most of them are subjects of other specialist publications and are avoided here.
- They are indigenous trees, bushes or grasses implicated in bush encroachment and veld degradation, which is covered in detail in the Chapter starting on p15.
- They are simply minor weeds and available space in this book is limited, so they are not covered in detail.
- The plant just happened to occur in a trial site, was formally identified and was included in the trial data used to apply for the registration. There is little further evidence of the plant being weedy.

Indigenous plants that appear on at least one herbicide label, but are not considered of sufficient importance to justify being covered in detail in this book include:

Aristea africana
Maagbossie

Triumfetta sp.
Burweed
Klitse

Cyanella lutea
Five fingers
Lady's hand
Vyfvingerblaar

Others are:

Cineraria lobata	
Cyperus distans	slender Cyperus * skraal Cyperus
Felicia muricata	wild aster * bloubossie * skaapbossie * karoo-aster
Ficinia filiformis	star grass * stergras * letjotjo (S)
Ficinia indica	star grass * swartkop-biesie * ystervarkgras
Fimbristylis dichotoma	biesie
Fimbristylis hispidula	slender sedge * fynbiesie
Galenia secunda	vanwyksbossie
Isolepis antarctica	sedge * biesie
Lasiochloa longifolia	haregrass * haasgras * witbiesie
Leucas martinicensis	tumbleweed * bobbin weed * tolbossie * waaibossie
Oncosiphon suffruticosum	(=*Pentzia suffruticosum*), bitter karoo bush * wurmbossie
Pentaschistis thunbergii	dune grass * duingras * haasgras
Pentzia globosa	bitter karoo bush * bitterkaroobossie * goedkaroo
Pentzia grandiflora	matricaria * stinkweed * stinkkruid
Triumfetta pilosa	burs * klitse

THE THREAT OF ORNAMENTAL AND CROP PLANTS

Ornamentals

Everyone loves a beautiful flower! Wherever man has gone, he has taken plants with him that remind him of home or for other ornamental purposes. There is, for example, a surviving letter from Darius the Great of Persia, who reigned from 521 to 486 B.C., commending an estate manager for deliberately introducing 'trees and plants from beyond the Euphrates'. Early colonists were encouraged to plant certain species as a 'footprint' and some were deliberately planted in a wild and desolate landscape, to indicate fresh water or a safe haven for those who followed. Anyone who has an interest in or appreciation of plants loves to cultivate and encourage beautiful, interesting and exotic species. It has been going on for centuries and ornamental intention is just one reason of many why plants are deliberately brought into a country.

However, it is only in relatively recent times that many of these imported ornamental plants have been accused of becoming part of the total threat that alien invasive plants are now known to pose. In many parts of the world, more than 50% of the serious pest plants are known to have come from gardening, landscaping or from some other amenity purpose. In Australia, 70% of 1 765 listed environmental weeds are invasive garden species (World Wildlife Fund – Australia 2005). Some 52% of all naturalised alien species in Europe are ornamental or horticultural introductions (National Environment Research Council 2008). Invasive garden plants also comprise 56% of the 36 land and aquatic plants on the list of the *World's Worst Invasive Species*. Furthermore it remains an unavoidable fact that the horticultural trade generally used to promote exotic species at the expense of indigenous ones. This may be changing, but it has been a fact of life for many years.

The problem has now been exposed and steps to correct it have been put in place. In recent years, international restrictions on the movement of plants and regulations controlling the trade of seeds and other propagules have been introduced worldwide. But often there are gaps and there are always individuals who, either through ignorance or deliberately, will avoid the regulations. How many holidaymakers, for example, have brought a plant home with them as a memento of their holiday? And this does not necessarily refer to an overseas holiday.

Not only do gardeners and collectors favour colourful and dramatic plants, but with selective breeding it is possible to produce hybrids or cultivars that are even more colourful and dramatic. These plants have formed the basis of commercial horticulture over the years and it is only fairly recently that a shift has occurred which now favours indigenous and local species for the garden. For example, gardeners wishing to attract butterflies and birds appreciate the role of indigenous species and will concentrate almost entirely on 'native' or indigenous gardening for this purpose.

Not all exotic plants pose a threat and there remains a very large number of alien species that for various reasons are still popular for planting in gardens, on verandas and in pots and have shown no tendency at all to propagate voluntarily. Many hybrids are completely sterile or produce very weak and uncompetitive offspring and pose no threat. Sometimes the plants are so fragile and sensitive that they can only survive with the care and attention of a dedicated gardener.

Sometimes the reverse is true. Some of our major alien weeds have a range of hybrids in their established populations. For example, nearly 50 different variants of *Lantana camara* have been identified in South Africa alone (650 worldwide), but they are all considered to be the same species. If it is a garden plant, this variety adds to the plant's attractiveness, but if it is a problem weed like lantana, then it greatly exacerbates the problem. This is because it can completely nullify all efforts at biological control and the continual variation may enhance the ability of lantana as a species to adapt to new environments.

Are sterile hybrids really sterile? Trailing, weeping or creeping lantana is marketed worldwide as an attractive and sterile hybrid or subspecies of Lantana and is often called *Lantana montevidensis*. But does it produce viable pollen? Can it cross-pollinate with *Lantana camara*? This photograph shows a berry on a 'sterile', yellow ornamental *L. camara* hybrid, which is growing right next to a wild and invasive *L. camara*.

Lantana camara hybrid

Where did the pollen come from that produced this berry? Will this berry be viable? Could the potential offspring be yet another highly invasive cultivar, perhaps exhibiting hybrid vigour with the ability to become an even worse invader than its parent? The problem is, we do not always know. This is just one example and it is clear the subject needs a lot more research. Meanwhile, it is being proposed that all seed-producing hybrids of *Lantana* that are non-indigenous should be totally banned.

There are many other species of exotic ornamental plants that are under suspicion for various reasons and need to be watched. Some very popular and common garden plants can already be found regenerating in the wild, but for various reasons may not have developed significant populations. Some may be recent introductions or recently developed hybrids and their invasiveness or sterility has not yet been determined. Sometimes it is possible that cross-pollination with another alien or even indigenous species has created a hybrid that will exhibit hybrid vigour and will have the ability to tolerate a new environment and allow it to flourish and threaten other plants in its habitat.

Crop plants

In exactly the same way, many alien weeds were originally introduced and are still grown for food, fodder or fibre (e.g. *Prosopis glandulosa* or the guava, *Psidium guajava*). There are also a lot of hybrids of these plants and some can even be ornamental; but as with the ornamentals, many are falling under suspicion as they can be found in places where they were not planted and their invasive status needs to be researched.

Many such ornamental and crop plants fell into Category 3 of the 1983 CARA Regulations, but are now being proposed as Category 1b on the new CARA and NEMBA lists. Some of these plants have not been formally classified, but will need investigating.

Nevertheless, exotic ornamentals and crop plants will always remain with us but it is hoped that those that have shown a strong propensity to escape and become invasive will be eliminated. These are just some ornamental and crop plants that are falling under suspicion and have not been covered elsewhere in this book. There are many more and we all need to stay alert.

Colocasia esculenta
Elephant's ear * madumbe * olifantsoor
Tropical Asia

Bougainvillea sp.
Bougainvillea
South America

Cyathea cooperi
Australian tree fern * Australiese boomvaring
Australia

Coffea arabica
Coffee tree * koffieboom
Subtropical Africa

Crataegus monogyna
English hawthorn * meidoring
Europe

Eugenia uniflora
Surinam cherry * pitanga
Tropical America

Pterocarya fraxinifolia, Caucasian wingnut * Kaukasiese vleuelneut, Eurasia

Mangifera indica, Mango * veselperske
India

Quercus robur, English oak * steeleik
Europe

Ficus carica, Edible fig * makvy, Southwest Asia

Ulmus parviflora, Chinese elm * Chinese iep, East Asia

APPENDIX: LIST OF DECLARED WEEDS AND INVADER PLANTS

Regulation 15 of the Conservation of Agricultural Resources Act, 1983 (Act 43 of 1983)

Botanical name	Common name	Type	CARA	PROPOSED CARA/NEMBA
Acacia baileyana	Bailey's wattle / Bailey-se-wattel	Invader	3	3
Acacia cyclops	Redeye / Rooikrans	Invader	2	2
Acacia dealbata	Silver wattle / Silwerwattel	Weed	1 in WC 2 in rest of SA	1b 1b
Acacia decurrens	Green wattle / Groenwattel	Invader	2	2 in KZN, MP, EC
Acacia elata	Pepper tree wattle / Peperboomwattel	Invader	3	1b
Acacia implexa	Screw-pod wattle	Weed	1	1a
Acacia longifolia	Long-leaved wattle / Langblaarwattel	Weed	1	1b
Acacia mearnsii	Black wattle / Swartwattel	Invader	2	2
Acacia melanoxylon	Australian blackwood / Australiese swarthout	Invader	2	2 in WC, EC 1b in rest of SA
Acacia paradoxa	Kangaroo wattle	Weed	1	1a
Acacia pendula	Weeping myall		X3	3
Acacia podalyriifolia	Pearl acacia / Vaalmimosa	Invader	3	1b
Acacia pycnantha	Golden wattle / Gouewattel	Weed	1	1b
Acacia saligna	Port Jackson willow / Port Jackson	Invader	2	1b
Acacia stricta	Hop wattle			1a
Acer buergerianum	Chinese maple / Chinese ahorn		X3	3
Acer negundo	Box elder / Essenblaarhorn		X3	3
Achyranthes aspera	Burweed / Grootklits	Weed	1	
Agave americana	Spreading century plant / Garingboom		X2	1b in WC
Agave sisalana	Sisal hemp / Garingboom	Invader	2	2
Ageratina adenophora	Crofton weed	Weed	1	1b
Ageratina riparia	Mistflower / Misblom	Weed	1	1b
Ageratum conyzoides	Invading ageratum / Indringerageratum	Weed	1	1b
Ageratum houstonianum excluding cultivars	Mexican ageratum / Meksikaanse ageratum	Weed	1	1b
Agrimonia procera	Scented agrimony / Geelklits		X3	1b
Ailanthus altissima	Tree of heaven / Hemelboom	Invader	3	1b
Albizia julibrissin	Silk tree / Syboom		X3	

Botanical name	Common name	Type	CARA	PROPOSED CARA/NEMBA
Albizia lebbeck	Lebbeck tree / Lebbeckboom	Weed	1	1b
Albizia procera	False lebbeck / Basterlebbeck	Weed	1	1b
Alhagi maurorum	Camel thorn bush / Kameeldoringbos	Weed	1	1b
Alisma plantago-aquatica	Mud plantain / Wateralisma			1b
Alnus glutinosa	Black alder / Swartels		X3	3
Alpinia zerumbet	Shell ginger / Skulpgemmer			3
Ammophila arenaria	Marram grass		X2	2
Anredera cordifolia	Madeira vine / Madeira-ranker	Weed	1	1b
Antigonon leptopus	Coral creeper / Koraalklimop			1b
Araujia sericifera	Moth catcher / Motvanger	Weed	1	1b
Ardisia crenata	Coralberry tree / Koraalbessieboom	Weed	1 in NP, KZN, MP only	1b
Ardisia elliptica	Shoebutton ardisia			1b
Argemone mexicana	Yellow-flowered Mexican poppy / Geelblom–bloudissel	Weed	1	1b
Argemone ochroleuca subsp. *ochroleuca*	White-flowered Mexican poppy / Witblom–bloudissel	Weed	1	1b
Aristolochia elegans	Dutchman's pipe / Sisblom			1b
Arundo donax	Spanish reed / Spaanse riet	Weed	1	1b
Atriplex inflata	Sponge-fruit saltbush / Blasiesoutbos			1b
Atriplex lindleyi subsp. *inflata*	Sponge-fruit saltbush / Blasiesoutbos	Invader	3	
Atriplex nummularia subsp. *nummularia*	Old man saltbush / Oumansoutbos	Invader	2	2
Azolla filiculoides	Azolla / Rooiwatervaring	Weed	1	1b
Azolla pinnata	Mosquito fern			1b
Bartlettina sordida	Bartlettina			1b
Bauhinia purpurea	Butterfly orchid tree / Skoenlapperorgideëboom	Invader	3	1b in KZN, LP, MP, EC. 3 in rest of SA
Bauhinia variegata	Orchid tree / Orgideëboom	Invader	3	1b in KZN, LP, MP, EC. 3 in rest of SA
Berberis thunbergii	Japanese barberry / Japanese berberis		X3	3
Billardiera heterophylla	Bluebell creeper			1a
Brachychiton populneus	Bottle tree / Koerajong		X3	

Botanical name	Common name	Type	CARA	PROPOSED CARA/NEMBA
Bryophyllum delagoense	Chandelier plant / Kandelaarplant	Weed	1	1b
Bryophyllum pinnatum	Cathedral bells			1b
Bryophyllum proliferum	Green mother of millions			1b
Buddleja davidii	Chinese sagewood / Chinese saliehout			3
Buddleja madagascariensis	Madascar sagewood / Madagaskar saliehout			3
Cabomba caroliniana	Carolina fanwort / Cabomba			1a
Caesalpinia decapetala	Mauritius thorn / Kraaldoring	Weed	1	1b
Caesalpinia gilliesii	Bird-of-paradise flower / Paradysvoëlblom			1b
Callisia repens	Creeping inch plant			1b
Callistemon citrinus	Lemon bottlebrush / Lemoenperdestert			3
Callistemon rigidus	Stiff-leaved bottlebrush / Perdestert		X3	1b in WC, EC. 3 in rest of SA
Callistemon citrinus	Lemon bottlebrush			1b in KZN, MP, LP, EC. 3 in rest of SA
Calotropis procera	Giant milkweed / Calotropis			1b
Campuloclinium macrocephalum	Pompom weed / Pompombossie	Weed	1	1b in GP, NW, LP, MP. 1a in rest of SA
Canna indica excluding hybrid cultivars	Indian shot / Indiese kanna	Weed	1	1b
Cardiospermum grandiflorum	Balloon vine / Blaasklimop	Weed	1	1b
Cardiospermum halicacabum	Lesser balloon vine / Blaasklimop		X3	3
Casuarina cunninghamiana	Beefwood / Kasuarisboom	Invader	2 not for use in dune stabilisation	2 within 100 m of natural ecosystem
Casuarina equisetifolia	Horsetail tree / Perdestertboom	Invader	2 not for use in dune stabilisation	2
Catharanthus roseus	Madagscar periwinkle / Begraafplaasblom			3
Celtis australis	Nettle tree / Netelboom		X3	3

Botanical name	Common name	Type	CARA	PROPOSED CARA/NEMBA
Celtis occidentalis	Common hackberry / Valswit-stinkhout		X3	3
Celtis sinensis	Chinese nettle tree / Chinese netelboom		X3	
Cereus jamacara	Queen of the Night / Nagblom	Weed	1	1b
Cestrum aurantiacum	Yellow or Orange cestrum / Oranjesestrum	Weed	1	1b
Cestrum elegans	Crimson cestrum / Karmosyn-sestrum	Weed	1	1b
Cestrum laevigatum	Inkberry / Inkbessie	Weed	1	1b
Cestrum parqui	Chilean cestrum / Chileense inkbessie	Weed	1	1b
Cestrum spp.	Cestrum			1b
Chondrilla juncea	Skeleton weed			1a
Chromolaena odorata	Triffid weed / Chromolaena	Weed	1	1b in KZN, MP, LP, EC. 1a in rest of SA
Cichorium intybus	Chicory / Sigorei			2
Cinnamomum camphora	Camphor tree / Kanferboom	Weed	1 in NP, KZN, MP only	1b in KZN, MP, EC, Southern Cape
Cirsium japonicum	Japanese thistle			1b
Cirsium vulgare	Scotch thistle / Skotse dissel	Weed	1	1b
Coffea arabica	Coffee tree / Koffieboom		X2	3
Convolvulus arvensis	Field bindweed / Akkerwinde	Weed	1	1b
Coreopsis lanceolata	Tickseed / Coreopsis		X3	3
Cortaderia jubata	Pampas grass / Pampasgras	Weed	1	1b in GP. 1a in rest of SA
Cortaderia selloana excluding sterile cultivars	Pampas grass / Pampasgras	Weed	1	1b in WC
Cotoneaster franchetii	Cotoneaster / Dwergmispel	Invader	3	1b
Cotoneaster glaucophyllus	Late cotoneaster / Bloudwerg-mispel			1b
Cotoneaster pannosus	Silver-leaf cotoneaster / Silwer-dwergmispel	Invader	3	1b
Cotoneaster salicifolius	Willow-leaved showberry			1b
Cotoneaster simonsii	Himalayan cotoneaster			1b
Crataegus pubescens	Mexican hawthorn / Meksi-kaanse meidoring		X3	
Crotalaria agatiflora	Canarybird bush / Voëltjiebos		X3	1a

Botanical name	Common name	Type	CARA	PROPOSED CARA/NEMBA
Cryptostegia grandiflora	Rubber vine / Rubberklimop			1a
Cuscuta campestris	Common dodder / Gewone dodder	Weed	1	1b
Cuscuta suaveolens	Lucerne dodder / Luserndodder	Weed	1	1b
Cynodon dactylon	Couch grass / Kweek		X2	
Cytisus monspessulanus	Montpellier broom / Montpellierbrem	Weed	1	
Cytisus scoparius	Scotch broom / Skotse brem	Weed	1	1a
Datura ferox	Large thorn apple / Grootstinkblaar	Weed	1	1b
Datura innoxia	Downy thorn apple / Harige stinkblaar	Weed	1	1b
Datura stramonium	Common thorn apple / Gewone stinkblaar	Weed	1	1b
Diplocyclos palmatus	Lollipop-climber			1a
Duchesnea indica	Wild strawberry / Wilde-aarbei			1b
Duranta erecta	Pigeon berry / Vergeet-my-nie-boom		X3	3
Echinodorus cordifolius	Creeping burhead			1b
Echinodorus tenellus	Amazon sword plant			1b
Echinopsis spachiana	Torch cactus / Orrelkaktus	Weed	1	1b
Echium plantagineum	Patterson's curse / Pers-echium	Weed	1	1b
Echium vulgare	Blue echium / Blou-echium	Weed	1	1b
Egeria densa	Dense water weed / Waterpes	Weed	1	1b
Eichhornia crassipes	Water hyacinth / Waterhiasint	Weed	1	1b
Elodea canadensis	Canadian water weed / Kanadese waterpes	Weed	1	1b
Equisetum hyemale	Rough horsetail			1a
Eriobotrya japonica	Loquat / Lukwart	Invader	3	3
Eucalyptus camaldulensis	Red river gum / Rooibloekom	Invader	2	1b/2
Eucalyptus cladocalyx	Sugar gum / Suikerbloekom	Invader	2	1b/2
Eucalyptus diversicolor	Karri / Karie	Invader	2	1b/2
Eucalyptus grandis	Saligna gum / Salignabloekom	Invader	2	1b/2
Eucalyptus lehmannii	Spider gum / Spinnekopbloekom	Weed	1 in WC 2 in rest of SA	1b/2
Eucalyptus paniculata	Grey ironbark / Grysysterbasbloekom	Invader	2	

Botanical name	Common name	Type	CARA	PROPOSED CARA/NEMBA
Eucalyptus sideroxylon	Black ironbark / Swartysterbasbloekom	Invader	2	
Eucalyptus tereticornis	Forest red gum / Bosrooibloekom			1b/2
Eugenia uniflora	Pitanga	Weed	1 in NP, KZN, MP 3 in rest of SA	1b
Euphorbia leucocephala	White poinsettia			1b
Fallopia sachalinensis	Giant knotweed			1a
Flaveria bidentis	Smelter's bush / Smelterbossie			1b
Foeniculum vulgare	Fennel / Vinkel			2 in WC
Fraxinus americana	American ash / Amerikaanse esseboom		X3	3
Fraxinus angustifolia	Algerian ash / Algeriese esseboom			3
Galium tricornutum	Tree-horned bedstraw			1b
Gaura coccinea	Scarlet gaura			3
Genista monspessulana	Montpellier broom / Montpellierbrem	Weed	1	1a
Gleditsia triacanthos excluding sterile cultivars	Honey locust / Amerikaanse driedoring	Invader	2	1b
Glyceria maxima	Reed meadow grass			2
Grevillea banksii	Australian crimson oak / Australiese rooi-eik			1b
Grevillea robusta	Australian silky oak / Australiese silwereik	Invader	3	1b
Hakea drupacea (=*H. suaveolens*)	Sweet hakea / Soethakea	Weed	1	1a
Hakea gibbosa	Rock hakea / Harige hakea	Weed	1	1b
Hakea salicifolia	Willow hakea / Wilgerhakea		X3	
Hakea sericea	Silky hakea / Syerige hakea	Weed	1	1b
Harrisia martinii	Moon cactus / Toukaktus	Weed	1	1b
Hedera helix subsp. *canariensis*	Canary ivy / Madeiraklimop		X3	3
Hedera helix subsp. *helix*	English ivy / Engelse hedera		X3	3
Hedychium coccineum	Red ginger lily / Rooigemmerlelie	Weed	1	1b
Hedychium coronarium	White ginger lily / Witgemmerlelie	Weed	1	1b

Botanical name	Common name	Type	CARA	PROPOSED CARA/NEMBA
Hedychium flavescens	Yellow ginger lily / Geelgemmerlelie	Weed	1	1b
Hedychium gardnerianum	Kahili ginger lily / Kahiligemmerlelie	Weed	1	1b
Homalanthus populifolius	Bleeding heart tree / Gebrokenhartjieboom			1b
Houttuynia cordata	Chameleon plant			3
Hydrilla verticillata	Hydrilla			1a
Hydrocleys nymphoides	Water poppy			1a
Hylocereus undatus	Night-blooming cereus			1b
Hypericum androsaemum	Tutsan			1b
Hypericum perforatum	St. John's wort / Johanneskruid	Invader	2 Controlled cultivation	2
Ipomoea alba	Moonflower / Maanblom	Weed	1 in NP, KZN, MP 3 in rest of SA	1b
Ipomoea carnea	Morning glory bush			1b
Ipomoea indica	Morning glory / Purperwinde	Weed	1 in NP, KZN, MP 3 in rest of SA	1b
Ipomoea purpurea	Morning glory / Purperwinde	Invader	3	1b
Iris pseudacorus	Yellow flag / Geel iris			1a
Jacaranda mimosifolia excluding sterile cultivar 'Alba'	Jacaranda / Jakaranda	Invader	3	1b in KZN, MP, LP. 2 in GP
Jatropha curcas	Physic nut / Purgeerboontjie			2
Jatropha gossypiifolia	Cotton-leaf physic nut			1b
Juniperus virginiana	Red cedar / Rooiseder		X3	
Kunzea ericoides	White tea tree / Burgan			1a
Lantana spp. All exotic, seed-producing species or seed-producing hybrids	Lantana	Weed	1	1b
Lavatera arborea	Tree mallow / Mak-kiesieblaar			1b
Lepidium draba	Pepper-cress / Peperbossie	Weed	1	1b
Leptospermum laevigatum	Australian myrtle / Australiese mirt	Weed	1	1b
Leptospermum scoparium	Manuka myrtle / Mankamirt		X3	
Leucaena leucocephala	Leucaena / Reusewattel	Weed	1 in WC 2 in rest of SA	1a in WC 2 in rest of SA

Botanical name	Common name	Type	CARA	PROPOSED CARA/NEMBA
Ligustrum japonicum	Japanese wax-leaved privet / Japanese liguster	Invader	3	1b in KZN, MP, LP, EC, WC, GP, NW. 3 in FS, NC
Ligustrum lucidum	Chinese wax-leaved privet / Chinese liguster	Invader	3 Only for us as rootstock if authorised by the Executive Official in terms of regulation 15C(5)	1b in KZN, MP, LP, EC, WC, GP, NW. 3 in FS, NC
Ligustrum ovalifolium	Californian privet / Kaliforniese liguster	Invader	3	1b in KZN, MP, LP, EC, WC, GP, NW. 3 in FS, NC
Ligustrum sinense	Chinese privet / Chinese liguster	Invader	3	1b in KZN, MP, LP, EC, WC, GP, NW. 3 in FS, NC
Ligustrum vulgare	Common privet / Gewone liguster	Invader	3	1b in KZN, MP, LP, EC, WC, GP, NW. 3 in FS, NC
Lilium formosanum	St Joseph's lily / Sintjosefslelie	Invader	3	1b
Limonium sinuatum	Statice / Papierblom			1b in WC, NC
Linaria dalmatica	Dalmatian toadflax			1b
Linaria vulgaris	Common toadflax			1b
Litsea glutinosa	Indian laurel / Indiese lourier	Weed	1	1b
Lolium multiflorum	Italian ryegrass / Italiaanse raaigras		X2	
Lolium perenne	Perennial ryegrass / Meerjarige raaigras		X2	
Lonicera japonica	Japanese honeysuckle / Japanse kanferfoelie		X3	3
Ludwigia peruviana	Water primrose			1a
Lythrum hyssopifolia	Hyssop loosestrife			1b
Lythrum salicaria	Purple loosestrife	Weed	1	1a
Macfadyena unguis-cati	Cat's claw creeper / Katteklouranker	Weed	1	1b
Malva verticillata	Mallow / Kiesieblaar			1b
Malvastrum coromandelianum	Prickly malvastrum			1b
Melaleuca hypericifolia	Red-flowering tea tree			1a

Botanical name	Common name	Type	CARA	PROPOSED CARA/NEMBA
Melia azedarach	Syringa / Maksering	Invader	3	1b in KZN, MP, LP, EC, NW, GP. 3 in rest of SA
Metrosideros excelsa	New Zealand Christmas tree / Nieu-Seelandse perdestert	Invader	3	1a in Overstrand District
Mimosa pigra	Giant sensitive plant / Raak-my-nie	Invader	3	1b
Mirabilis jalapa	Four o'clock / Vieruurtjie		X3	1b
Montanoa hibiscifolia	Tree daisy / Montanoa	Weed	1	1b
Morus alba excluding cultivar 'Pendula'	White mulberry / Witmoerbei	Invader	3 Only for us as rootstock if authorised by the Executive Official in terms of regulation 15C(5)	2
Morus nigra	Black mulberry / Swartmoerbei		X3	
Murraya paniculata	Orange Jessamine / Oranjejasmyn			1b in KZN, MP, LP, EC
Myoporum insulare	Manatoka		X3	3
Myoporum laetum	New Zealand manitoka / Nieu-Seelandse manitoka		X3	3
Myoporum tenuifolium subsp. montanum	Manitoka	Invader	3	1b
Myriophyllum aquaticum	Parrot's feather / Waterduisendblaar	Weed	1	1b
Myriophyllum spicatum	Spiked water-milfoil	Weed	1	1b
Nassella tenuissima	White tussock / Witpolgras	Weed	1	1b
Nassella trichotoma	Nassella tussock / Nassella polgras	Weed	1	1b
Nasturtium officinale	Watercress / Bronkors	Invader	2	2
Nephrolepis cordifolia	Erect sword fern			1b in KZN, MP, LP, EC, WC. 3 in rest of SA
Nephrolepis exaltata excluding cultivars	Sword fern / Swaardvaring	Invader	3	1b in KZN, MP, LP, EC, WC. 3 in rest of SA
Nerium oleander excluding sterile, double-flowered cultivars	Oleander / Selonsroos	Weed	1	1b
Nicandra physalodes	Apple-of-Peru / Basterappelliefie			1b

Botanical name	Common name	Type	CARA	PROPOSED CARA/NEMBA
Nicotiana glauca	Wild tobacco / Wildetabak	Weed	1	1b
Nymphaea mexicana	Yellow water lilies / Geelwaterlelies			1b
Nymphoides peltata	Gringed water lily			1a
Oenothera indecora	Evening primrose / Nagblom			
Oenothera rosea	Pink evening primrose / Pienkaandblom		X3	
Oenothera stricta	Sweet sundrop / Soetnagblom		X3	
Oenothera tetraptera	White evening primrose / Witnagblom		X3	
Opuntia aurantiaca	Jointed cactus / Litjieskaktus	Weed	1	1b
Opuntia exaltata	Long spine cactus / Langdoringkaktus	Weed	1	1b
Opuntia ficus-indica excluding all spineless cactus pear cultivars and selections	Sweet prickly pear / Boereturksvy	Weed	1	1b
Opuntia fulgida	Rosea cactus / Roseakaktus	Weed	1	1b
Opuntia humifusa	Large-flowered prickly pear	Weed	1	1b
Opuntia imbricata	Imbricate cactus / Kabelturksvy	Weed	1	1b
Opuntia lindheimeri	Small round-leaved prickly pear / Klein rondeblaarturksvy		1	1b
Opuntia microdasys	Yellow bunny ears			1b
Opuntia monacantha	Cochineal prickly pear / Suurturksvy	Weed	1	1b
Opuntia robusta	Blue-leaf cactus / Robusta turksvy			2
Opuntia spinulifera	Large round-leaved prickly pear / Grootrondeblaarturksvy	Weed	1	1b
Opuntia stricta	Pest pear of Australia / Suurturksvy	Weed	1	1b
Orobanche minor	Lesser broomrape / Klawerbesemraap	Weed	1	1b
Orobanche ramosa	Blue broomrape / Blouduiwel			1b
Paraserianthes lophantha	Stink bean / Stinkboon	Weed	1	1b
Parkinsonia aculeata	Jerusalem thorn/Mexikaanse groenhaarboom			1b
Parthenium hysterophorus	Parthenium	Weed	1	1b
Paspalum hysterophorus	Tussock paspalum			1a

Botanical name	Common name	Type	CARA	PROPOSED CARA/NEMBA
Passiflora caerulea	Blue passion flower / Siergrenadella	Weed	1	1b
Passiflora edulis	Grenadilla		X2	2 in KZN, MP, LP, EC
Passiflora mollissima	Bananadilla / Piesangdilla	Weed	1	1b
Passiflora suberosa	Indigo berry	Weed	1	1b
Passiflora subpeltata	Granadina	Weed	1	1b
Paulownia tomentosa	Empress tree / Keiserinboom			1a
Pennisetum clandestinum	Kikuyu / Kikoejoegras		X2	
Pennisetum purpureum	Napier grass / Olifantsgras		X2	1b
Pennisetum setaceum excluding sterile cultivar 'Rubrum'	Fountain grass / Pronkgras	Weed	1	1b
Pennisetum villosum	Feathertop / Veergras	Weed	1	1b
Pereskia aculeata	Barbados gooseberry / Pereskia	Weed	1	1b
Persicaria capitata	Knotweed / Knoopkruid			1b
Phytolacca americana	Poke weed / Inkbos			1b
Phytolacca dioica	Belhambra / Bobbejaandruifboom	Invader	3	3
Phytolacca octandra	Forest inkberry / Inkbessie		X1	1b
Pinus canariensis	Canary pine / Kanariese den	Invader	2	3
Pinus elliottii	Slash pine / Basden	Invader	2	2
Pinus halepensis	Aleppo pine / Aleppoden	Invader	2	2
Pinus patula	Patula pine / Treurden	Invader	2	2
Pinus pinaster	Cluster pine / Trosden	Invader	2	2
Pinus pinea	Stone pine / Sambreelden		3	
Pinus radiata	Radiata pine / Radiataden	Invader	2	2
Pinus roxburghii	Chir pine / Tjirden	Invader	2	2
Pinus taeda	Loblolly pine / Loblollyden	Invader	2	2
Pistia stratiotes	Water lettuce / Waterslaai	Weed	1	1b
Pittosporum crassifolium	Karo / Styweblaarkasuur		X3	3
Pittosporum undulatum	Australian cheesewood / Australiese kasuur	Weed	1	1b
Plectranthus comosus	Woolly plectranthus / 'Abessiniese' coleus	Invader	3	1b
Polypodium aureum	Rabbit's foot fern / Haaspootvaring		X3	
Pontederia cordata	Pickerel weed / Jongsnoekkruid	Invader	3	1b

Botanical name	Common name	Type	CARA	PROPOSED CARA/NEMBA
Populus alba	White poplar / Witpopulier	Invader	2	2
Populus x canescens	Grey poplar / Vaalpopulier	Invader	2	2
Populus deltoides	Match poplar / Vuurhoutjiepopulier		X3	
Populus nigra var. italica	Lombardy poplar / Italiaanse populier		X2	
Populus simonii	Simon poplar / Simon populier		X3	
Prosopis glandulosa and hybrids	Honey mesquite / Heuningprosopis	Invader	2	1b in NW, FS, EC, WC. 2 in NC
Prosopis velutina and hybrids	Velvet mesquite / Fluweelprosopis	Invader	2	1b in NW, FS, EC, WC. 2 in NC
Prunus cerasifera	Cherry plum / Kersiepruim		X3	
Prunus serotina	Black cherry / Swartkersie			1b
Psidium cattleianum	Strawberry guava / Aarbeikoejawel	Invader	3	1b
Psidium guajava and hybrids	Guava / Koejawel	Invader	2	2 in KZN, MP, LP, EC
Psidium guineense	Brazilian guava / Brasiliaanse koejawel	Invader	3	1b
Psidium x durbanensis	Durban guava / Durbanse koejawel	Weed	1	1b
Pueraria montana var. lobata	Kudzu vine / Kudzuranker	Weed	1	1a
Pyracantha angustifolia excluding cultivars	Yellow firethorn / Geelbranddoring	Invader	3	1b
Pyracantha coccinea	Red firethorn / Rooibranddoring	Invader		1b
Pyracantha crenatoserrata	Chinese firethorn			1b
Pyracantha crenulata	Himalayan firethorn / Rooivuurdoring	Invader	3	1b
Pyracantha koidzumii	Formosa firethorn	Invader		1b
Pyracantha rogersiana	Firethorn			1b
Rhus glabra	Scarlet sumac / Gladde sumak		X3	3
Rhus succedanea	Wax tree / Wasboom	Weed	1	
Ricinus communis	Castor-oil plant / Kasterolieboom	Invader	2	1b
Rivina humilis	Bloodberry / Bloedbessie	Weed	1	1a

Botanical name	Common name	Type	CARA	PROPOSED CARA/NEMBA
Robinia pseudoacacia	Black locust / Witakasia	Invader	2 Only for us as rootstock if authorised by the Executive Official in terms of regulation 15B(10)	1b
Rosa canina	Dog-rose / Hondsroos		X3	
Rosa rubiginosa (= *R. eglanteria*)	Eglantine / Wilderoos	Weed	1	1b
Rubus cuneifolius and hybrid *R. x proteus*	American bramble / Amerikaanse braam	Weed	1	1b
Rubus flagellaris	Bramble / Braam		X1	1b
Rubus fruticosus	European blackberry / Braam	Invader	2	2
Rubus niveus	Ceylon raspberry			1b
Rumex crispus	Curly dock / Weeblaar		X3	
Rumex usambarensis	East African dock / Oos-Afrikaanse tongblaar			1b
Salix babylonica	Weeping willow / Treurwilger	Invader	2	
Salix fragilis	Crack willow / Brittle willow	Invader	2	
Salsola kali	Tumbleweed / Rolbossie			1b
Salsola tragus	Russian tumbleweed / Russiese rolbossie			1b
Salvia tiliifolia	Lindenleaf sage			1b
Salvinia molesta and other species of the family *Salviniaceae*	Kariba weed / Watervaring	Weed	1	1b
Sambucus canadensis	Canadian elder / Kanadese vlier		X3	1b
Sambucus nigra	European elder / Europese vlier			1b
Sasa ramosa	Dwarf yellow-striped bamboo/ Dwerggeelstreepbamboes			3
Schefflera actinophylla	Australian cabbage tree / Australiese kiepersol		X3	1b in KZN, MP, LP, EC
Schefflera arboricola	Dwarf umbrella tree/Hawaiiese dwerg			1b in KZN, MP, LP, EC
Schefflera elegantissima	False aralia / Vals-aralia			1b in KZN, MP, LP, EC
Schinus molle	Peper tree / Peperboom		X3	
Schinus terebinthifolius	Brazilian pepper tree / Brasiliaanse peperboom	Weed	1 in KZN 3 in rest of SA	1b in KZN, MP, LP, EC. 3 in rest of SA

Botanical name	Common name	Type	CARA	PROPOSED CARA/NEMBA
Senna bicapsularis	Rambling cassia	Invader	3	1b
Senna didymobotrya	Peanut butter cassia / Grond-boontjiebotterkassia	Invader	3	1b in KZN, MP, LP, EC. 3 in rest of SA
Senna hirsuta	Woolly senna			1b
Senna occidentalis	Stinking weed			1b
Senna pendula var. glabrata		Invader	3	1b
Senna septemtrionalis			X3	1b
Sesbania punicea	Red sesbania / Rooi-sesbania	Weed	1	1b
Solanum betaceum	Tree tomato / Boomtamatie		X3	3 in KZN, MP, LP, EC
Solanum chrysotrichum	Giant devil's fig			1b
Solanum elaeagnifolium	Silver-leaf bitter apple / Satansbos	Weed	1	1b
Solanum mauritianum	Bugweed / Luisboom	Weed	1	1b
Solanum pseudocapsicum	Jerusalem cherry / Jerusalemkersie		X3	1b
Solanum seaforthianum	Potato creeper / Aartappelranker	Weed	1	1b
Solanum sisymbriifolium	Dense-thorned bitter apple / Doringtamatie	Weed	1	1b
Sorghum halepense	Johnson grass / Johnsongras	Invader	2	2
Spartium junceum	Spanish broom / Spaanse besem	Weed	1	1b in WC. 3 in rest of SA
Spathodea campanulata	African flame tree / Afrikaanse vlamboom			3 in KZN, MP, LP, EC.
Stachytarpheta spp.	Snakeweeds			3
Syngonium spp.	Arrow-head vines / Gansvoete			1b in KZN, MP, LP, EC
Syzygium cuminii	Jambolan	Invader	3	1b in KZN, MP, LP. 1a in rest of SA
Syzygium jambos	Rose apple / Jamboes	Invader	3	3
Syzygium paniculatum	Australian water pear / Australiese waterpeer		X3	
Tamarix aphylla	Athel tree / Woestyntamarisk		X3	1b
Tamarix chinensis	Chinese tamarisk / Chinese tamarisk	Weed	1 in NC, WC, EC 3 in rest of SA	1b
Tamarix gallica	French tamarisk / Franse tamarisk			1b
Tamarix ramosissima	Pink tamarisk / Perstamarisk	Weed	1 in NC, WC, EC 3 in rest of SA	1b

Botanical name	Common name	Type	CARA	PROPOSED CARA/NEMBA
Tecoma stans	Yellow bells / Geelklokkies	Weed	1	1b
Tephrocactus articulatus	Pine cone cactus			1a
Thelechitonia trilobata	Singapore daisy / Singapoermadeliefie	Weed	1 in KZN	
Thevetia peruviana	Yellow oleander / Geel-oleander	Weed	1	1b
Tipuana tipu	Tipu tree / Tipoeboom	Invader	3	3
Tithonia diversifolia	Mexican sunflower / Mexikaanse sonneblom	Weed	1	1b
Tithonia rotundifolia	Red sunflower / Rooisonneblom	Weed	1	1b
Toona ciliata	Toon tree / Toonboom	Invader	3	1b
Tradescantia fluminensis	Wandering Jew / Wandelende Jood			1b
Tradescantia zebrina	Wandering Jew / Wandelende Jood			1b
Triplaris americana	Ant tree / Triplaris	Weed	1	1b
Tropaeolum speciosum	Chilean flame creeper			3
Ulex europaeus	European gorse / Gaspeldoring	Weed	1	1b
Ulmus parviflora	Chinese elm / Chinese iep		X3	
Verbena bonariensis	Wild verbena / Blouwaterbossie			1b
Verbena brasiliensis	Brazilian verbena			1b
Vinca major	Greater periwinkle / Gewone opklim			1b
Vinca minor	Lesser periwinkle			1b
Vitex trifolia	Indian three-leaf vitax			1b
Wigandia urens	Wigandia			3
Xanthium spinosum	Spiny cocklebur / Boetebos	Weed	1	1b
Xanthium strumarium	Large cocklebur / Kankerroos	Weed	1	1b

- EC: Eastern Cape Province
- FS: Free State Province
- GP: Gauteng Province
- KZN: KwaZulu-Natal Province
- LP: Limpopo Province (formerly NP: Northern Province)
- MP: Mpumalanga Province
- NC: Northern Cape Province
- NW: North West Province
- WC: Western Cape Province

CARA: 1983/2001

Category 1 Plants are prohibited and must be controlled.
Category 2 Plants (commercially used plants) may be grown in demarcated areas, providing that there is a permit and that steps are taken to prevent their spread.
Category 3 Plants (ornamentally used plants) may no longer be planted; existing plants may remain, except within the flood line of watercourses and wetlands, as long as all reasonable steps are taken to prevent their spread.
Proposed Weeds and Invaders: X (and proposed category)

CARA/NEMBA (Proposed 2009)

Category 1a Plants are high-priority emerging species requiring compulsory control. All breeding, growing, moving and selling are banned.
Category 1b Plants are widespread invasive species controlled by a management programme.
Category 2 Plants are invasive species controlled by area. Can be grown under permit conditions in a demarcated area. All breeding growing, moving, selling banned without a permit.
Category 3 Plants are ornamental and other species that are permitted on a property but may no longer be planted or sold.

Please note:

The pictograms on the main species pages use only the lowest category number in the above list. Please refer to this list for other categories and provincial restrictions.

GRASSES AND SEDGES

CYPERACEAE

Cyperus esculentus
yellow nutsedge * geeluintjie

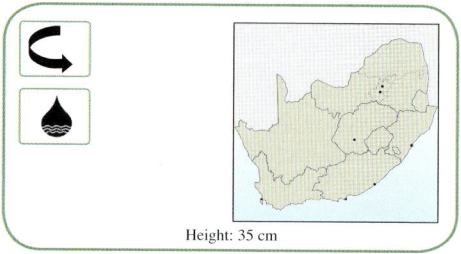

Height: 35 cm

Origin and description: This weed is of uncertain origin, but it is thought to be exotic. Although sometimes mistakenly referred to as a grass, *C. esculentus* is not a grass but a sedge. It reproduces from tubers or 'nuts' that are produced in vast numbers underground at the end of the rhizomes. One tuber can produce 1 900 plants and in turn nearly 7 000 tubers, covering an area of 2 m² in one year. The tubers are edible, with a faint nutty taste (unlike the bitter taste of *C. rotundus*), and are easily washed to new areas by storm water. If chewed, these nuts are said to be an efficient relief of indigestion and heartburn. The plant can also reproduce from shallow-germinating seeds, which will spread it further afield.

Impact: *Cyperus* species are serious and competitive weeds of many crops, not only because they are widespread, difficult to control and aggressive, but also because they can give off a toxin that can suppress the growth of other plants. This phenomenon is known as allelopathy. The rhizomes can cause serious damage to crops such as potatoes. They are able to grow right through a developing potato and 'nuts' are often found inside the potato tuber.

Other common names
earth almond * edible galingale * water-grass * yellow nut-grass * patrysuintjie * hoenderuintjie * chufa (Sh) * indawo (Z) * manakalali (S)

Cyperus esculentus

Control: Of the two *Cyperus* species, *C. esculentus* is easier to control as it is susceptible to the acetanelides, bendioxide and thiocarbamates like EPTC. However, specific conditions are required for these chemicals to be effective. Mechanical control is usually not very effective, as the tubers are not easily killed by desiccation.

Cyperus esculentus

The main differences between *C. esculentus* and *C. rotundus* (see overleaf)

	C. esculentus	*C. rotundus*
Leaf colour	pale	darker
Growth habit	usually erect	flatter
Height	30–40 cm	10–20 cm
Stem base	no hard lump	hard lump
Leaf shape	tapered	pointed
Nuts	round, end of rhizomes, nutty taste	irregular, in strings, bitter taste
Distribution	everywhere	warmer areas
Selective herbicides	acetanelides, bendioxide, EPTC, MSMA, halosulfuron	EPTC, MSMA, halosulfuron
Desiccation	less effective	more effective

CYPERACEAE

Cyperus rotundus subsp. *rotundus*
purple nutsedge * rooi-uintjie

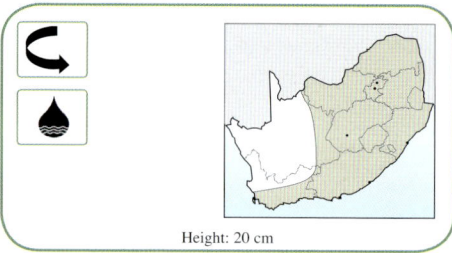

Height: 20 cm

Origin and description: *C. rotundus* subsp. *rotundus*, also of uncertain origin, is not as widespread as *C. esculentus*, being confined to the warmer, frost-free areas. It is a major weed worldwide, having been recorded as a weed in 52 crops in 92 countries. Like *C. esculentus*, it produces massive numbers of tubers or 'nuts' (up to 8 700 per m^2) but unlike *C. esculentus*, it can only produce a few viable seeds. Its growth habit is flatter than that of *C. esculentus* and the 'nuts' have a bitter taste. *C. rotundus* subsp. *tuberosa* is similar but considerably taller (75 cm) and the leaf tip is more pointed. The swellings, or basal tubers, on the underground rhizomes of these sedges will produce further rhizomes and roots, which will grow horizontally and create a chain of active rhizomes and tubers. They can then produce shoots and an interconnected mat of plants.

Impact: *Cyperus* species are serious and competitive weeds of many crops, not only because they are widespread, difficult to control and aggressive, but also because they give off a toxin that can suppress the growth of other plants. The rhizomes can cause serious damage to crops such as potatoes.

Other common names
coco grass * red nut-grass * eendjiesgras * knoppiesgras

Similar species
C. rotundus subsp. *tuberosa*: common north of Durban
C. natalensis: common in the sandy soils of Zululand

Control: *C. rotundus* subsp. *rotundus* is much more difficult to control as it resists more selective herbicides, except thiocarbamates like EPTC as well as MSMA. It is therefore very important to identify the weeds accurately before expensive control programmes are initiated. Furthermore, the taller growth habit of *C. rotundus* subsp. *tuberosa* means that it may escape any foliar herbicide applications intended for *Cyperus* control. Tubers of *C. rotundus* are more likely to be killed by desiccation and exposure than those of *C. esculentus*.

Cyperus rotundus

POACEAE

Agrostis montevidensis
fog grass

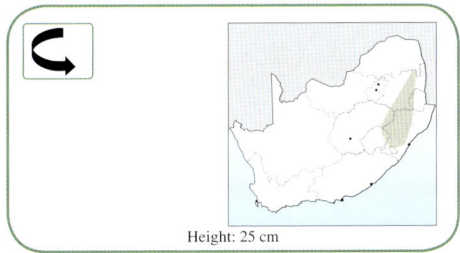

Height: 25 cm

Origin and description: Introduced into South Africa from South America, this annual grass is now common along the eastern escarpment. It is found on roadsides, in waste areas and in gardens. It has slender stems and a fine, feathery inflorescence. An infestation of this grass can be seen from a distance, especially in the early morning when there is heavy dew and the dew-covered, almost white inflorescences look like fog lying on the ground (hence the common name). There are no known Afrikaans names.

Impact: It is a particular nuisance in gardens, and can become dense and competitive on occasions, especially in disturbed areas. It replaces indigenous species.

Control: There are no specific recommendations for its control, but it can be easily removed by cultivation and responds to normal industrial herbicides that are used on roadsides.

Agrostis montevidensis

POACEAE

Arundo donax
giant reed * Spaanse riet

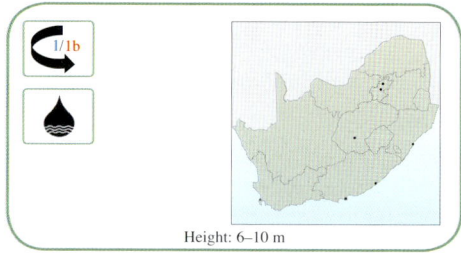

Height: 6–10 m

Other common names
bamboo reed * Spanish reed

Origin and description: This perennial plant from Eurasia is now a serious weed in the eastern parts of South Africa, especially the Mpumalanga lowveld and coastal region of southern KwaZulu-Natal. It was probably introduced for ornamental purposes. The giant reed must not be confused with the indigenous *Phragmites australis* (common reed / gewone fluitjiesriet) or *Phragmites mauritianus* (lowveld reed / laeveldfluitjiesriet). The latter two species of reed are much less robust and lack the large leaf lobes of *A. donux* (see page 98). The biggest difference apart from the relative heights, however, is the large and compact inflorescences of *A. donax*, which for some reason are not often produced above 1 000 m altitude.

The giant reed favours moist, but not wet places (as opposed to common reeds). It has strong underground rhizomes and can take root if the stems are cut and used as stakes.

Impact: *A. donax* can be beneficial in the right

Arundo donax

place, acting as an erosion control agent, filtering muddy floodwaters and being a haven for a wide range of wildlife. However, it competes vigorously with indigenous species, destroying biodiversity and forms dense stands in riverbeds and on riverbanks. These stands obstruct access to watercourses, increase siltation and will block rivers in time of flood. On roadsides these plants obstruct the vision of motorists.

Control: All reeds are difficult to control. Physical methods must include total removal of the rhizomes as the plants can re-grow from bits left in the soil, even when under 1–2 m of water. It is possible to control A. donax with chemicals. The plant should be cut down to ground level, stacked and preferably burnt – do not use cut stems as stakes! The lush regrowth must be sprayed with a systemic herbicide when the plant has reached a height of 1–2 m, usually about 6–8 weeks later. Good coverage with the recommended herbicide is essential for optimum results. Thorough follow-up treatments are required for effective long-term control. Observe any restrictions on the herbicide label with regard to subsequent water use.

Arundo donax

POACEAE

Avena fatua
common wild oats * gewone wildehawer

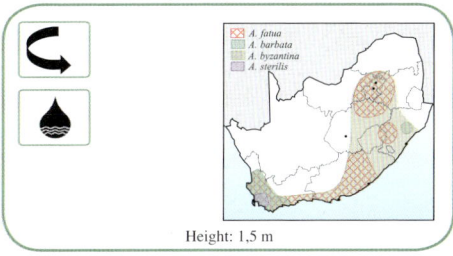

Height: 1,5 m

Origin and description: *A. fatua* is a major grass weed, occurring all over the world. It was introduced into South Africa from Europe or Asia. It is prevalent in most wheat-growing areas, especially where 'wheat on wheat' methods are followed, for example in the Free State and Western Cape. *A. fatua* can be identified at an early stage in wheat fields by its flatter growth habit, relatively long ligules and hairs at the base of the leaf. It has a distinctive seed and must not be confused with volunteer commercial oats (*A. sativa*). *A. fatua* is usually spread by contaminated wheat seed and

Similar species:
A. barbata (slender wild oats * wildebaardhawer) from the Mediterranean region
A. byzantina (red oats * rooihawer) from Eurasia
A. sterilis (tall wild oats * groot wildehawer) from Eurasia

by machines such as combine harvesters.

Impact: *A. fatua* becomes highly competitive and severely reduces the grain yield in infested fields. Wild oats is difficult to control chemically.

Control: *A. fatua* seed can lie dormant in the soil for about nine years and this makes the weeds extremely difficult to control by means of crop rotation. Selective post-emergence herbicides must be used for winter wheat under dryland conditions. Since *A. fatua* is closely related to cereal crops, the registered herbicides (of which there are several) have to be highly selective and specialised. In the winter-rainfall region, winter ploughing helps to reduce the weeds but this can only be done in the absence of winter wheat.

Avena fatua

POACEAE

Bambusa balcooa
common bamboo * gewone bamboes

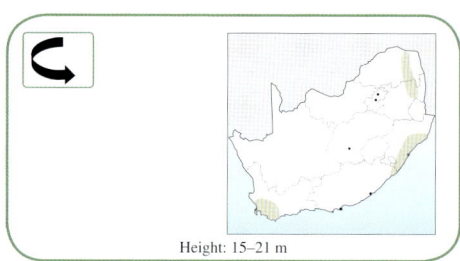

Height: 15–21 m

Similar species:
B. glaucescens ⊂ (hedge bamboo * oriental hedge) from China

Origin and description: *B. balcooa* is originally from India and was introduced as an ornamental and for the use of the strong straight stems as poles and for crafts. The shoots are also popular in Eastern cuisine.

Impact: Bamboo can become very invasive, replacing indigenous vegetation. A small plant in a garden can soon completely take over and can escape into neighbouring gardens, hedges, roadsides and neglected areas.

Control: Repeated cutting of the stems to ground level will eventually exhaust the plant. The stems should not be allowed to get taller than about 30 cm. Faster results can be achieved if the cut stumps are given a vertical chop with an axe and then painted with a suitable systemic herbicide.

Photo by: Derek Ramsay

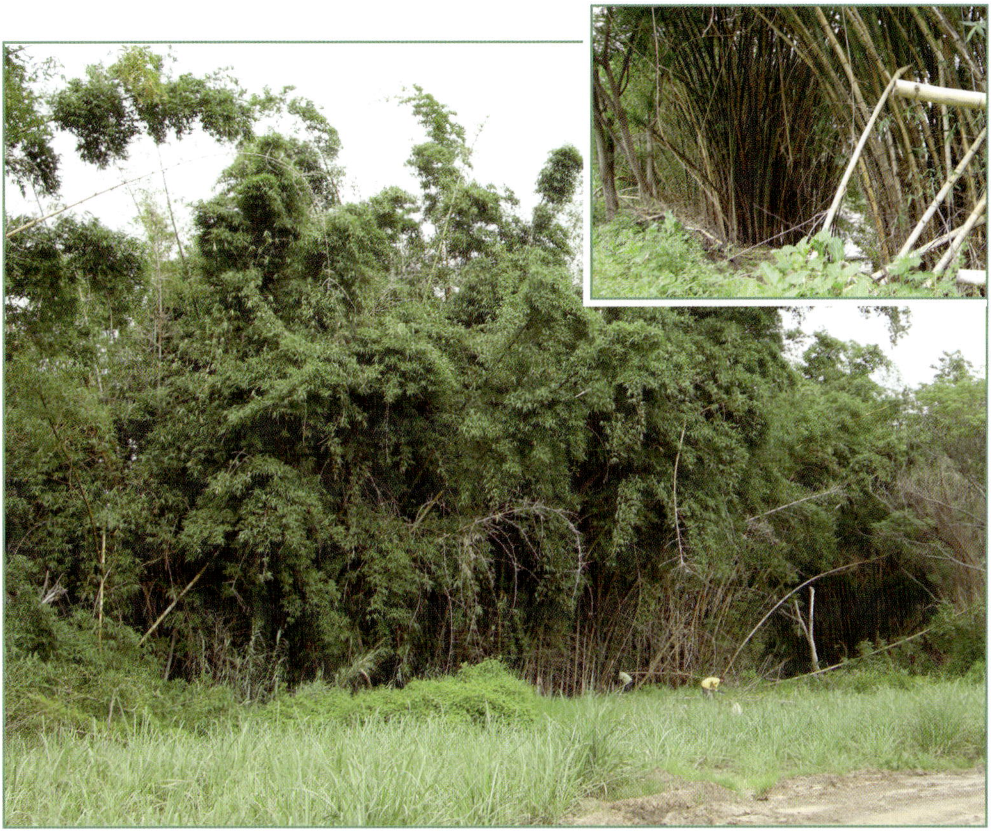

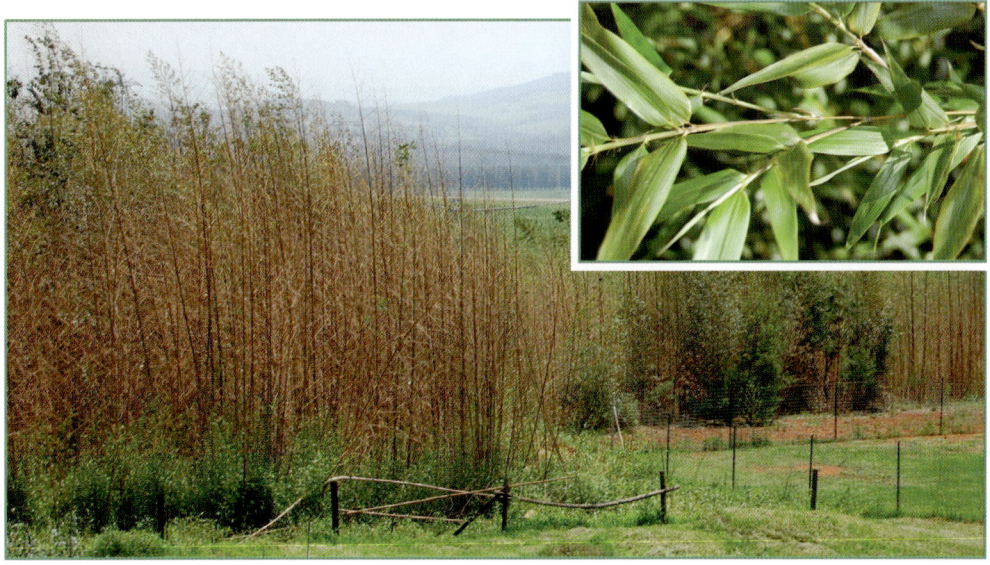

Bambusa balcooa

Bambusa glaucescens

POACEAE

Brachiaria deflexa *(=Pseudobrachiaria deflexa)*
false signal grass * bastersinjaalgras

Brachiaria eruciformis
sweet signal grass * litjiesinjaalgras

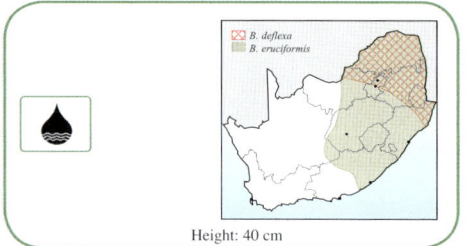
Height: 40 cm

Origin and description: Indigenous. These *Brachiaria* species are fairly widespread, common annual grasses and are just two of 20 *Brachiaria* species that occur in South Africa. *B. eruciformis* tends to restrict itself to damper sites and turf soils and is more common in Mpumalanga and the Free State. *B. deflexa* with its characteristic broad leaves, is found mainly in the Limpopo Province and Northern Cape in moist, shady areas. It is somewhat less weedy than *B. eruciformis*.
Impact: *Brachiaria* species are common weeds of lands and gardens but seldom become a serious problem.
Control: These grasses are susceptible to many pre-emergence and post-emergence grass killers and feature on many such herbicide labels. They are also controlled by shallow cultivation during the seedling stage.

Other common names
B. eruciformis: khlane (S) * umfisane (Z)

Brachiaria deflexa

Brachiaria eruciformis

POACEAE

Briza maxima
large quaking grass * bewertjies

Briza minor
small quaking grass * kleinbewertjies

Origin and description: 'Quaking grass' represents two species of annual grass introduced into South Africa from the Mediterranean region, probably as ornamentals, having attractive flower heads that 'quake' in a breeze and are used in flower arrangements. The quaking grasses have become naturalised and are now found throughout the southern Cape, where they are common.
Impact: Both species are weeds of roadsides, orchards and gardens. They occasionally appear in irrigated crops. *Briza* species rarely become very competitive unless left unattended.
Control: The plants are susceptible to many herbicides and will succumb to cultivation, especially during the seedling stage.

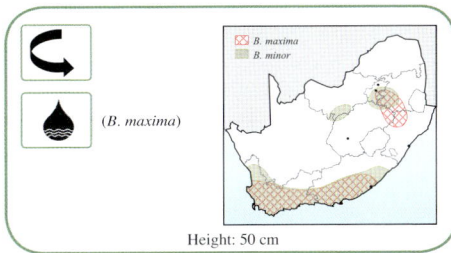
(*B. maxima*)
Height: 50 cm

Other common names
fairy bells * lady's heart grass * trilgras

Briza maxima

Briza maxima

Briza minor

POACEAE

Bromus catharticus *(=B. unioloides/ B. willdenowii)*
rescue grass * reddingsgras

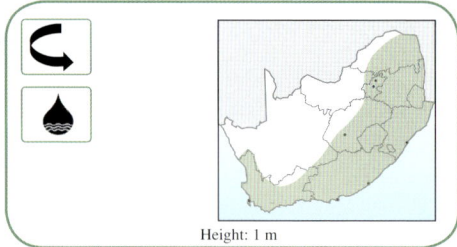
Height: 1 m

Origin and description: *B. catharticus* is a common and widespread annual or perennial grass weed from South America. It grows vigorously in winter but requires moisture and low grazing pressure to flourish. It is also a weed in Europe and North America.

Impact: It has invaded orchards, vineyards, roadsides, waste places and pastures, where although palatable, it does not compare to ryegrass in performance, with the result that it is undesirable in pastures. It can also be a problem in winter grains such as wheat and plays an important role in the life cycle and survival of the Russian wheat aphid in the Free State. It is an untidy weed in all the areas mentioned above.

Other common names
brome grass * broncho grass * orchard grass * prairie grass * rescue brome * beesgras * Vandermerwegras * wintergras

Control: In pastures or grass crops such as wheat, highly selective herbicides are required to control this grass. In other areas rescue grass is susceptible to the usual chemicals. It is also easily removed by cultivation.

Bromus catharticus

POACEAE

Bromus diandrus
ripgut brome * predikantsluis

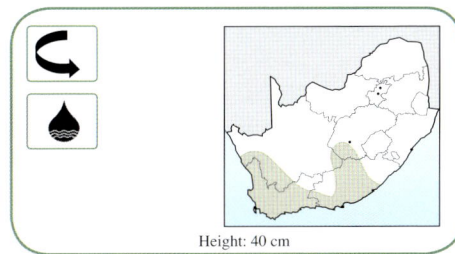

Height: 40 cm

Origin and description: *B. diandrus* was introduced into South Africa from the Mediterranean region in contaminated feed grain. It has spread rapidly, especially in the southwestern Cape. The species has hybridised with a close relative, creating what is referred to as an 'aggregate'. The seedlings have distinctive striped leaf sheaths and the leaves are hairier than those of the other common grass weeds of wheat in the Western Cape.

Impact: When mature, this grass is avoided by stock as the long, barbed awns get into their nostrils and mouths. The spikes also contaminate clothing and sheep's wool. *B. diandrus* can be a competitive weed and is a host for cereal root diseases. Contaminated grain also blocks grading screens. It is a major weed of wheat in the Sandveld and Swartland areas. Wheat monoculture and reduced tillage systems have encouraged this weed.

Control: Control is especially difficult in wheat as *B. diandrus* is not susceptible to the selective grass killers normally used to control wild oats and other grasses. Suppression can be achieved

Other common names
broncho grass * great brome * stick grass * bronkhorstgras * langnaaldbromus
Similar species
B. pectinatus (from Eurasia)

with winter fallow ploughing. A suitable crop rotation such as lupins, canola or lucerne, in which the cyclohexenone grass killers can be used, is also recommended. If wheat is grown on wheat, sowing can be delayed until after the first rains have caused mass germination of seedlings. The seedlings can then be destroyed with non-selective herbicides or tilling. However, not all the seeds germinate at the same time. Stubble burning in April also assists in suppressing this grass.

Bromus diandrus

POACEAE

Cenchrus brownii
burgrass * knopklitsgras

Cenchrus incertus
mat sandbur * dubbeltjiegras

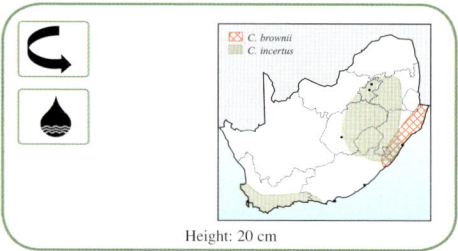

Height: 20 cm

Origin and description: Introduced from tropical America, *C. brownii* and *C. incertus* are troublesome and unpleasant annual weeds occurring in many parts of South Africa. *C. brownii* is thought to have arrived on a tramp steamer in 1945 which was harboured at the Bluff, Durban, and was first seen in nearby oil installations. Realising the potential danger, the authorities proclaimed it a weed in 1946 and formed a Burgrass Eradication Committee, which soon had the main infestation under control. However, by 1957 the weed had spread along the KwaZulu-Natal south coast and to several sites further inland.

C. incertus is more widespread than *C. brownii*, and is found in all provinces. Its route of introduction, however, is uncertain. Both these weeds are usually found in disturbed veld, waste areas and fallow land. They are only occasionally found in croplands such as orchards.

Impact: The burs, especially the more robust ones of *C. incertus*, can injure the feet and mouths of grazing animals as well as humans and domestic pets. They can also stick to clothing and contaminate sheep's wool.

Control: These two species of *Cenchrus* are not controlled effectively by pre-emergence herbicides. Effective control is achieved by post-emergence herbicides or by physical removal whilst they are seedlings.

Other common names
C. incertus: spiny bur grass * sand bur grass * sandklitsgras

Cenchrus incertus

POACEAE

Chloris gayana (=*C. abyssinica*)
Rhodes grass * Rhodesgras

Chloris pycnothrix
spiderweb chloris * spinnerak-chloris

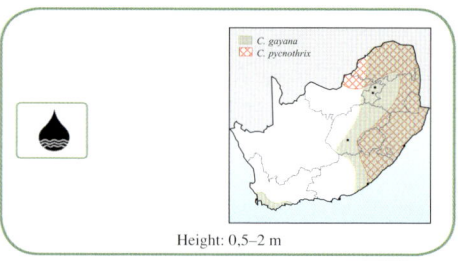

Height: 0,5–2 m

Origin and description: These species of *Chloris* are thought to be indigenous. *C. gayana* is a tall, tufted, stoloniferous perennial grass common on roadsides and disturbed soil in general, but favouring the shade under trees and bushes. Originally recommended as a pasture grass by Cecil John Rhodes, it has now lost much of its popularity, mainly due to its lack of persistence and unsuitability for silage. The plant's inflorescence of up to 20 spikes is quite distinctive. The origin of *C. gayana* is uncertain. Even though it is indigenous to Central Africa, there is a possibility that early introductions into South Africa were made from India.

C. pycnothrix is similar to *C. gayana*, but less widespread and smaller (up to 500 mm tall). It is recognisable by the rounded leaf tips. It is an annual and occurs mainly in gardens and on pavements and road verges.

Impact: These grasses can become a nuisance along roadsides, for example, where they may have to be controlled to improve visibility and tidiness and to reduce the fire hazard, etc. *C. gayana* can be a weed of cultivation, but will usually die out after 4–5 years if not disturbed or fertilised.

Control: These species are relatively easy to control by cultivation during the seedling stage and with pre-emergence grass herbicides including atrazine. They are more tolerant once they have become established.

Other common names
C. gayana: hunyani grass * Rhodesian blue grass * rooiklosgras * bruinvingergras * nyankomo (Z)
C. pycnothrix: orchard grass * radiate finger grass

Chloris gayana

Chloris pycnothrix

POACEAE

Chloris virgata
feathertop chloris * witpluim-chloris

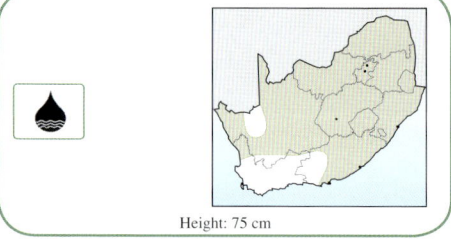

Height: 75 cm

Origin and description: *C. virgata* is a widespread indigenous grass but it is only a serious arable problem in parts of the western Free State and western Mpumalanga. Stems can vary in height from only a few centimetres to 90 cm, even on the same plant. Roots may develop from the lower nodes where they touch the ground. *C. virgata* is easily identified by its characteristic feathery white spikes in the inflorescence. The spikes in each inflorescence vary from four to 15 but are usually closer to four. The Xhosa use a decoction of this grass or its roots in a bath for the treatment of colds and rheumatism. It can behave as a perennial but, when weedy, acts as an annual. It is also a weed in overgrazed veld in the southern Free State and the Karoo.

Impact: This grass is a strong competitor and can severely reduce yields when it occurs in crops.

Control: *C. virgata* is relatively easy to control by cultivation during the seedling stage and with pre-emergence grass herbicides of which several are registered. It is more tolerant once it has become established.

Other common names
hay grass * oldlandgrass * sweetgrass * white grass * kwasgras * paardgras * amafusine (Z) * sehabane (S) * umadolwana (X)

Chloris virgata

POACEAE

Cortaderia selloana
pampas grass * pampasgras

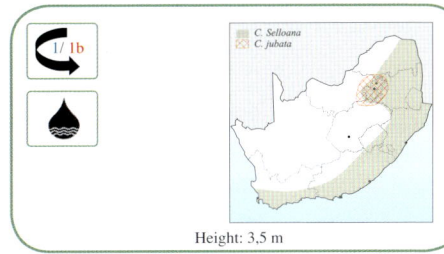

Height: 3,5 m

Origin and description: *C. selloana* is a robust evergreen grass that was introduced into South Africa from southern South America as an ornamental and to assist in the stabilisation of mine dumps. It has escaped from these places and is now a common sight on many roadsides and in waste areas. It can tolerate very harsh growing conditions hence its use on mine dumps.

Impact: It is a large, untidy plant that replaces indigenous vegetation and hampers roadside maintenance. The leaves have sharp edges, and although the seed heads are used in floral arrangements, they can irritate the eyes and nose. The dead leaves stay on the plant, smother other plants and provide a haven for vermin. A fertile plant can produce over 1 million wind-borne seeds during its lifetime, so the sterility or non-invasiveness of cultivars has to be investigated.

Control: Because of its large rhizomatous roots, pampas grass is very difficult to destroy by fire. Repeated applications of a systemic herbicide will be translocated to the roots and will eventually kill the entire plant. If removed by hand, protective clothing should be worn and all rhizomes must be carefully removed and destroyed.

Other common names
silwergras

Similar species
C. jubata 1/1b (purple pampas grass) also from South America, but with a less compact and often purple inflorescence

Cortaderia selloana

POACEAE

Cynodon dactylon
couch grass * kweek

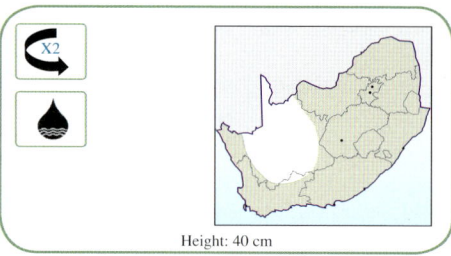

Height: 40 cm

Origin and description: Thought to have originated in tropical Africa or Asia, *C. dactylon* is now a widespread and troublesome weed. It is considered to be the most widely distributed grass weed in the world. In South Africa its widespread and troublesome nature is demonstrated by the fact that it has 65 English and Afrikaans common names as well as more than 14 recorded names in the various indigenous languages. It is a creeping perennial and is spread by means of an extensive system of stolons and underground rhizomes. It also disperses further afield on animal hooves and cultivator tines. The flower produces viable seed.

Impact: *C. dactylon* is a vigorous grower, making it highly competitive in crops such as sugarcane. It is capable of breaking up tarred or concrete surfaces. Its tough growth habit, however, also makes it a valuable grass for combating erosion. Several cultivars have been developed for use in lawns, golf greens and sports fields.

Control: Because of its extensive underground systems, this weed is extremely difficult to eradicate. Repeated winter ploughing and harrowing will give a fair degree of control by breaking up the runners and exposing them to the elements. This reduces root reserves and increases the

Other common names
quickgrass * Bermuda grass * twitch grass * doobgras * mohloa (S) * mothowa (T) * ngwengwe (Z) * uqaqaqa (X) * uqethu (Z)

Cynodon dactylon

efficiency of the systemic post-emergence grass killers when they are applied to the regrowth. It can also help if the grass is mowed or heavily grazed and even watered and fertilised before spraying. This will encourage fresh, green growth that is more capable of absorbing and translocating a herbicide. Although this grass is not susceptible to most pre-emergence herbicides, it can be controlled pre-emergent on roadsides and in industrial situations with some industrial herbicides.

Cynodon dactylon

POACEAE

Cynodon nlemfuensis
star grass * stergras

Origin and description: *C. nlemfuensis* was introduced from Kenya in 1919 and is now naturalised in parts of the warmer areas of South Africa, especially the KwaZulu-Natal coastal regions and northern and eastern Mpumalanga. It is a vigorous grower, palatable and nutritious.

Impact: *C. nlemfuensis* is planted for pasture and used for hay but once established, easily spreads into other areas where it becomes a nuisance. It occurs as a weed in patches in open and dense bushveld, cattle kraals, on roadsides, as well as in orchards and agricultural land, particularly in moist places.

Control: *C. nlemfuensis* is a perennial plant with a strong system of stolons. It is therefore extremely difficult to control. If one attempts to remove the plant mechanically it just tends to break up the runners and spread them around. Systemic herbicides can be used, but the grass should first be mowed or heavily grazed and even watered and fertilised. This will encourage fresh, green growth that is more capable of absorbing and translocating the herbicide.

Height: 1 m

Other common names
East African couch * robust stargrass * gifgras * reusekweekgras * sterkgras

Similar species
C. aethiopicus
C. plectostachyus
(Both are exotics from elsewhere in Africa.)

Cynodon nlemfuensis

POACEAE

Dactylis glomerata
cocksfoot * koksvoetgras

Origin and description: Introduced into South Africa from Eurasia for hay and fodder, *D. glomerata* has escaped into the wild and is now widespread. It can often be found in the shade and in disturbed areas such as roadsides especially where water accumulates, even though it does not tolerate being waterlogged. A large number

Height: 80 cm

Other common names
akaroa * orchard grass * Australiese gras * kropaar

of varieties of *D. glomerata* are available commercially. They vary in their flowering time, leafiness and seasonal spread of foliar production. The plants are frost tolerant and are used as winter pastures in much the same way as ryegrass. It is a perennial grass reproducing only by means of seeds.

Impact: This grass replaces indigenous species.

Control: *D. glomerata* is probably only a commercial problem on roadsides, where it is easily controlled by industrial herbicides.

Dactylis glomerata

POACEAE

Dactyloctenium aegyptium
crowfoot (grass) * hoenderspoor

Dactyloctenium giganteum
giant crowfoot * reusehoenderspoor

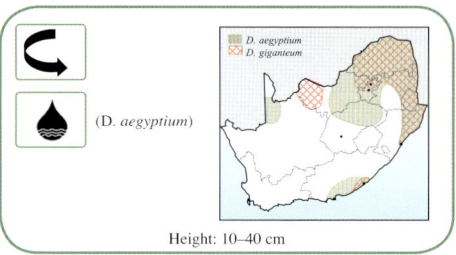

Height: 10–40 cm

Origin and description: *D. aegyptium*, being of uncertain origin but probably exotic, is an annual spreading grass. It is found mainly in the summer-rainfall region of South Africa, both in the temperate and subtropical regions, but is more common in the latter region. *D. giganteum* is indigenous and widespread in the subtropical parts of the summer-rainfall region. It grows much taller than

Other common names
coast (duck) grass * coast button grass * duck grass * starfish grass * Natalkweek * ungwengwe (Z)
Similar species
Eleusine coracana subsp. africana

D. aegyptium. Dactyloctenium species spread by means of stem-suckers formed where nodes on the stem touch the ground and take root. The plant produces an abundance of seed.
Impact: It is found along roadsides and in waste places and is frequently a pest in subtropical fruit orchards. It is also found in lawns and gardens.
Control: Although *Dactyloctenium* species can be strong competitors, they can be controlled effectively by conventional grass herbicides and shallow cultivation.

Dactyloctenium giganteum

Dactyloctenium aegyptium

POACEAE

Digitaria sanguinalis
crab finger-grass * kruisvingergras

Height: 60 cm

Origin and description: *D. sanguinalis* is of uncertain origin, but it was probably introduced from Europe. It has a relatively flattened growth habit, growing to only about 60 cm high, putting down roots from nodes where they touch the ground. Often referred to by farmers as one of the four 'landgrasses', along with *Eleusine coracana*, *Urochloa panicoides* and *Panicum schinzii*, because they are common and, as seedlings, appear similar. There are also several other very similar, closely related species of *Digitaria*.

Impact: It is a major weed of crops and gardens in most areas, particularly the KwaZulu-Natal midlands, the highveld and the Eastern Cape. *D. sanguinalis* is a serious problem in maize, where it competes vigorously for available moisture, especially later in the season.

Other common names
manna * wild millet * kopersaadgras * kruisgras * moqopshoe (S)
Similar species
D. eriantha (common finger grass, indigenous)

Control: It is controlled effectively by the acetanilide group of herbicides. Prior to the introduction of these chemicals, however, this weed was increasing in importance, as it did not respond well to thiocarbamates like EPTC. The seeds only germinate in soil temperatures exceeding 34°C. This means that in most of the maize-producing areas, plants emerge late in early planted crops. For this reason chemicals with a long residual effect are required. It is often advantageous to use a 'split application' of a pre-emergence residual grass herbicide, using some early and saving the rest for a later application when temperatures have become higher.

Digitaria sanguinalis

POACEAE

Echinochloa colona
marsh grass * kleinwatergras

Echinochloa crus-galli
barnyard grass * hanepootmanna

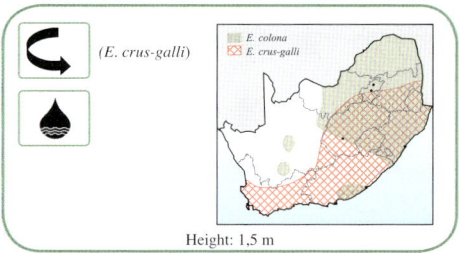

Height: 1,5 m

Origin and description: *E. colona* is probably indigenous whereas *E. crusgalli* is thought to be from Eurasia. They are reported to be weeds of more than 35 crops in over 60 countries worldwide. Despite their different origins, these annual grasses are very similar in appearance. *E. crus-galli* tends to have a more purple coloration and the awns on the seeds are longer than that of *E. colona*. They are both widespread and common, favouring moist places and often grow in standing water.

Impact: Both species are weeds of ditches and irrigated crops, especially rice. *E. crus-galli* in

Other common names
E. colona: jungle rice * shama millet * moerasgras
E. crus-galli: barnyard millet * cockspur grass * blousaadgras * tuinmanna * joang-ba-masimo (S)

particular is often a serious problem in irrigated crops. It is sometimes planted as forage. *E. colona* is of less economic importance, preferring even moister conditions than *E. crus-galli*. It is commonly found in gardens, on roadsides and in waste places. *E. crus-galli* is considered one of the world's worst weeds, causing severe crop losses as it can remove up to 80% of available soil nitrogen. Also, this accumulated nitrogen can be harmful to livestock.

Control: These grasses are effectively controlled by shallow cultivation and conventional pre- or post-emergence grass herbicides.

Echinochloa colona

Echinochloa crus-galli

POACEAE

Ehrharta longiflora
oat-seed grass * hawersaadgras

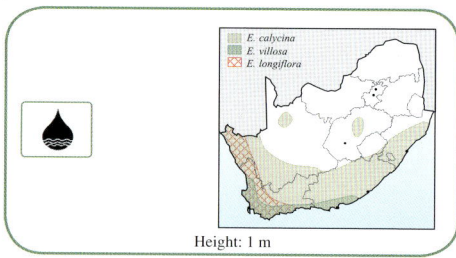

Height: 1 m

Ehrharta longiflora

Other common names
annual veld grass * veldgras
Similar species
E. brevifolia
E. calycina (common ehrharta)
E. villosa (pipe grass)

Origin and description: There are over 30 indigenous species of *Ehrharta*. Several of them are considered weedy, although they are of relatively minor importance. *E. longiflora* is an annual and, like most of the genus, is found mainly in the southern Cape regions. It must not be confused with any of the wild oat species. These grasses often grow together, especially on roadsides.
Impact: It occurs in disturbed areas such as roadsides and orchards and can become dense and competitive.
Control: *E. longiflora* can be controlled by shallow cultivation during the seedling stage and is susceptible to pre- and post-emergence herbicides registered for use in orchards.

POACEAE

Eleusine coracana subsp. *africana* (=*E. indica* subsp. *africana*/=*E. africana*)
African goosegrass * jongosgras

Origin and description: Goosegrass is probably indigenous and is considered by some authorities to be the most common grass weed of cultivated land in South Africa. *E. coracana* is often referred to by farmers as one of the four 'landgrasses', along with *Digitaria sanguinalis*, *Urochloa panicoides* and *Panicum schinzii*. As seedlings the four species look very similar.

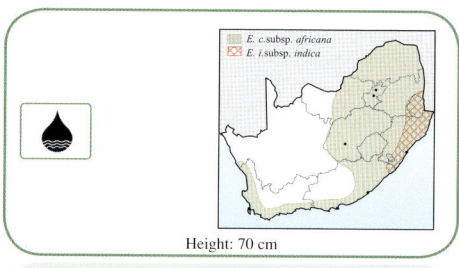

Height: 70 cm

Other common names
rapoko grass * osgras * makha (Sh) * maseka (T) * moseli (S) * pokwana (N) * unyankomo (Z)

Impact: This is a widespread and troublesome annual grass weed of crops and gardens. Because of its extended and vigorous root system it is a severe competitor in dryland summer crops. *E. coracana* subsp. *africana* is difficult to pull up by hand because of these strong roots, and it is not easy to cut in lawns because of its tough stems. It is not a very palatable grazing grass, but in times of famine the seed is ground into flour and eaten.

Control: *E. coracana* subsp. *africana* will continue to germinate throughout summer; therefore cultivation is not an efficient method of control. Fortunately it is very susceptible to grass herbicides and even to some pre-emergence so-called 'broadleaf-weed' herbicides. However, because of its highly competitive nature, the use of specialised grass killers is often essential.

> **Similar species**
> *E. indica* subsp. *indica* ⟲ (Indian goosegrass, probably exotic)

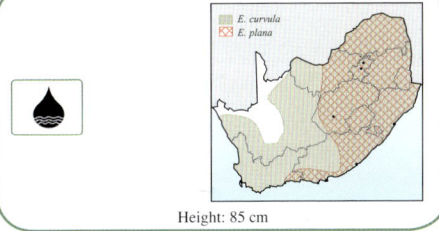

Eleusine coracana subsp. *africana*

POACEAE

Eragrostis curvula
weeping love grass * oulandsgras

Origin and description: There are many indigenous species of *Eragrostis*, and several of them are considered weeds. Zulus and farmers use the term 'mtshiki' to describe grasses like this or the veld in which they occur.

Impact: *E. curvula* is used as a hay crop and in permanent pastures. It is versatile and widely distributed. Unfortunately, it is a problem as a volunteer, as it invades some perennial crops. It is also found in waste areas and old lands.

Control: *E. curvula* is effectively controlled by pre-emergence herbicides. Post-emergence grass killers are generally less effective, especially once the grass has passed the seedling stage or it has tillered.

Height: 85 cm

> **Other common names**
> Boer love grass * weeping grass * wire grass * fyngras * renostergras * matolo (S) * seritsoana (Z)
>
> **Similar species**
> *E. plana* (fan love grass)
> *E. ciliaris* (stink love grass * woolly love grass) reported as a weed of sugarcane
>
> Many other Eragrostis species are referred to as love grass and are occasionally considered as weeds.

Eragrostis curvula

POACEAE

Hordeum murinum subsp. *murinum*
wild barley * wildegars

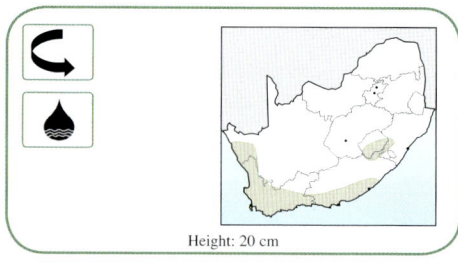

Height: 20 cm

Origin and description: Of European origin, this grass is now an annual weed causing serious problems in parts of the southern and southwestern Cape. It is only rarely seen outside of these areas.

Impact: *H. murinum* subsp. *murinum* is commonly found on roadsides, in gardens, waste places and in croplands, especially on headlands in wheat fields that have been cultivated but not planted. It frequently grows in lucerne where it is cut along with the lucerne and eaten by livestock. It has sharp awns that can cause ulceration of the mucous membranes of the animals' mouths. Apparently the mouths of horses are particularly

Other common names
false barley * mouse barley * muiswildegars * kruipgras

sensitive to this grass. Ostriches can die from its effects.

Control: *H. murinum* subsp. *murinum* is controlled effectively by the selective post-emergence herbicides registered for use in lucerne.

Hordeum murinum subsp. *murinum*

POACEAE

Hyparrhenia tamba
blue thatching grass * bloutamboekiegras

Origin and description: There are many species of *Hyparrhenia* in South Africa. All of them are indigenous, perennial and collectively referred to as 'thatching' grass. Although representatives of this genus can be found throughout the country in a wide range of sites, they are especially common on roadsides.

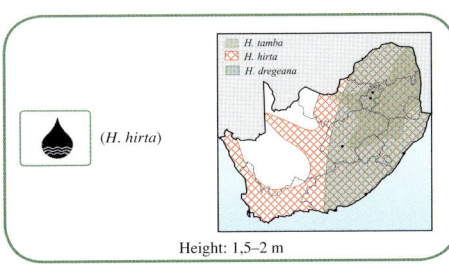

Height: 1,5–2 m

Other common names
H. hirta: Boesmansgras * intunga (Z) * mofula-tsephe (S) * mohlomo (S)
Similar species
H. hirta (common thatching grass)
H. dregeana (tambuki grass)

Impact: Although these grasses are widely used for thatching purposes, they become a nuisance on roadsides, as they are relatively tall, thereby obstructing visibility.
Control: Several industrial herbicides have been registered for use where these grasses have to be controlled on roadsides and in industrial areas.

Hyparrhenia tamba

POACEAE

Imperata cylindrica
cotton wool grass * donsgras

Origin and description: This grass is indigenous to many regions of the world including eastern and southern Africa, India and Australia. It is also a weed of many regions of the world and is said to be one of the most troublesome grass weeds of some tropical countries. It is a serious alien invasive species in the USA. *I. cylindrica* occurs in moist places throughout South Africa, tending to take over once it has established itself.

Impact: It is an unpalatable climax species and is often a serious problem on roadsides and in industrial areas. It is a spreading perennial grass with a system of underground rhizomes, which makes it very difficult to control. However, *I. cylindrica* can sometimes be of value in vleis and eroded watercourses as it may be the only plant able to survive the waterlogged conditions and

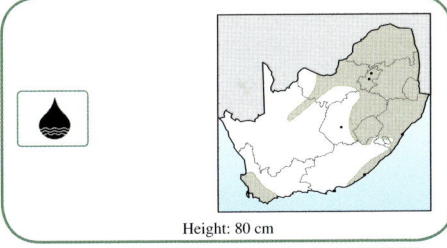

Height: 80 cm

Other common names
silver spike * sword grass * cogongrass (USA) * beddinggras * silweraargras * mohlorumo (S) * umthente (Z)

act as a soil binder, thus preventing erosion. This grass is a potential problem in tropical plantation crops such as coffee.
Control: When this weed occurs on roadsides and in industrial areas it has considerable economic impact, as it requires the use of expensive systemic weed-killers or non-selective industrial herbicides.

Imperata cylindrica

POACEAE

Lagurus ovatus
hare's tail (grass) * haasstert(gras)

Origin and description: Of Mediterranean origin, this annual grass is now a weed in parts of South Africa, mainly the southern Cape. It is also found to the east as far as Port Elizabeth and has even been recorded in Pretoria. It was probably introduced as an ornamental because it is used in flower arrangements.

Impact: *L. ovatus* flourishes on roadsides, in waste places and in gardens. It replaces indigenous species. It is especially noticeable in the dry summer months in the Cape when other vegetation has gone brown.

Control: No specific control measures are recommended for this weed, but it is easily removed by cultivation and would be susceptible to the usual herbicides. In lawns it will have to be removed by hand.

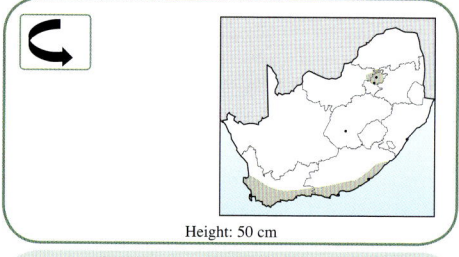
Height: 50 cm

Other common names
bunnie's tails * hare's foot * klossiesgras * pluisiesgras

Lagurus ovatus

POACEAE

Lolium multiflorum
Italian ryegrass * Italiaanse raaigras

Lolium temulentum
darnel (ryegrass) * drabok(raaigras)

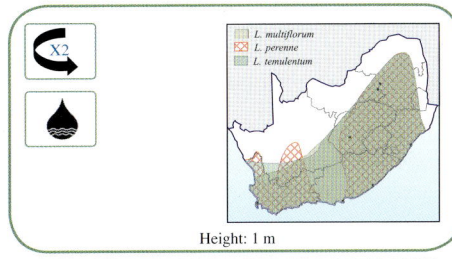

Height: 1 m

Origin and description: These closely related species from Europe are difficult to tell apart even for botanists, and are now naturalised throughout South Africa. *L. temulentum* is a wild type and was probably introduced by early settlers as a contaminant of grain seed. It was the first weed to receive official attention and in 1659 it became the subject of a 'plakkart' or regulation posted in public places by Van Riebeeck's council. It is now widespread in South Africa in cultivated lands, gardens and other disturbed places. When the seed is milled with wheat it is said to cause the flour to become grey, bitter and even poisonous. It also forms hybrids with the other species.

Impact: The ryegrasses have been hybridised and bred as high-performing pasture grasses, but often escape into the wild and become troublesome volunteer crops. *L. multiflorum* can become infested with the nematode *Anguina agrostis* and a bacterium, *Corynebacterium rathayi*, which together can cause fatal poisoning of livestock. The first danger sign that the ryegrass is infected, is the presence of a yellow bacterial slime, usually occurring on the inflorescences in September.

Control: The ryegrasses are difficult to control when they occur in cereal crops. They are sensitive, however, to most of the selective grass and wild oats herbicides, even if it seems that there

Other common names
L. multiflorum: annual ryegrass * roggras
Similar species
L. perenne (perennial ryegrass)

is some variation in herbicide sensitivity between the different species and varieties. Unfortunately, there is now evidence that these grasses have developed resistance to certain herbicides in the Cape. Where practicable, they can easily be removed by cultivation during the seedling stage.

Lolium multiflorum

Grasses and Sedges

POACEAE

Melinis repens* subsp. *repens *(=Rynche-lytrum repens)*
Natal red-top * Natalse rooipluim

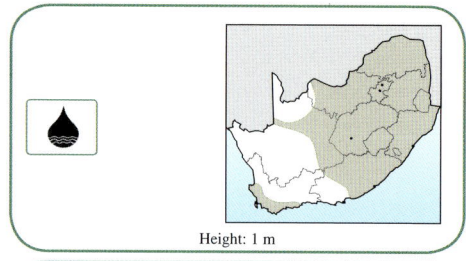
Height: 1 m

Origin and description: Although the name of this weed suggests that it is restricted to KwaZulu-Natal, it is widely distributed throughout South Africa and indeed, the world. In fact, it is possibly not even indigenous as its origin is uncertain. It reproduces only by means of seeds, but frequently takes root at the lower nodes.

Impact: *M. repens* subsp. *repens* is a common annual or short-lived perennial weed of such places as roadsides and waste places. It is capable of invading fallow lands but is seldom common in undisturbed veld. Animals do not find it palatable, but it does provide cover in disturbed

Other common names
fairy grass * bergrooigras * blinkgras * ferweelgras * wolgras

areas, thereby reducing erosion. Used in flower arrangements.

Control: Shallow cultivation will control this grass in the seedling stage and it is susceptible to many pre-emergence grass herbicides. On roadsides it is controlled effectively by conventional industrial herbicides.

Melinis repens subsp. *repens*

POACEAE

Nassella neesiana (=*Stipa neesiana*)
spear grass * pylgras

Nassella tenuissima (=*Stipa tenuissima*)
white tussock * witpolgras

Nassella trichotoma (=*Stipa trichotoma*)
nassella tussock * nassella-polgras

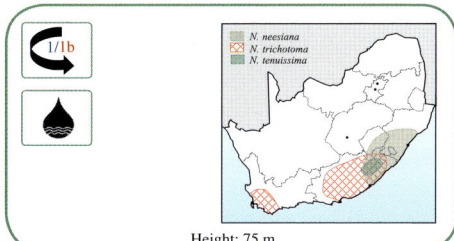

Height: 75 m

Origin and description: These species of *Nassella*, which are serious perennial weeds in South Africa, all originate from South America. *N. neesiana* is found in the Eastern Cape and southern KwaZulu-Natal. *N. tenuissima* occurs mainly around Barkley East near an infestation of *N. trichotoma*. This suggests a common origin which is thought to have been 130 000 tons of fodder brought into Port Elizabeth and East London for military purposes during the Anglo-Boer War. *N. trichotoma* has also been found at the Rhodes Memorial in Cape Town and near Stellenbosch and Swellendam.

Impact: Once *N. trichotoma* has become established, it forms dense stands, totally displacing the naturally occurring species. Before flowering,

Nassella neesiana

which occurs from November to December, sheep and cattle will graze this grass, but after it has flowered the animals will not touch it. At maturity the whole inflorescence breaks off and blows away. At the peak of the invasion during the 1960s, these inflorescences formed large drifts against fences. The long awns on the seeds of all three species adhere to clothing and hair, which is a further very efficient means of dispersal. *N. trichotoma* tussock is a declared weed with the potential to force farmers off their land if it is not contained.

Control: Tetrapion is registered for the selective control of *N. trichotoma* seedlings. Glyphosate and proprop are also registered, but effective long-term eradication requires a comprehensive programme of chemical or physical removal over many years.

Nassella trichotoma

POACEAE

Panicum maximum
common buffalo grass * gewone buffelsgras

Origin and description: This grass is indigenous to South Africa and is a widespread weed except

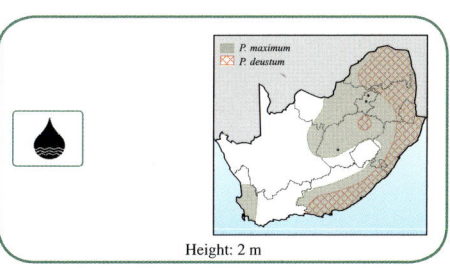

Height: 2 m

in the Western Cape. As it is a palatable plant, some varieties have been selected and bred for commercial use as pasture grasses and to make hay. It is a vigorous grower and can develop large perennial clumps that can produce stems over 2 m tall. In extreme cases, stems of up to 4 m have been recorded. The leaves and stems are hairy, unlike that of *P. schinzii* (see below) with which it can be confused, and it does not usually exhibit such strong purple coloration.

Impact: Despite its very good qualities, *P. maximum* is the principal weed of sugarcane and many other crops, and as such it is a source of great concern in agriculture.

Control: The seedlings are very weak and slow growing and high mortality rates are experienced during drought periods. It is important to control this grass early, as plants that are not controlled at an early stage, develop large perennial clumps or 'stools'. These clumps are tolerant of even the strongest herbicides and must usually be removed by hand. Seedlings are susceptible to conventional pre-emergence grass killers and are also easily controlled by post-emergence chemicals or shallow cultivation. Care must be taken to remove all late-germinating individuals as escaped plants become very competitive.

Other common names
guinea grass * blousaadsoetgras * lehola (S) * mphaga (T) * ubabe (Z) * umhatji (N)

Similar species
P. schinzii (sweet buffalo grass)
P. deustum (broad-leaved panicum)

Panicum maximum

POACEAE

Panicum schinzii *(=P. laevifolium)*
sweet buffalo grass * blousaad(buffelsgras)

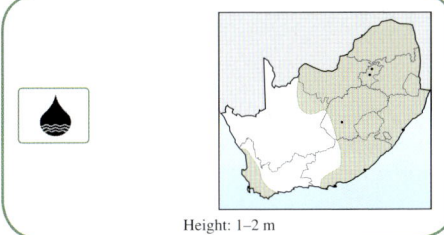

Height: 1–2 m

Origin and description: *P. schinzii* is a widespread and indigenous annual grass weed, especially in the cooler and/or moister areas. It can grow to 2 m, with hollow stems that are usually purple or red. The stems are hairless, except at the ligule. This characteristic makes it easy to differentiate from *P. maximum* (see above). It is a highly palatable grass and excellent for hay or silage although its high moisture content and thick hollow stems make it rather difficult to make into hay, especially in the summer-rainfall areas. Being an annual, unlike *Eragrostis* for example, it does not regrow after cutting. *P. subalbidum* is very similar in appearance, is native to Africa but has recently been found infesting sugarcane in KwaZulu-Natal. It does not respond well to herbicides.

Other common names
blue panic * vlei-panicum * soetgras * vleibuffelsgras * mofantsoe-o-moholo (S)

Similar species
P. maximum (common buffalo grass)
P. subalbidum (elbow buffalo grass * elmboogbuffelsgras)

Impact: *P. schinzii* is a common weed of crops and gardens in most areas. In the high-rainfall regions it is often the dominant grass weed. It competes vigorously for moisture in crops such as maize and sugarcane. It is one of the four species commonly referred to as 'landgrasses' because their seedlings are similar and they occur frequently. The others are: *Digitaria sanguinalis*, *Eleusine coracana* and *Urochloa panicoides*.

Control: *P. schinzii* is susceptible to conventional pre-emergence grass herbicides. At a young stage they can be killed with certain post-emergence chemicals and by means of shallow cultivation.

Panicum schinzii

Panicum schinzii

POACEAE

Paspalum dilatatum
common paspalum * gewone paspalum

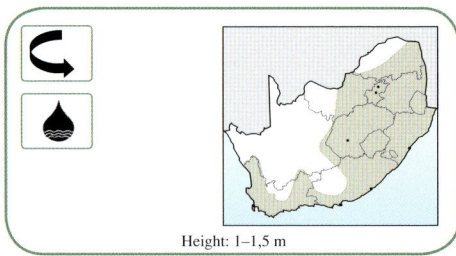

Height: 1–1,5 m

Origin and description: *P. dilatatum* is from South America. It is a perennial grass, being spread by seeds or rhizomes. It was originally used as a pasture grass and is also suitable for hay and silage. However, it had to be cut before flowering to avoid the hay being spoiled by ergot-infected seed heads. Although it can grow to 150 cm, it usually has a flattened tuft of leaves with only the seed heads standing erect. This habit helps it survive mowing and slashing.

Impact: *P. dilatatum* is especially difficult to control once established. It is an important weed of fruit orchards and pineapple fields, often interfering with micro-irrigation systems, and it is also a significant weed in sugarcane, where it is highly competitive and sometimes difficult to tell apart from the crop. It is also a common weed of lawns and turf where it can only be effectively controlled by physical removal.

Other common names
dallis grass * giant paspalum * golden crown grass * large water grass * mupunganini (To)

Similar species
P. urvillei (giant paspalum). This species is taller and lacks the distinctive hairs on the inflorescence and at the base of the plant; it grows from a dense tuft. (See page 94)

Control: Although this weed should be controlled before it becomes established, it is susceptible to the systemic grass killers, where these chemicals are suitable.

Paspalum dilatatum

POACEAE

Paspalum distichum *(=P. paspalodes)*
couch paspalum * kweekpaspalum

Paspalum notatum
lawn paspalum * bahiagras

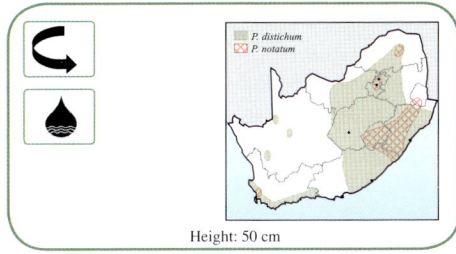

Height: 50 cm

Other common names
P. distichum: buffalo quick paspalum * buffelskweekpaspalum
P. notatum: bahia paspalum

Origin and description: *P. distichum* is thought by some authorities to be indigenous, but by others to be a native of South America. *P. notatum* is unanimously considered to be from South America and is now naturalised in South Africa. *P. distichum* has light green flowers, often with contrasting black female parts (stigmas). Despite favouring moister areas, it can withstand drought conditions for a reasonable period. It is, however, a palatable species on which sheep may do well in winter, provided they are able to get to the succulent underground stolons which remain nutritious even after frost has occurred.

Impact: *P. distichum* is a widespread perennial weed in South Africa, distributed in moist sites such as waterways, drainage channels and in moister crop lands. *P. notatum* is an aggressive invader of lawns, but can often make a good surface for some kinds of sports fields. It generally does not require much cutting, but the flowering stems shoot up rapidly and make the turf unsightly if they are not trimmed frequently.

Control: *P. distichum* has an underground system of creeping stems that makes it very difficult to eradicate. Ploughing and discing merely tend to spread it. High rates of systemic grass killers are required to control this weed successfully. *P. notatum* also has a system of underground rhizomes, which necessitates the use of systemic herbicides for effective control.

Paspalum notatum

POACEAE

Paspalum urvillei
tall paspalum * langbeenpaspalum

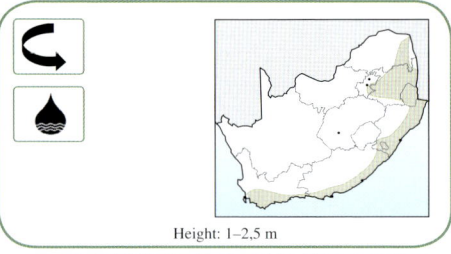
Height: 1–2,5 m

Origin and description: This grass is from South America and is a widespread weed in South Africa, being particularly well known to sugarcane and pineapple growers. *P. urvillei* only reproduces by seed but can grow into a large, tufted, perennial clump which is tolerant to many herbicides.

Impact: It is a major weed of forestry, pineapples and sugarcane, but is also common in waste places, along roadsides and moist areas in general. It is palatable and nutritious and in the right place, a useful pasture grass.

Control: Control of *P. urvillei* should be implemented early, as it is difficult to control once mature. When young and growing vigorously, it can be controlled by the systemic grass killers and some industrial herbicides. Before germination it is susceptible to the conventional pre-emergence grass herbicides.

Other common names
casey paspalum * upright paspalum * vasey grass * langbeenwatergras
Similar species
P. dilatatum (*P. urvillei* is taller, with a more erect but tufted growth habit.)

Paspalum urvillei

POACEAE

Pennisetum clandestinum
kikuyu * kikoejoe

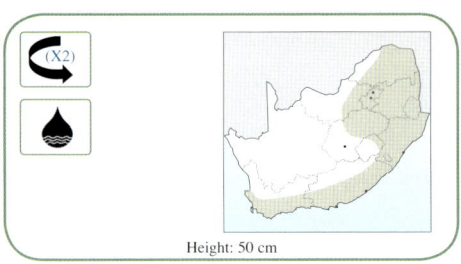
Height: 50 cm

Origin and description: Introduced from East Africa, *P. clandestinum* is now widely distributed in the summer-rainfall regions, both as a cultivated plant and where it has escaped, as a weed. It is a robust, perennial, creeping plant and a valuable summer grazing grass widely used in lawns and sports turfs.

Impact: The fine thread-like anthers on a *P. clandestinum* sward are a common sight, but the local strains have always produced little or no seed. This made pasture establishment difficult but, at the same time, it also reduced the likelihood of escape. The advent of seeding varieties may make this grass more troublesome as a weed. Its rapid and vigorous growth habit helps it swamp and then eliminate other plants and it can even choke ponds and waterways. It can also damage paving.

Control: *P. clandestinum* is particularly sensitive to glyphosate, which can be used for its

Other common names
tajoe (S)

general as well as for its selective control in *Cynodon* turf. Otherwise, as this grass is frost-sensitive, winter ploughing and discing give a fair degree of suppression. It should be borne in mind though, that any remaining runners will spread rapidly under suitable conditions. *P. clandestinum* is not susceptible to pre-emergence herbicides unless it is growing from seed.

Pennisetum clandestinum

POACEAE

Pennisetum purpureum
napier grass * olifantsgras

Origin and description: Native to the tropical grasslands of Africa where it is a popular food for elephants, *P. purpureum* is a tall, erect perennial grass growing well in moist soil and along water courses, although it does not tolerate water logging. It is planted in the same way as sugarcane by burying sections of stem possessing one or two nodes and is popularly planted as fodder, as screening, as a windbreak, and to prevent erosion.

Impact: However, *P. purpureum* is a vigorous

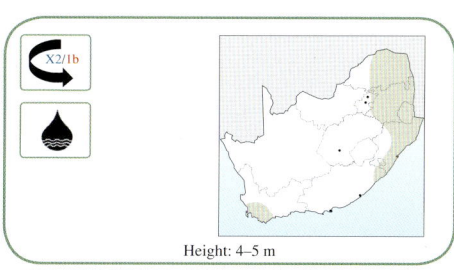

Height: 4–5 m

Other common names
elephant grass * napier fodder * Uganda grass * mufufu (Shona)

and prolific grower, making it highly competitive where it is not wanted and it easily replaces indigenous vegetation. It narrows

watercourses, increases siltation and obstructs access. It invades forest margins and any area where the soil is moist.
Control: Because of its extensive underground systems, this weed is difficult to eradicate once established. Systemic grass herbicides should be applied to fresh green growth for maximum effect. Where possible, indigenous alternatives such as reeds or turpentine grasses (*Cymbopogon* species) should be used instead.

Pennisetum purpureum

POACEAE

Pennisetum setaceum
fountain grass * pronkgras

Pennisetum villosum
feathertop * haarwurmgras

Origin and description: *P. setaceum* and *P. villosum* are both from North Africa. They were introduced as ornamentals and have escaped into the wild.
Impact: They are now common perennial weeds of roadsides and waste places. They are attractive in flower arrangements, but on roadsides they are obstructive and untidy. *P. setaceum* in particular has a strong pioneering nature and can invade industrial areas. They replace indigenous vegetation.
Control: Both species can be controlled by the usual industrial herbicides used on roadsides.

Pennisetum setaceum

Pennisetum villosum

POACEAE

Phalaris canariensis
common canary grass * gewone kanariegras

Phalaris minor
little-seeded canary grass * kleinsaad-kanariegras

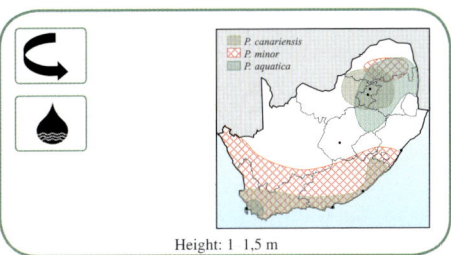

Origin and description: Introduced from the Mediterranean region, probably as a fodder crop, these annual grasses are now widespread in South Africa. The seedlings can be identified quite easily as they will exude a red juice from the stem and roots when broken.

Although the two species are very similar in appearance, *P. minor* is somewhat smaller than *P. canariensis*. This grass is grown commercially in some parts of the world for birdseed, hence the name.
Impact: They are particularly troublesome in cereals in the Western Cape, where they are highly competitive.
Control: These grasses are easy to control in cereals, as they are susceptible to the highly effective range of selective grass herbicides registered for use in these crops.

Other common names
P. canariensis: birdseed grass * kwarrelgras
P. minor: small canary grass * lesser canary grass
Similar species
P. aquatica is a perennial and not as widespread. It is also from the European region but is usually found only on roadsides and in waste

Phalaris minor

Phalaris canariensis

POACEAE

Phragmites australis
common reed * fluitjiesriet

Typha capensis
common bulrush * gewone papkuil

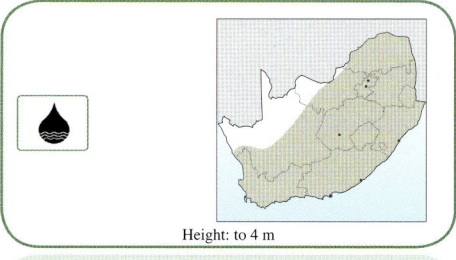
Height: to 4 m

Origin and description: These are indigenous and widespread perennial plants commonly found growing in damp places and in water.
Impact: Reeds and bulrushes are usually beneficial species, acting as erosion control agents, filtering muddy floodwaters and being a haven for a wide range of wildlife. However, they are sometimes unwanted as they can hinder access to lakes and dams and can block drainage channels. Do not confuse them with the highly invasive giant reed, *Arundo donax*.
Control: All reeds are difficult to control. Glyphosate and imazapyr are registered as foliar sprays

Other common names
P. australis: carrizo * sonquasriet * vaderlandsriet * vinkriet * vlakkiesriet
T. capensis: cat's tail * cossack asparagus * poker plant * reedmace * matjiesgoed * palmiet * papies * motsitla (SS) * umkhanzi (X) * ibhuma (Z)
Similar species
P. mauritianus: lowveld reed * dekriet * laeveld-fluitjiesriet * olumbungu

and are best applied to the vigorous regrowth that appears after the main plant is slashed. Treatment should be done before flowering and avoiding water that is to be used for irrigation.

Phragmites australis

Typha capensis

Phragmites australis

POACEAE

Poa annua
wintergrass * wintergras

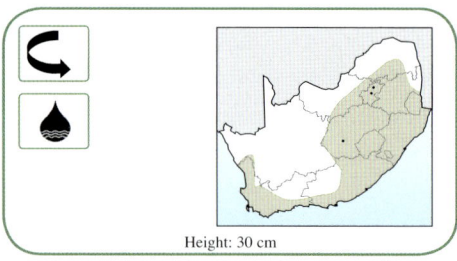
Height: 30 cm

Origin and description: This grass is a relatively small, bright green annual plant that is a native of Europe. It is now distributed worldwide and is common in all areas of South Africa. It grows throughout the year in damp and shady places but is particularly noticeable in winter.
Impact: It can be a troublesome weed in gardens, lawns, golf courses and bowling greens where lit-

Other common names
annual bluegrass * dwarf meadow grass * low spear grass * six weeks grass * eenjarige blougras * straatgras * joang-ba-lintja (S)

Poa annua

tle green tufts remain when the cultivated grass has frosted off. It is also found near places such as leaking taps, where the soil may be too damp for other species.
Control: During winter, when golf courses or lawns are brown and dormant, this weed can be controlled by means of non-selective contact herbicides. However, in the summer months, selective post-emergence chemicals will be required. Courses with 'bent grass' (usually *Agrostis* species) greens can use various stress elimination techniques to promote the growth of the desired grass. Pre-emergence herbicides can prevent regeneration of *P. annua* from seed.

POACEAE

Polypogon monspeliensis
rabbit's foot * brakbaardgras

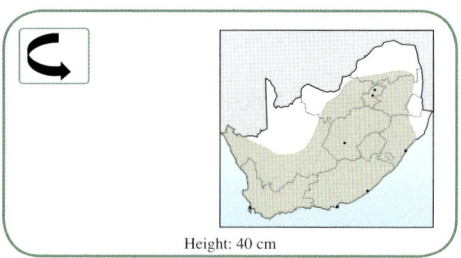
Height: 40 cm

Origin and description: Originally from Europe and Asia, this grass was probably introduced as an ornamental and is now naturalised in many parts of South Africa, especially coastal areas. It grows well on 'brak' or brackish soils but despite its lush growth, it is not a good grazing species. It favours moist places.
Impact: It replaces indigenous vegetation.
Control: Although it is an annual, *P. monspeliensis* appears to be relatively tolerant to many herbicides. It is probably best controlled by shal-

Other common names
beard grass * brakgras * brakbaardgras

low cultivation in the early stages. No herbicides have been registered for controlling this weed.

Polypogon monspeliensis

POACEAE

***Rottboellia cochinchinensis** (=R. exaltata)*
itch grass * tarentaalgras

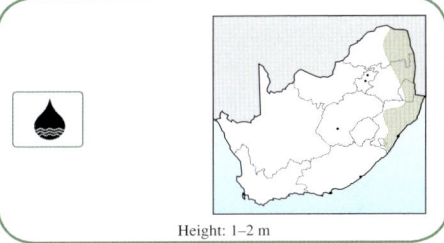

Height: 1–2 m

Other common names
guinea-fowl grass * kokoma grass * Raoul grass * shamva grass

Origin and description: This grass is probably indigenous, but is also native to India and Australia and is a major annual grass weed worldwide. It is a problem plant in several crops, especially sugarcane in the subtropical areas of the Philippines, Colombia, Cuba, the USA, as well as in southern Africa. *R. cochinchinensis* only grows from its barrel-shaped seeds, which are large and mainly dispersed by water, wind and human activities. Guinea-fowl, in particular, find the seeds very palatable.

Impact: The stiff, sharp hairs on the basal leaf sheath of this grass are easily dislodged and cause skin irritations. For this reason, it is internationally known as itch grass and is very unpleasant to have to remove by hand. It is found in many areas in South Africa where sugarcane is grown, particularly northern Zululand. It also occurs in maize at places such as Normandien and Colenso in KwaZulu-Natal and at various sites in

Rottboellia cochinchinensis

Mpumalanga, especially in the eastern lowveld. It becomes dense and competitive and is a serious weed in sugarcane.

Control: *R. cochinchinensis* is a deep-rooted, large-seeded grass and not selectively controlled by pre-emergence grass herbicides in sugarcane and maize. It germinates at variable depths, which results in extended emergence periods and makes effective control even more difficult. It is best controlled at the seedling stage with post-emergence herbicides or by cultivation. MSMA is frequently used in mixtures to assist in sugarcane.

Rottboellia cochinchinensis

POACEAE

Setaria megaphylla *(=S. chevalieri)*
ribbon bristle grass * breëblaarborselgras

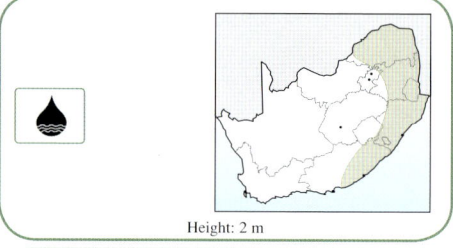

Height: 2 m

Other common names
broad-leaved setaria * bush grass * forest buffalo grass * macopo grass * solitz grass * breëblaarsetaria * mufhafha (V)

Origin and description: *S. megaphylla* is indigenous and widespread in the eastern half of South Africa's summer-rainfall region, but also throughout tropical Africa. It is a large and robust perennial grass that favours moist and shady areas such as those under trees. It is planted in gardens as an ornamental and plays a major role in water purification as it absorbs excess nutrients from the water.

Impact: It is considered a major weed of silviculture. It hampers regeneration and in the first few years of stand development, it competes vigorously with the young trees, causing a loss of increment. It is difficult to eradicate once it has become established, since the underground rhizomes can resist most systemic chemicals.

Control: *S. megaphylla* can be controlled in forestry with relatively high rates of systemic herbicides like glyphosate, but this is laborious and expensive on large plants. Control should be initiated when the plants are still small, certainly before flowering. Some foresters have had success eradicating the weed even before clear felling. This ensures that the rapid regrowth of this grass is prevented when the trees are removed.

Setaria megaphylla

Setaria megaphylla

POACEAE

Setaria pallide-fusca
red bristle grass * rooiborselgras

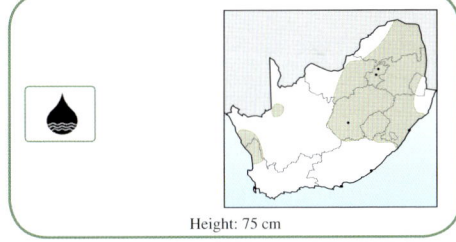

Height: 75 cm

Origin and description: There are many species of *Setaria* that are troublesome in South Africa, including this one. They are all probably indigenous. The bright orange seed heads of *S. pallide-fusca* make it stand out from other weeds. The seedlings of this grass are characterised by their upright growth habit and bright red stem bases.

Impact: *S. pallide-fusca* is an annual agricultural weed causing serious problems in most areas of the summer-rainfall region. As it often appears in maize late in the season, it escapes regular pre-emergence herbicides. It subsequently presents a fire hazard in winter.

Control: This grass is controlled effectively by pre- and post-emergence herbicides when applied timeously. The seedlings are easily destroyed by cultivation.

Other common names
cat's tail * garden setaria * horse grass * perdesoetgras * tuinsetaria * lehola-la-lipere (S)

Setaria pallide-fusca

POACEAE

Setaria verticillata
sticky bristle grass * klitsborselgras

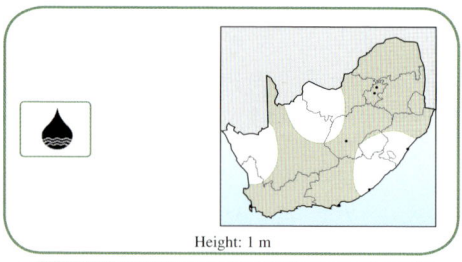
Height: 1 m

Origin and description: This is one of the many species of *Setaria* that are troublesome in southern Africa. *S. verticillata* is native to Europe and is now naturalised in South Africa. It is an annual grass that is widely distributed in South Africa, except in KwaZulu-Natal, where it is uncommon.

Setaria verticillata

Impact: It is an annual species and is found in gardens and other disturbed places, especially where it is damp or shaded. The bristles characteristically adhere to clothing or the hair of animals, which is an efficient means of dispersal. It contaminates sheep's

Other common names
bristly foxtail * bur bristle grass * cat's tail * hooked bristlegrass (USA) * love grass * klawergras * siergras * steekgras

wool. In parts of Africa the seed head is used to cover stored grain to deter rats, as their fur becomes entangled in the barbed bristles.

Control: This grass is controlled effectively by pre- and post-emergence herbicides when applied timeously. The seedlings are easily destroyed by cultivation.

POACEAE

Sorghum bicolor subsp. *arundinaceum* (=*S. verticilliflorum*)
common wild sorghum * gewone wildesorghum

Sorghum bicolor subsp. *drummondii* (=*S. almum*)
wild grain sorghum * wildegraansorghum

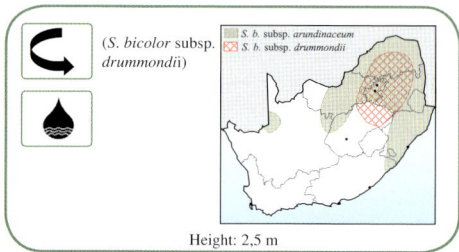
Height: 2,5 m

Origin and description: *S. bicolor* subsp. *arundinaceum* is an indigenous grass. It reproduces only by means of seeds (no rhizome), but is capable of producing large perennial clumps. It is common throughout KwaZulu-Natal and Mpumalanga. *S. bicolor* subsp. *drummondii* is an annual from elsewhere in Africa. It resembles grain sorghum and has only become a problem since the 1970s. Its distribution is centred around Lichtenburg in North West Province and Bronkhorstspruit in eastern Gauteng.

Impact: These closely related grasses cause serious problems in agriculture. The pollen of *S.*

Sorghum bicolor subsp. *arundinaceum*

bicolor subsp. *arundinaceum* is known to cause hay fever and, like all *Sorghum* species, it can cause prussic acid poisoning in animals. It is a particularly problematic weed in subtropical crops in the KwaZulu-Natal coastal belt and the Mpumalanga lowveld. In sugarcane, for example, this grass can cause major problems. It can grow as tall as the cane and is difficult to locate because its broad leaves are very similar to that of the sugarcane. It is also a problem in maize, where it can only be controlled effectively with expensive, specialist herbicides.

Control: If allowed to establish itself, *Sorghum* species can only be controlled effectively by hand. Pre-emergence control tends to be erratic. As seedlings, these grasses are relatively susceptible to pre- or early post-emergence grass herbicides. Effective control of these weeds therefore depends on the timeous and efficient application of these herbicides. Once these grasses become established, they are extremely difficult to eradicate.

Sorghum bicolor subsp *drummondii*

Sorghum bicolor subsp *arundinaceum*

POACEAE

Sorghum halepense
Johnson grass * Johnsongras

Sorghum versicolor
black seed wild sorghum * swartsaadwildesorghum

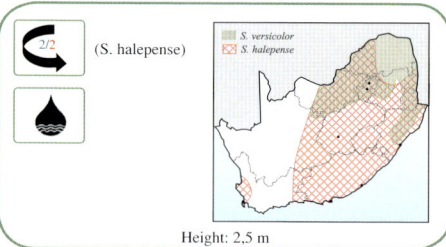

Height: 2,5 m

Origin and description: *S. halepense*, which was introduced from Europe, is perhaps potentially the most serious *Sorghum* weed. At present though in South Africa, it is not as common as *S. bicolor* subsp. *arundinaceum* and although quite widespread, it is most common in Mpumalanga. *S. halepense* differs from the other *Sorghum* species by having rhizomes from which it can reproduce and grow large perennial 'stools'. *S. versicolor* is an indigenous annual grass, characterised by hairy nodes and inflorescences that are partly black. It prefers black turf soils in Mpumalanga and northern KwaZulu-Natal.

Impact: These are large and competitive grasses and will out-perform most crops.

Control: On account of the robust rhizomes, *S. halepense* is extremely difficult to eradicate once it has become established. High rates of nonselective herbicides are required for chemical control, or the total removal of the plant by manual means.

Other common names
S. halepense: aleppo grass * Egyptian millet * evergreen millet * iquangoboto (N)

Sorghum halepense

POACEAE

Sporobolus africanus
rat's tail dropseed * rotstertfynsaadgras

Origin and description: 'Dropseeds' are indigenous, perennial grasses that are members of the group referred to as 'mtshiki' by Zulus and many farmers, because of their similar, upright growth habit. Other species in this group are *Eragrostis curvula* and *E. plana*.

Impact: *S. africanus* is the most common of the 'dropseeds', being widespread in South Africa and found in disturbed ground and compacted areas. It is a nuisance in lawns on account of its tough, wiry nature. *S. fimbriatus* and *S. pyramidalis* are relatively less common, but also occur as weeds in disturbed or compacted areas. They are not usually weeds of annual crops.

Control: Control of these grasses with herbicides is variable and usually the higher rates are recommended. They are probably most effectively controlled by cultivation when still in the seedling stage.

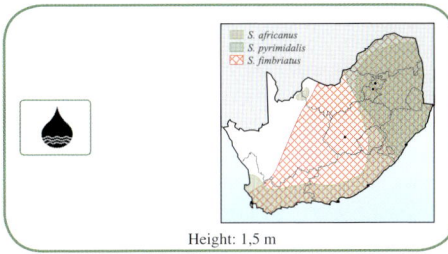
Height: 1,5 m

Other common names
rush grass * tough dropseed * taaipol * mtshiki (Z)

Similar species
S. fimbriatus (common dropseed)
S. pyrimidalis (cat's tail * dropseed)

Sporobolus pyrimidalis   *Sporobolus africanus*

POACEAE

Stenotaphrum secundatum
St Augustine grass * Augustinuskweek

Origin and description: *S. secundatum* is a creeping perennial grass of uncertain origin, but is thought by some authorities to have originated in North America, the West Indies and Australia. It is common in the eastern coastal areas of southern Africa.

Impact: It is widely used in lawns on sandy soils where it forms a dense, rather coarse mat. It does

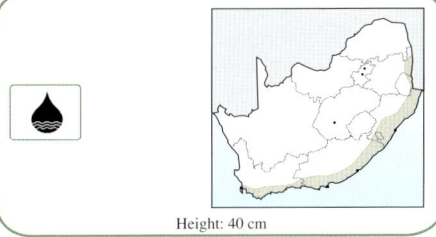
Height: 40 cm

Other common names
Cape kweek * carpet grass * coastal buffalo grass * mission grass *

not often invade crops, but is usually found as a minor weed on roadsides and in waste places, especially in coastal areas where it can tolerate the salty environment and can out-compete more desirable species. Several varieties are available for commercial use.

Control: *S. secundatum* is controlled effectively by industrial and systemic grass killers.

Other common names
ramsammy grass * Charleston grass (USA) * buffelskweek * strandbuffelsgras * marotlo-a-mafubelu (S) * umtombo (Z)

Stenotaphrum secundatum

POACEAE

Tragus berteronianus
small carrot-seed grass * kleinwortelsaadgras

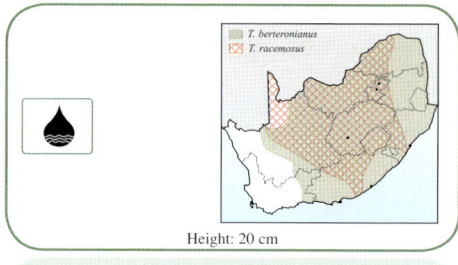

Height: 20 cm

Origin and description: This indigenous, annual grass is widespread in South Africa, except in the southwestern Cape.

Impact: As a weed of crops, *T. berteronianus* occurs mainly in the lowveld and northern areas of Mpumalanga where it appears in many locations. It is also common in KwaZulu-Natal as a

Other common names
burgrass * haasgras * kousgras * kousklits * luisgras * wolgras * wortelsaadgras * bore-ba-ntja (S)

weed of sugarcane. Elsewhere it usually grows in sandy places, in waste areas and on roadsides, often in pure stands and forming loose mats. It is a prolific producer of viable seeds and therefore spreads rapidly. The hooked spikelets often cling to the wool of sheep and to clothing.

Control: *T. berteronianus* is susceptible to normal pre- and post-emergence grass herbicides.

Similar species
T. racemosus (large carrot-seed grass), very similar but with larger seeds.

Tragus berteronianus

POACEAE

Urochloa mosambicensis
bushveld herringbone grass * bosveldbeesgras

Urochloa panicoides
herringbone grass * beesgras

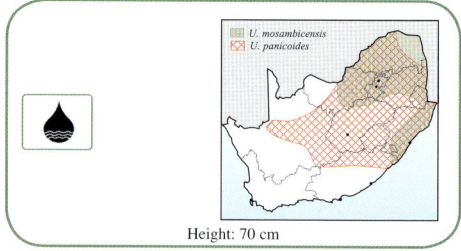

Height: 70 cm

Other common names
U. panicoides: garden urochloa * kuri-millet * poke weed * tuinurochloa * bore-ba-ntja (S)
U. mosambicensis: bushveld signal grass

Origin and description: *U. panicoides* is a common and often serious indigenous annual grass weed. It is most common in the northwestern Free State, the highveld and eastern areas of Mpumalanga. *U. mosambicensis* is similar and also indigenous but is found only in the warmer, eastern regions of South Africa. It is a perennial, but seldom occurs as such in cultivated lands as it grows readily from seeds which are produced in quantity and are used by rural people to make porridge. *U. panicoides* is one of the four species of grasses commonly referred to as 'landgrasses', because they are common and as seedlings, similar to annual grass weeds of croplands. The other 'landgrasses' are *Digitaria sanguinalis*, *Eleusine coracana* and *Panicum schinzii*.

U. panicoides is easily distinguished from the other 'landgrasses' by its crinkly leaf margins and by being noticeably hairy. When established, it often roots from the lower nodes.

Impact: *U. panicoides* is a weed of most crops in most areas, but usually on heavier soils. *U. mosambicensis* also invades veld and can produce underground runners, especially under heavy grazing. It is, in fact, very palatable and in such instances, desirable.

Control: As germination takes place over a short period in the spring, timely cultivation can give good control. Furthermore, *U. panicoides* is well controlled by most of the grass herbicides although it seems to be relatively tolerant to the triazines. *U. mosambicensis* is very susceptible to the normal grass herbicides.

Urochloa mosambicensis

Urochloa panicoides

POACEAE

Vulpia myuros
rat's tail fescue * langbaardswenkgras

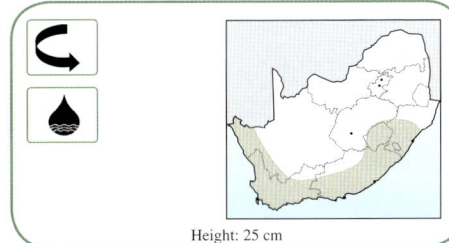

Height: 25 cm

Vulpia myuros

Other common names
wildegars
Similar species
V. bromoides, a very similar species, can be found as far north as KwaZulu-Natal, where it tends to favour comparatively drier areas

Origin and description: Introduced from Europe, *V. myuros* is now a serious annual grass weed in the southern regions of South Africa. Although this grass is restricted mainly to the southern Cape and Eastern Cape, it is common in these areas.
Impact: *V. myuros* is a common weed of roadsides, waste places and gardens, especially in lawns, and in lucerne.
Control: This weed is difficult to control once it has become established and does not respond to most selective grass killers. Control should be initiated when the plants are small.

TREES

Within the section, the plants are arranged alphabetically according to family and genus.

SPECIES	Compound leaves	Phyllodes (stalkless leaves)	Needle-like leaves	Thorny stems	Deciduous or semideciduous	Red flowers	Yellow flowers	Purple flowers	White flowers	Green or indistinct flowers
Ailanthus altissima, p. 158	X				X					X
Albizia spp., p. 127	X							X	X	
Gleditsia triacanthos, p. 129	X			X	X		X			
Jacaranda mimosifolia, p. 118	X				X			X		
Melia azedarach, p. 137	X				X			X		
Paraserianthes lophantha, p. 131	X								X	
Phytolacca dioica, p. 146	X									
Prosopis glandulosa, p. 132	X			X			X			
Rhus succedanea, p. 115	X				X					X
Robinia pseudoacacia, p. 133	X			X	X				X	
Schefflera actinophylla, p. 117	X					X				
Schinus spp., p. 116	X									X
Tipuana tipu, p. 134	X				X		X			
Acacia spp., p. 121–127		X					X			
Hakea spp., p. 152			X						X	
Pinus patula, p.148			X							
Pinus pinaster, p. 149			X							
Acer spp., p. 114					X	X				X
Bauhinia spp., p. 129					X	X		X	X	
Morus spp., p. 138					X					X
Salix spp., p. 157					X					X
Populus x canescens, p. 155					X	X				
Callistemon spp., p. 139						X				
Eucalyptus spp., p. 140, 141						X	X			
Prunus persica, p. 155						X			X	
Triplaris americana, p. 150						X				
Grevillea spp., p. 151						X				

Trees

SPECIES	Compound leaves	Phyllodes (stalkless leaves)	Needle-like leaves	Thorny stems	Deciduous or semideciduous	Red flowers	Yellow flowers	Purple flowers	White flowers	Green or indistinct flowers
Cestrum laevigatum, p. 158							■			
Litsea glutinosa, p. 136							■			■
Nicotiana glauca, p. 160							■			
Solanum mauritianum, p. 161								■		
Cinnamomum camphora, p. 135									■	
Eriobotrya japonica, p. 154									■	
Leptospermum laevigatum, p. 143									■	
Myoporum spp., p. 138									■	
Pittosporum undulatum, p. 150									■	
Psidium guajava, p. 144									■	
Sambucus nigra, p. 120									■	
Syzygium spp., p. 145									■	

ACERACEAE

Acer buergerianum
Chinese maple * Chinese ahorn
Acer negundo
ash-leaved maple * essenblaarahorn

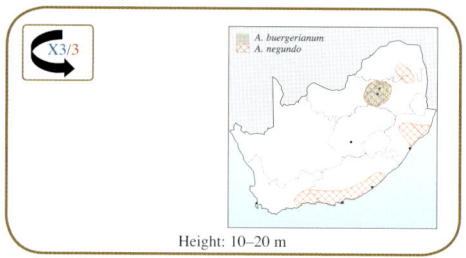

Height: 10–20 m

Origin and description: *A. buergerianum* comes from eastern China, whereas *A. negundo* comes from North America. Both were introduced for shade trees and as ornamentals. *A. buergerianum* has typical three-lobed maple leaves, whereas *A. negundo*, unlike most maples, has compound leaves. They both produce large numbers of two-winged fruits called samaras that festoon the trees in summer and autumn.

Impact: There are many members of the *Acer* genus that are planted as ornamentals throughout the world, including South Africa. However, these two can be found regenerating from the large numbers of seeds and can invade such places as roadsides, graveyards and urban waste areas. Usually found in urban areas close to the parent tree. *A. buergerianum* has been noticed as a significant invader in parts of Gauteng and Mpumalanga, whereas *A. negundo* is recognised as an emerging weed along the Garden Route, in parts of KwaZulu-Natal and as far north as Pilgrim's Rest.

Other common names
A. buergerianum: trident maple * Chinese esdoring
A. negundo: box elder * Kaliforniese esdoring

Acer negundo

Control: Unwanted plants should be physically removed.

Acer buergerianum

ANACARDIACEAE

Rhus succedanea *(=Toxicodendron succedaneum)*
wax tree * wasboom

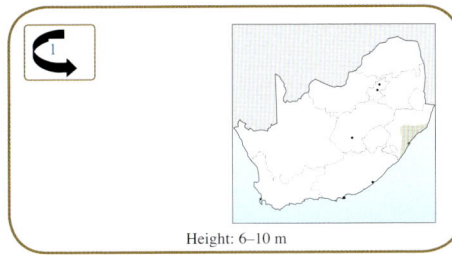

Height: 6–10 m

Origin and description: A deciduous, small, spreading tree or shrub with soft, fuzzy branches and shoots. Originally from eastern Asia, it is grown primarily as an ornamental and a street tree but elsewhere wax is extracted from the berries and resin from the stems. The leaves change colour dramatically in the autumn. There are many indigenous species of *Rhus* (now classified in the genus *Searsia*) and some of them have also been accused of being weedy.

Impact: The wax tree can cause severe dermatitis, with a rash, itching and blisters where the skin has touched the plant. The sap causes the worst reaction, but contact with any part of the tree can result in some symptoms. Birds that eat the fruit have spread this plant into the wild, especially in the moist soils of KwaZulu-Natal, and it is also spread in soil that is moved around and between gardens. It is not a pleasant tree to come into contact with and its cultivation should be discouraged.

Control: Unwanted trees should be removed and replaced with indigenous ones. Protective clothing should be worn if the tree is to be removed manually and contact with the sap should be avoided at all costs. The branches should not be mulched or chipped for garden use, as the toxic resin remains active for many months.

Similar species
Rhus glabra (smooth sumac * scarlet sumac * gladdesumak), related but is nontoxic 3/3

Rhus succedanea

ANACARDIACEAE

Schinus terebinthifolius
Brazilian pepper tree * Brasiliaanse peperboom

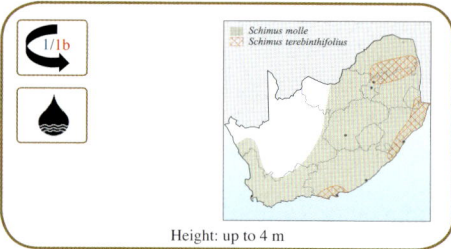

Height: up to 4 m

Origin and description: Introduced from South America as an ornamental plant, *S. terebinthifolius* is an evergreen, perennial shrub or tree that can reach a height of more than 6 m. It was used as a firebreak between sugarcane plantations, is still planted as an ornamental and is commonly used as a hedge in the centre of highways. It has escaped and become well established along the coastal roads and in the coastal bush of KwaZulu-Natal.

Impact: The flowers of *S. terebinthifolius* are a source of nectar for honeybees and the bright red berries are a familiar sight in late summer. It is, however, a vigorous invader and easily replaces indigenous vegetation by shading them out. It is particularly invasive of wetland or riverine areas. It has been nominated as among 100 of the 'World's Worst' invaders.

Control: Although large and established root systems are difficult to remove, this plant responds well to physical methods of control. If mature trees are cut down, care should be taken not to disperse seeds with the trash. Triclopyr is registered for application as a basal stem treatment, but it can take several weeks to be effective.

Other common names
Brazilian holly * Christmas berry tree * pepper hedge * South American pepper

Similar species
S. molle (pepper tree * peperboom), also from South America, is a widespread ornamental tree X3

Schinus molle

Schinus terebinthifolius

ARALIACEAE

Schefflera actinophylla *(=Brassaia actinophylla)*
umbrella tree * Australiese kiepersol

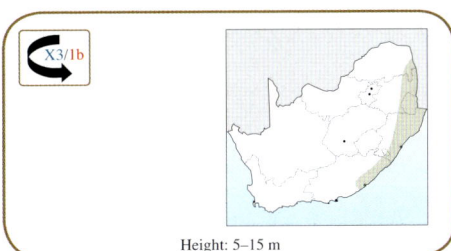
Height: 5–15 m

Origin and description: *S. actinophylla* is a decorative, soft-wooded evergreen tree from tropical Australasia and is a common ornamental in subtropical towns and cities such as Durban. It is also favoured as an indoor and patio plant. It is usually multi-trunked and the octopus-like flowers develop at the top of the tree. The other species are also popular ornamentals, with a wide range of cultivars.

Impact: Unfortunately, *S. actinophylla* is an aggressive plant and its roots can dominate surrounding soil. It can also grow on another tree, eventually strangling and killing it. It has already been declared an invasive alien weed in Florida and Hawaii. All these species are under suspicion and are on the CARA proposed lists.

Control: Unwanted plants should be removed and care should be taken that garden specimens do not spread elsewhere.

Other common names
S. actinophylla: amate * Australian cabbage tree * octopus tree * Queensland umbrella tree
Similar species
S. arboricola X 3 (=*Heptapleurum arboricolum*) dwarf umbrella tree * Hawaiise dwerg, from Taiwan
S. elegantissima X 3 (=*Dizygotheca elegantissima*) falsearalia * vals-aralia, from New Caledonia

Schefflera actinophylla

117

BIGNONIACEAE

Jacaranda mimosifolia
jacaranda * jakaranda

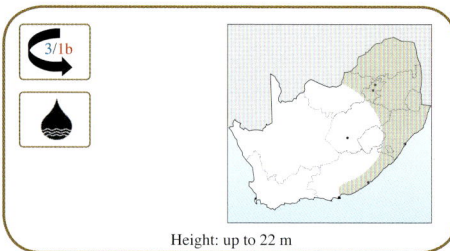

Height: up to 22 m

Origin and description: Introduced from Central and South America in 1888, *J. mimosifolia*, apart from being used as an ornamental, is also a widespread and invading weed. It was first planted in Pretoria in a garden at what is now Sunnyside Primary School and is now planted all over Pretoria, with nearly 75 000 specimens, earning Pretoria the label of 'Jacaranda City'. There are also about 100 white jacarandas in Pretoria; these are a 'sterile' hybrid. Many other towns and cities in Mpumalanga, Gauteng and KwaZulu-Natal have streets lined with jacarandas. It is also a serious invader in Australia.

Impact: *J. mimosifolia* is being planted in towns and gardens throughout South Africa and has invaded natural vegetation, especially along roads and watercourses. In these areas it is a highly undesirable plant because of its aggressive growth habit, replacing indigenous vegetation and consuming valuable soil moisture.

Control: *J. mimosifolia* is very difficult to eradicate once established. Large trees must be ring-barked, cut below ground level or cut down and the regrowth treated with systemic herbicides such as imazapyr.

Other common names
blue Brazilian * blue jacaranda * Brazilian rosewood * fern tree

Jacaranda mimosifolia

Jacaranda mimosifolia

CAPRIFOLIACEAE

Sambucus canadensis
Canadian elder * Kanadese vlier

Sambucus nigra
European elder * Europese vlier

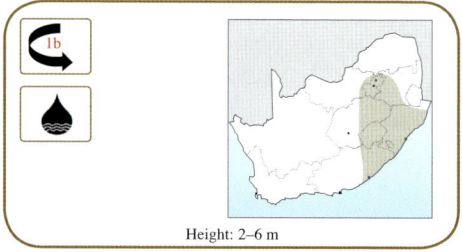
Height: 2–6 m

Other common names
black elder * elder bush * elderberry

Origin and description: There are up to 30 species of *Sambucus* or 'elder' that occur naturally throughout the world, but mainly in the northern hemisphere. *S. nigra*, from Europe, flowers in the spring and then produces clusters of red to black berries. These berries, which are mildly toxic when immature, are much favoured by birds and are used to make a passable wine and even a brandy. The leaves have a strong smell and were used in the past, tied to a horse's mane, to keep the flies away. *S. canadensis*, from North America is very similar and is thought by some authorities to be a subspecies of *S. nigra*. Some authorities place *Sambucus* in the Adoxaceae, while SANBI has now classified it in the Caprifoliaceae.

Impact: *S. nigra* can easily escape from gardens and establish itself in waste areas. It can be untidy and can replace indigenous vegetation. Although having been present in South Africa for some time, these plants are only now becoming significant and noticeable invaders.

Control: Unwanted plants should be cut down and removed.

Sambucus nigra

FABACEAE

Acacia baileyana
Bailey's wattle * Bailey-se-wattel

Acacia podalyriifolia
pearl acacia * vaalmimosa

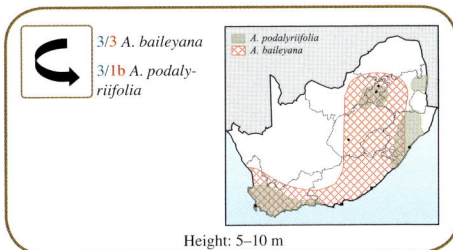
Height: 5–10 m

Origin and description: Another two *Acacia* species brought in from Australia, but as ornamentals. The bright yellow blossoms of these trees are a beautiful contrast with the silver-grey leaves and are much favoured in flower arranging. Unfortunately they seed down easily and can rapidly establish themselves as weeds in gardens, along roadsides and in open urban spaces. The seeds are believed to be dispersed by birds and ants and reach peak germination after fire. *A. podalyriifolia* is found mainly in and around the large conurbations of Johannesburg, Durban and Cape Town, whereas *A. baileyana* is more widespread.

Impact: They replace indigenous species and, although not as aggressive as many of the other acacias, they are generally invasive and threaten biodiversity.

Control: These species respond to the various control techniques in the same way as the other Australian acacias.

Other common names
A. baileyana: Cootamundra wattle
A. podalyriifolia: golden ball wattle * Queensland silver wattle * pêrel-akasia

Acacia baileyana

Acacia podalyriifolia

FABACEAE

Acacia cyclops
redeye * rooikrans
Acacia melanoxylon
blackwood * swarthout

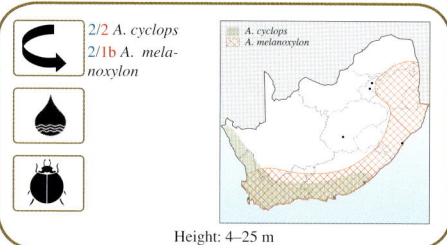

Height: 4–25 m

Origin and description: Like the other invasive wattles, these weeds are from Australia. The seeds of both these species are black with bright red seed stalks. The seeds of *A. melanoxylon* are smaller whereas the plant is considerably taller than *A. cyclops*.

   A. cyclops was introduced and used with *A. saligna* (see page 126) in the second half of the 19th century to stabilise the shifting sands on the Cape Flats. Many farmers, especially in dry areas, value this plant as fodder. *A. cyclops* is well established in Mountain and Lowland Fynbos vegetation types throughout the Cape coastal region. It can be a tall tree or a shrub, swamping indigenous vegetation. *A. melanoxylon* can grow up to 35 m and is found from the southwestern Cape, through KwaZulu-Natal and into the high-rainfall escarpment areas of Mpumalanga. It produces excellent timber and was originally introduced and planted as a forest replacement species in the Knysna forest around 1856. *A. cyclops* rarely coppices but *A. melanoxylon* can regenerate from vigorous root suckers. They both produce large quantities of long-lived seeds that make a massive seed bank in the soil. Apart from birds and other animals, the seeds are spread in sand taken from infested areas and used for building.

Impact: Both these species form large mono-

Other common names
A. cyclops: redwreath acacia * baaibos * hoenderboom * rooipit
A. melanoxylon: stinkboontjie * stinkpeul

Acacia cyclops

Acacia melanoxylon

specific stands that eliminate indigenous flora and totally transform the landscape. In certain areas, however, these plants can still play a useful role as stabilising agents and as a source of firewood.

Control: Successful long-term control requires a co-ordinated programme of physical, chemical and cultural techniques. Biological control by means of a beetle from Australia is progressing.

FABACEAE

Acacia dealbata
silver wattle * silwerwattel

Acacia decurrens
green wattle * groenwattel

Acacia mearnsii
black wattle * swartwattel

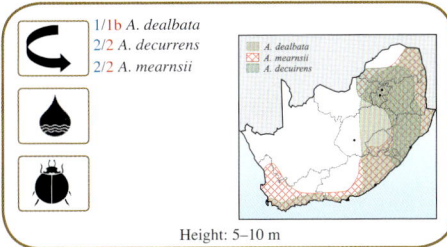

Height: 5–10 m

Other common names
A. dealbata: blue wattle * bloubasboom * blouwattelboom * vaalwattel * uwatela (Z)
A. decurrens: Australian wattle * Australiese basboom * groenbasboom

Origin and description: These are three species of 'wattle' that originate from Australia. *A. mearnsii* was introduced into present-day KwaZulu-Natal in the 19th century by an immigrant named John Vanderplank, who planted seeds from Tasmania on his farm at Camperdown. Shortly afterwards, commercial planting of *A. mearnsii* began for the production of tannic acid. This is used in the leather industry and the timber is of value for pulp, firewood and the mining industry. *A. mearnsii* and *A. dealbata* are the more serious invaders and may be difficult to tell apart; the glands on the leaf rachis of *A. dealbata* are evenly distributed whereas those of *A. mearnsii* are irregular. The overall appearance of *A. dealbata* is a light grey-green colour, whereas *A. mearnsii* and *A. decurrens* have darker leaves. *A. decurrens* is not very common and, along with *A. dealbata*, was probably introduced by mistake. Neither of them is planted commercially, unlike *A. mearnsii*.
Impact: All three species are now serious invaders of veld, fynbos, indigenous bush, watercourses, roadsides and, occasionally, perennial crops such as sugarcane. They severely threaten biodiversity. *A. mearnsii* is the most prominent invader by far in South Africa (Henderson 2007) and is No. 1 on the list of the *World's Worst Invasive Species*.
Control: Long-term control of wattles is difficult as they coppice easily and produce large numbers of seeds that can remain dormant for

Acacia dealbata

well over 50 years. These seeds are efficiently dispersed by water and their germination is stimulated by fire. Plants should not be felled, bulldozed or burnt without immediate follow-up with herbicides. A combination of chemical, mechanical and management techniques, including the use of competitive cover crops, is usually required for effective long-term control. There are soil- and foliar-applied herbicides registered. Insects that attack the seeds are under investigation and were released in 1994. The effects are still being monitored.

Acacia dealbata

Acacia mearnsii

FABACEAE

Acacia longifolia
long-leaved wattle * langblaarwattel

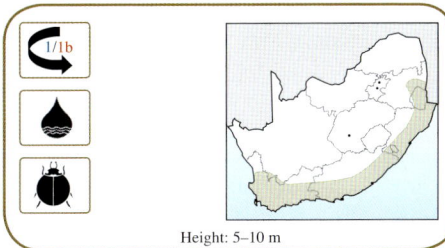

Height: 5–10 m

Origin and description: Introduced from Australia, *A. longifolia* is now widespread with isolated populations occurring throughout the eastern parts of the country. It was first introduced in 1827 but it was not until 1945 that it was recorded as being a problem plant alongside rivers at Houwhoek and Mitchell's Pass.

Other common names
golden wattle * Port Jackson acacia * sallow wattle * Sydney golden wattle

Impact: It favours moist sites and is the only invasive wattle with finger-like inflorescences. It does not coppice easily. It rapidly becomes dense; it is competitive and can totally transform the landscape.

Control: Control of this invasive acacia can only be achieved with a combination of techniques of which the use of herbicides is only one aspect. Chopping, burning and cultural techniques must all play a part, so it is advisable to consult an expert before embarking on a control campaign. In coastal areas, *A. longifolia* is being successfully controlled with a gall insect that reduces seed set. (The galls are clearly visible in the photograph.) Unfortunately, it seems that this method of control is much less effective in Mountain Fynbos, and where there has been success, the areas are being re-colonised by other exotic species.

Acacia longifolia

FABACEAE

Acacia pycnantha
golden wattle * goue wattel

Acacia elata (=*A. terminalis*)
pepper tree wattle * elataboom

Height: 5–15 m

Origin and description: Both these wattles were introduced from Australia. *A. pycnantha* was planted in 1893 in an attempt to reclaim dunes at Port Elizabeth and the Cape Flats. These sites are still the foci of modern infestations. *A. elata* is also planted as an ornamental and is established in parts of the Western Cape. *A. pycnantha* has a high tannin content but is a relatively slow grower

Other common names
A. pycnantha: broadleaf wattle
A. elata: cedar wattle * mountain cedar wattle *peperboomwattel

compared to other species grown commercially for their tannin.

Impact: Neither of these species is as invasive as other wattles and they are less widely distributed. *A. pycnantha* was introduced as a stabilising agent and therefore has a strong pioneering nature. *A. elata* is used as an ornamental or as a shade tree and, being less aggressive, it is now mainly found on roadsides and in urban areas. They are, however, large trees and replace indigenous vegetation.

Control: Control of these plants requires a combination of chemical, physical and cultural techniques.

Acacia elata

FABACEAE

Acacia saligna (=*A. cyanophylla*)
Port Jackson * goudwilger

Origin and description: *A. saligna* is from Western Australia and was first planted in 1848 to stabilise the loose sand that threatened to cover the new road from Cape Town to Bellville. It is now found around the South African coast, from the Orange River in the west to Kosi Bay in the east and has more recently spread inland. This *Acacia* also thrives in Zambia, especially on mine dumps. It can grow rapidly in soil with a low level of nutrients and produces large quantities of seed, which are dispersed by ants, birds, other animals and by human activities. Germination is stimulated by the passage through the digestive tract. *A. saligna* seeds are also transported in sand used for roads and building purposes and these seeds have even germinated in painted concrete walls.

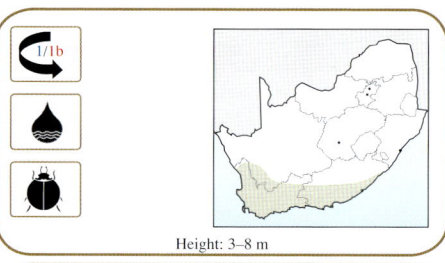

Height: 3–8 m

Other common names
golden willow * Port Jackson willow * saligna wattle * portjeksonwilg

Acacia saligna

Impact: Veld fires stimulate germination of the long-lived Port Jackson seeds and give the species an advantage over destroyed indigenous vegetation. It coppices relatively easily, even from burned stumps and forms large monospecific stands, which eliminate indigenous flora and totally transform the landscape. It increases erosion, prevents coastal dune movement and reduces the diversity of vertebrates. In certain areas, however, this plant can still play a useful role as a stabilising agent and as a source of firewood.

Control: Control of exotic *Acacia* species requires a combination of chemical, physical and cultural techniques. The introduction in 1987 of the acacia gall rust fungus (*Uromycladium tepperianum*) from its native land is proving very effective. A seed weevil was also introduced in 2001 and the benefit of this will be seen soon.

Acacia saligna

FABACEAE

Albizia lebbeck
lebbeck tree * lebbeckboom

Albizia procera
false lebbeck * basterlebbeck

Albizia julibrissin
silk tree

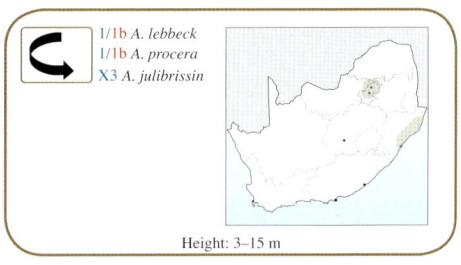

1/1b *A. lebbeck*
1/1b *A. procera*
X3 *A. julibrissin*

Height: 3–15 m

Origin and description: From tropical Asia, these species of *Albizia* have been introduced as ornamental and shade trees. Both *A. lebbeck* and *A. procera* are Category 1 plants, being the most potentially invasive, while *A. julibrissin* poses less of a threat. *A. lebbeck* is sometimes called 'the mother-in-law tree' on account of the constant rattling of the pods as they blow in the wind. There are many beautiful indigenous species of *Albizia*.

Impact: These trees can replace indigenous species. They are generally invasive and threaten biodiversity. *A. lebbeck* is now naturalised along the KwaZulu-Natal north coast.

Control: Unwanted plants should be physically removed and replaced with an indigenous species.

Albizia lebbeck

Albizia procera

Albizia julibrissin

FABACEAE

Bauhinia variegata var. *variegata*
orchid tree * orgideeboom

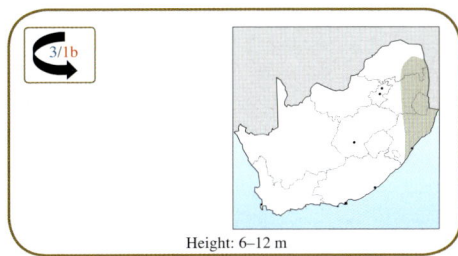
Height: 6–12 m

Origin and description: *B. variegata* is from East Asia and is the floral emblem of Hong Kong. It was originally introduced as an ornamental. Although pink is the most common flower colour, many cultivars with different colours have been developed since; many of them are sterile. The tree is semi-evergreen.
Impact: It is found in urban open spaces, coastal bush and riverbanks in the subtropical regions. It is competitive and replaces indigenous vegetation.
Control: Unwanted plants should be cut down and the stump either removed or treated with herbicide.

Other common names
butterfly tree * mountain ebony
Similar species
Bauhinia purpurea 3/1b (butterfly orchid tree * skoenlapper-orgideeboom)

There are several indigenous species of *Bauhinia* in South Africa, all with the characteristic bi-lobed leaves. Some, such as *B. galpinii* (pride of De Kaap * vlam-van-die-vlakte) and *B. petersiana* subsp. *macrantha* (camel's foot * kameelspoor), have also been recorded as weeds after being cultivated as ornamentals.

Bauhinia variegata var. *variegata*

FABACEAE

Gleditsia triacanthos
honey locust * soetpeulboom

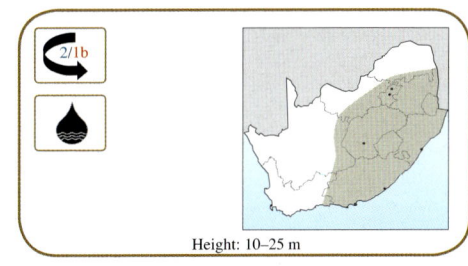
Height: 10–25 m

Origin and description: Originally from eastern North America, the honey locust is a deciduous tree that was introduced mainly to counter erosion, for fodder, and also for honey, although this is not a significant use. Thornless cultivars exist, but the wild-type, which is present in South Africa, is armed with enormous, often 3-branched spines. It has large, flat, twisted pods, which are up to 20 cm long and have sweet and edible flesh. This tree must therefore not be confused with the black locust (*Robinia pseudoacacia*), which is toxic. It is one of the first species to shed its leaves in autumn and one of the last to start budding in spring. It is dioecious, meaning that male and female flowers occur on different trees. Trees with male flowers will therefore never produce fruit.

Impact: The seeds are spread by seed-eating animals, and by the pods that float away in water. It will sucker freely when cut down. The plant now invades grassland, riverbanks, roadsides and waste areas. It is fast growing, forming dense thickets, which easily replace indigenous vegetation.

Control: This tree is not easily controlled by physical removal but it is susceptible to herbicides used as cut-stump or basal-bark treatments. The sterility or non-invasiveness of cultivars has to be investigated.

Other common names
honeyshuck * sweet locust * driedoringboom * soetpeul * springkaanboom * leoka (S)

Gleditsia triacanthos

FABACEAE

Paraserianthes lophantha **subsp. lophantha**
(=Albizia lophantha, P. lophantha)
stinkbean * stinkboon

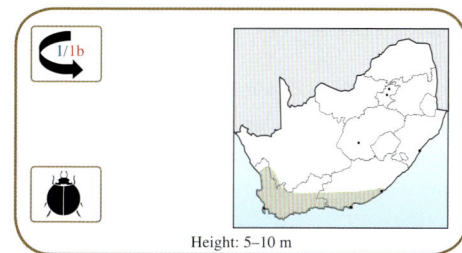

Height: 5–10 m

Origin and description: *P. lophantha* was first introduced from Australia by Baron Von Ludwig, who planted it in his garden in Cape Town in 1833. It escaped and is now found along most of the Cape coastal region and many inland areas. It favours the southern slopes of mountains and is dense in such places as the Du Toitskloof Pass. It is an adaptable species and the fact that it has successfully established itself in Cornwall, England, suggests that it has the potential to spread further in South Africa. When the seeds are broken and moistened, they produce an obnoxious odour, hence the common name 'stinkbean'. It produces copious quantities of seed, which are spread around by various means, including in sand transported for building purposes.

Impact: It forms monospecific stands and has altered the landscape in many areas where it occurs.

Control: *P. lophantha* does not coppice readily if burned or cut close to the ground. Regeneration after fire is usually from seed. Long-term control will therefore require repeated follow-up operations. Seed-feeding biocontrol agents were released in 1989, and even though they have established themselves, their effect is not dramatic.

Other common names
Australian albizia * Cape wattle * crested wattle * silk tree * sirus

Paraserianthes lophantha

Paraserianthes lophantha

FABACEAE

Prosopis glandulosa var. *torreyana*
mesquite * muskietboom

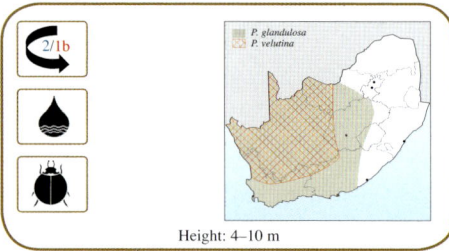

Height: 4–10 m

Origin and description: A native of northeastern Mexico and the southwestern United States, *P. glandulosa* was first introduced in 1897 in the Okahandja Experimental Garden in the then South West Africa. German settlers in the area planted it for shade and fodder and by 1912 it was recorded as having established itself in the wild. It was also recorded as being cultivated around Upington in 1900. It is now widespread in the Karoo and Kalahari thornveld. It hybridises with *P. velutina*, making identification difficult. There are several other weedy species of *Prosopis* from North America but they are of minor importance. *P. glandulosa* has drooping branches, feathery foliage and straight, paired spines on the twigs.

Impact: The plant is extremely tolerant of drought, high temperatures and overgrazing. It forms dense thickets, thereby excluding natural vegetation. Although it provides fodder, it has transformed the landscape in large areas of the Great Karoo. The IUCN considers this plant as one of the world's worst invasive species.

Control: Control is difficult because plants damaged by inadequate removal, resprout from dormant buds just below ground level, resulting

Other common names
honey mesquite * Suidwesdoring * heuningprosopis

Similar species also recorded as naturalised
P. chilensis (algaroba), from South America
P. pubescens (screw bean), from southwestern USA
P. velutina 2/1b (velvet mesquite * fluweelboontjie), from the deserts of Central America; very similar

in a dense multistemmed shrub. Cut-stump, foliar and soil-applied herbicide registrations exist, but with either chemical or physical means of control, follow-up treatments are always necessary. Releases of a beetle from Arizona since 1987 and other seed feeders, are achieving significant results.

Prosopis glandulosa

FABACEAE

Robinia pseudoacacia
black locust * witakasia

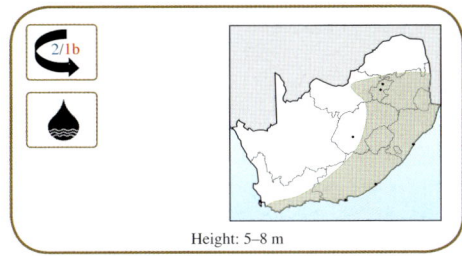
Height: 5–8 m

Origin and description: A deciduous, perennial plant from the Appalachian Mountains of North America, *R. pseudoacacia* was probably introduced as an ornamental and for soil stabilisation. It was the first North American woody plant to be introduced into Europe in about 1630. It suckers freely when the roots are disturbed and forms dense stands or clonal clusters (i.e. originating from one plant and interconnected by a common root system). The seedlings and sprouts grow rapidly and are easily identified by long, paired thorns. The fragrant flowers appear in drooping clusters and have a yellow blotch on the uppermost petal. Although parts of the plant are toxic, it is highly nutritious if grazed and is a source of nectar for honeybees. The timber makes excellent fence posts. Being a leguminous plant, it fixes nitrogen in the soil. It is found mainly in the summer-rainfall regions but it does occur occasionally in the southern and Western Cape.

Impact: It is commonly found on riverbanks and along roadsides. The seeds, inner bark and young shoots are poisonous and the flowers compete with native plants for pollinating bees. The dense clonal clusters replace other vegetation.

Control: Any attempt to cut out stems only stimulates sucker production from roots and stumps. These suckers can be up to several metres from the original stem. Because of the vigorous root system, this weed is not easy to remove. It does not respond well to herbicides, but systemic products are registered as foliar sprays and on the cut stump. Because plants that appear to have been killed can re-sprout even several years after treatment, annual monitoring should be conducted and follow-up treatments made where necessary. Physical removal requires care and persistence.

Other common names
false acacia * locust tree * yellow locust * valsakasia

Robinia pseudoacacia

FABACEAE

Tipuana tipu *(=T. speciosa)*
tipu tree * tipoeboom

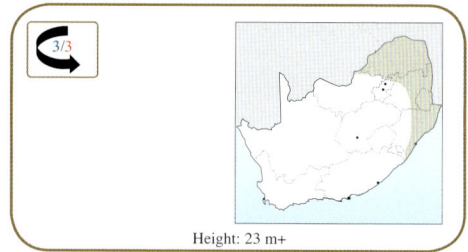
Height: 23 m+

Origin and description: A large and majestic tree from Bolivia and Brazil, *T. tipu* is widely planted for shade and as an ornament, especially as a street tree; unfortunately, it can also be found invading roadsides and urban open spaces. It is drought resistant, and frost and salt tolerant. These characteristics, in addition to its ability to produce many seeds and achieve high germination rates, make the tipu tree potentially a serious threat to native plants. It is a source of the 'rosewood' used in Victorian times for cabinet-making and is also sometimes planted as fodder.
Impact: Where it has become invasive, *T. tipu* can become competitive and replace indigenous vegetation. The leaves and seeds can clog drains and waterways.
Control: Care should be taken to ensure that all non-intended plants are physically removed.

Other common names
pride of Bolivia * racehorse tree * rosewood

Tipuana tipu

LAURACEAE

Cinnamomum camphora
camphor tree * kanferboom

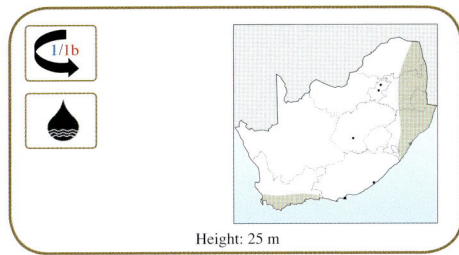

Height: 25 m

Other common names
camphor laurel * camphorwood

Origin and description: Native to East Asia where it is cultivated for camphor and timber production, the camphor tree has been widely planted in South Africa as an ornamental, for shade and as a source of honey. It can grow into a huge tree of over 30 m high. Six camphor trees planted about 1699 near Somerset West were declared National Monuments in 1942.

Impact: The leaves have a smell of camphor when crushed and this camphor inhibits the germination of other species, which ensures the plant's competitive success. The camphor tree is an evergreen, which makes it popular with landscapers and on golf courses. However, in urban areas its huge roots cause problems with drainage and sewage systems. This plant invades gardens and parks in the vicinity of existing specimens and is then spread further afield by birds where it invades riverbanks and coastal bush. In large quantities, the ripe berries are toxic to humans and the green fruits are high in chemicals known to cause sterility in birds. In other countries, camphor trees have been associated with fish kills and the absence of frogs in nearby wetlands.

Control: Seedlings can be hand-pulled and plants up to 3 m tall can be sprayed with a herbicide. Above that it should receive a basal-bark treatment or should be cut down and the stump chemically treated.

Cinnamomum camphora

LAURACEAE

Litsea glutinosa *(= L. sebifera)*
Indian laurel * Indiese lourier

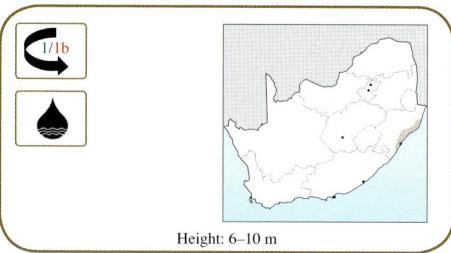

Height: 6–10 m

Origin and description: A native of tropical Southeast Asia and the Himalayas, *L. glutinosa* has spread rapidly as a weed in the subtropical areas of KwaZulu-Natal, particularly around Durban. It was probably introduced as an ornamental plant even though the numerous flowers and fruit are not particularly impressive.

Impact: Indian laurel invades the bush and forest margins and can quickly develop into a monospecific infestation by out-competing and eliminating other species. It produces large numbers of seeds, which are favoured and spread by indigenous birds, but this simply increases the threat of this plant to the indigenous plant communities. It can grow into a large tree and the ground beneath it is usually a mass of smaller plants at all stages of development.

Control: Large trees can be treated with a basal stem treatment with a registered herbicide, but this is less efficient and slower on larger trees and it may be better to cut them down first before they can produce viable seeds. Smaller plants can be pulled by hand and are also susceptible to herbicides, but care must be taken to remove the long and strong tap root. Frequent follow-up is critical on account of the large seed reserves usually found under these trees.

Other common names
ukwatyapheya (Z)

Litsea glutinosa

MELIACEAE

Melia azedarach
syringa * sering (boom)

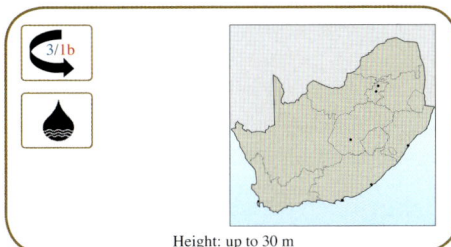

Height: up to 30 m

Origin and description: *M. azedarach* is native to a large area extending from India to Australia, with the type invading southern Africa being an Indian cultivar. It was first recorded as naturalised in 1894 in present-day KwaZulu-Natal. It is a deciduous tree that produces an abundance of marble-sized, pale yellow berries, which although poisonous to humans, are spread by birds, other animals, water and human activities. These berries, which are not poisonous to birds, often remain on the tree for a year or longer. It makes a very good timber.

Impact: *M. azedarach* is one of the most widespread of all alien invaders in South Africa and is commonly found along streams, on railway embankments and in waste areas. It establishes itself easily in such areas, where it replaces indigenous vegetation, blocks waterways and is generally unsightly. Very large specimens are found in the semi-arid, western areas of the country and in towns such as Kuruman, where this tree is used for roadside shade. The berries are one of the most common causes of human poisoning in South Africa and fatalities have been recorded.

Control: *M. azedarach* is fast growing and coppices strongly, even from stumps that are cut to ground level and burned. For this reason it is very difficult to control. Physical removal of the stump and roots is effective, but laborious and expensive. Ring-barking and bark-stripping usually stimulates coppicing and the development of root suckers. Trees should be cut well below ground level to prevent this. Triclopyr is registered for use as basal stem treatment; it should be mixed with diesel oil and painted onto the stem and stump immediately after cutting. In order to get maximum coverage, it is important not to cut the tree down too close to the ground. Foliar application will work well on trees less than 2,5 m tall. It is important to apply annual follow-up treatments in order to destroy escaped individuals and seedlings.

Other common names
bead tree * berry tree * Cape lilac * China berry * China tree * Persian lilac * white cedar * bessieboom * maksering * umsilinga (Z)

Similar species
Do not confuse with the indigenous syringas *Kirkia acuminata* and *K. wilmsii*. *Syringa* is in fact the lilac genus

Melia azedarach

MORACEAE

Morus alba **var.** *alba*
white mulberry * witmoerbei

Morus nigra
black mulberry * swartmoerbei

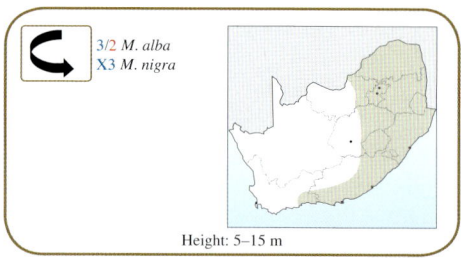

Height: 5–15 m

Origin and description: Originally from northern China, the mulberry was introduced for its fruit and as in many other parts of the world, as a food for silkworms. It is widely cultivated for this purpose in China and even parts of Europe. There are other cultivars with sweeter and more succulent fruit but they are not as invasive. It is a short-lived and fast-growing tree being evergreen in warmer regions, but deciduous in colder ones. The seeds are widely dispersed by birds. The fruit of *M. alba* are bland and insipid but those of *M. nigra* are very tasty when ripe.
Impact: The mulberry invades roadsides, riverbanks and urban open spaces. It out-competes and replaces indigenous vegetation.
Control: This tree should be physically removed or cut down and the stump treated with a herbicide.

Morus alba

MYOPORACEAE

Myoporum tenuifolium **subsp.** *montanum*
(=*M. montanum*)
manitoka

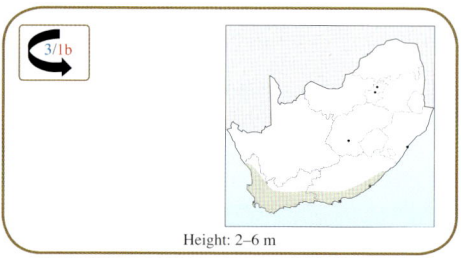

Height: 2–6 m

Origin and description: A native of Australia, *M. tenuifolium* is an evergreen shrub of the figwort family that has been planted throughout the Cape provinces as a shade tree, a windbreak and as an ornamental. Even though the whole plant is poisonous to humans and other mammals, the fleshy purple fruit is eaten safely by birds, which is how the species is being spread into the wild.
Impact: Manitoka has spread rapidly as a weed in the Cape provinces and can be found invading coastal fynbos, coastal dunes and riverbanks. It replaces indigenous vegetation.
Control: Unwanted plants should be cut down and removed. Planting of indigenous alternatives is encouraged.

Similar species cultivated as ornamentals
M. insulare X3/3 (boobialla), also from Australia
M. laetum X3/3 (New Zealand manitoka * ngaio * mousehole tree), from New Zealand

Myoporum tenuifolium

MYRTACEAE

Callistemon rigidus
stiff-leaved bottlebrush * perdestert

Callistemon citrinus
crimson bottlebrush * lemon bottlebrush

Callistemon viminalis
weeping bottlebrush

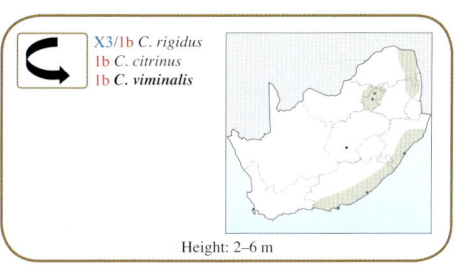

X3/1b *C. rigidus*
1b *C. citrinus*
1b *C. viminalis*

Height: 2–6 m

Similar species
Melaleuca spp. 1a (paperbarks), also from Australia

Origin and description: There are about 34 species of bottlebrush that are native to Australia and New Caledonia, with several of them being commonly planted throughout the world as ornamentals. The usual flower colour is red but there are cultivars with other colours. The characteristic beautiful red flowers have the shape of the familiar cylindrical 'bottlebrush'. Although *C. citrinus* has red flowers as well, the crushed leaves have a lemon aroma.

Impact: Although not usually very invasive, *C. rigidus* and *C. viminalis* are proposed Category 1b in certain provinces, with the latter species known to be becoming a problem further north into Zimbabwe. They are considered as an emerging threat in the Karoo and along the Garden Route. They invade watercourses, wetlands, roadsides, disturbed and burnt sites and can become competitive, thereby replacing indigenous vegetation.

Control: Unwanted plants should be cut down and removed. The sterility or non-invasiveness of cultivars has to be investigated.

Callistemon rigidus

Callistemon viminalis

MYRTACEAE

Eucalyptus camaldulensis
red river gum * rooibloekom

Eucalyptus cinerea
florist's gum * penny gum

Eucalyptus lehmannii
spider gum * spinnekopbloekom

Origin and description: Three of approximately 14 species of Australian *Eucalyptus* or gum tree, that are naturalised in South Africa.

Impact: Although of marginal use for timber, these species are planted as ornamentals and as windbreaks. *E. camaldulensis* is now a widespread invader and is particularly common along rivers in the Western Cape and the Free State. The penny gum is a popular windbreak in the high-altitude grasslands and is commonly planted along the main highways. It is clearly invasive in the southern Drakensberg. It is also popular with florists on account of its silvery coin-shaped, juvenile foliage. Do not confuse these juvenile leaves with those of *Acacia podalyriifolia*.

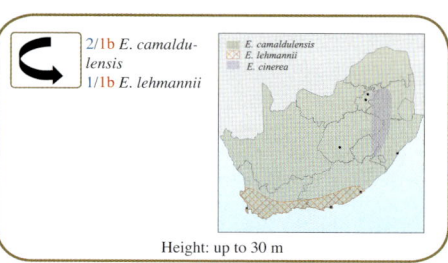
2/1b *E. camaldulensis*
1/1b *E. lehmannii*
Height: up to 30 m

Other common names
E. camaldulensis: Murray red gum * red gum * rostrata gum
E. cinerea: Argyle apple * mealy stringybark
E. lehmannii: bushy yate * Lehman's gum

Eucalyptus camaldulensis

E. lehmannii is common in the Cape where it is also planted as a windbreak and as an ornamental.
Control: As for *E. grandis*.

Eucalyptus lehmannii   *Eucalyptus cinerea*

Eucalyptus cinerea

MYRTACEAE

Eucalyptus grandis
saligna gum * salignabloekom

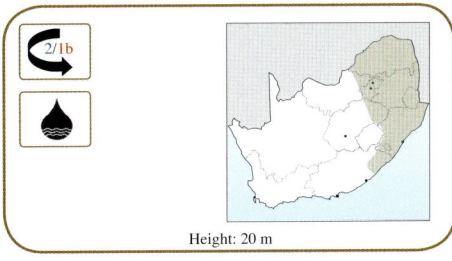

Height: 20 m

Origin and description: There are over 700 species of *Eucalyptus*, native mainly to Australia, but some also come from nearby Asian countries; eight have been declared as invasive weeds in South Africa and there are others that are the subject of at least one herbicide registration. Many of these species are planted throughout the world as ornamentals, for their timber and as firewood; they are also popular with bee-keepers. They are called gum trees because many of them (but not all) exude a sticky gum from breaks in the bark.

Impact: *Eucalyptus* trees, whether in formal plantations, woodlots or growing wild, use large amounts of water, thereby lowering the water table and threatening the water supply and the ecology of a region such as Zululand. Unwanted plants should therefore be identified and destroyed. *Eucalyptus* gum trees also invade watercourses, blocking rivers, impeding floodwater and recreational use. They can outcompete indigenous species, thereby threatening biodiversity. *E. grandis* is the most widely cultivated timber species and is particularly prevalent as an invasive species in the eastern parts of the country.

Control: Gum trees will quickly re-grow when cut down, so the stumps of unwanted plants should be treated with a suitable herbicide. There are also registrations for soil, foliar, frill and aerial applications. Seedlings can be sprayed or removed by hand and are also susceptible to fire.

Other common names
E. grandis: bluegum * rose gum * bloukom * saligna

Some similar species also naturalised in South Africa
E. cladocalyx 🗲 2/1b 💧 (sugar gum * suikerbloekom)
E. diversicolor 🗲 2/1b 💧 (karri * karie)
E. dunnii 🗲 💧 (Dunn's white gum * Killarney ash)
E. globulus (eurabbie * southern blue gum)
E. gomphocephala (tuart)
E. macarthurii 🗲 💧 (Camden woollybutt)
E. maculata 🗲 💧 (spotted gum * gevlekte bloekom)
E. paniculata 🗲 2 💧 (grey iron bark * grysysterbasbloekom)
E. sideroxylon 🗲 2 💧 (black iron bark * red iron bark * swartysterbasbloekom)
E. smithii 🗲 💧 (blackbutt peppermint * swartstampeperment)
E. tereticornis 🗲 1b (forest red gum)

Eucalyptus grandis

MYRTACEAE

Leptospermum laevigatum
Australian myrtle * Australiese mirteboom

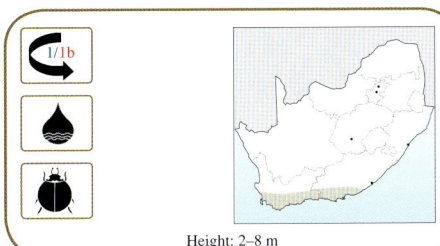

Height: 2–8 m

Origin and description: This weed was introduced from Australia as a hedge or windbreak, with the first recorded sighting having been made in 1850 from the White Sands plantation near Cape Town. Called 'tea tree' because it is said that Captain Cook used to make a tea from the leaves to prevent scurvy. Since that time *L. laevigatum* has become naturalised in South Africa and is found in patches on sandy soils from the southwestern Cape to as far east as Port Elizabeth. It is particularly common around Hermanus.

It has distinctive seed capsules and the mature stems are twisted, furrowed and with ribbony bark. It is killed by fire and does not coppice if it is cut at ground level. However, fire stimulates the release of the seeds, with a consequent massive germination after a fire. Since it takes about four years before seeds are produced, follow-up burns for the seedlings have to be done only after four years. It is tolerant of frosts and salty sea spray.

Impact: The plant has strong lateral roots that produce a mat of tiny rootlets in the top 5 cm of the soil. These rootlets are so efficient at extracting moisture from the soil surface that other plants cannot compete, even other invaders such as *Acacia saligna* (see page 126). *L. laevigatum* can quickly form dense thickets that threaten the flora and fauna of the Western Cape. No indigenous birds are known to nest in the shrub.

Control: Soil-applied herbicides have been registered for the control of this weed. These are applied at the base of standing plants and eliminate the need for chopping or using fire. However, because the plant is sensitive to fire, this remains an effective management tool. Considerable success is being achieved with biocontrol agents released since 1994.

Other common names
coast tea tree * small-leaved tea tree

Similar species
Leptospermum scoparium X3 (manuka myrtle * mankamirt), from New Zealand, also called 'tea tree' but with many colourful garden cultivars

Leptospermum laevigatum

MYRTACEAE

Psidium guajava
guava * koejawel

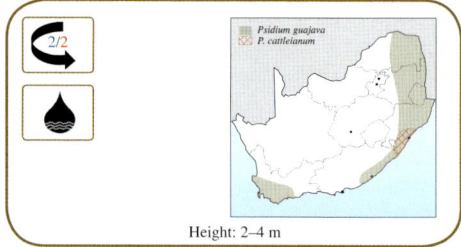

Height: 2–4 m

Origin and description: A native of South America, the guava was brought to South Africa to be grown primarily as an agricultural crop, but also for shade and as an ornamental. It has now established itself in the wild and has become a serious weed. It is grown as a subtropical fruit crop and is found growing wild in all subtropical areas, especially in Zululand and the Mpumalanga lowveld. It can be found growing wild wherever the soil has been disturbed. The fruit is eaten by people and a wide range of birds and animals, who disseminate the seeds and thereby assist in the spread of the weed. *P. x durbanensis* is a sterile South African hybrid which spreads by suckering.

Impact: Wild guavas are a host for fruit flies and act as a source of infestation of fruit flies to orchards of other fruit. They are very aggressive invaders and can replace indigenous vegetation, thereby transforming the landscape.

Control: Once established, wild guavas are extremely difficult to control. They are evergreen,

Similar species
P. cattleianum 3/1b (strawberry or cherry guava * aarbeikoejawel)
P. guineense 3/1b (Brazilian guava * Brasiliaanse koejawel)
P. x durbanensis 1/1b (Durban guava * Durbanse koejawel)

perennial plants with a strong root system, being able to withstand many foliar- and soil-applied herbicides. They coppice when cut and produce vigorous root suckers. Ring-barking, bark-stripping and felling can encourage root sucker development and thereby a greater density of the infestation. They can be controlled by efficient use of a registered herbicide, but will require repeated applications.

Psidium guajava

MYRTACEAE

Syzygium cuminii
jambolan

Syzygium jambos
rose apple * jamboes

Syzygium paniculatum
Australian water pear * Australiese waterpeer

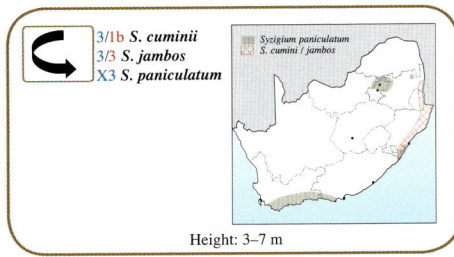

Height: 3–7 m

Other common names
S. paniculatum: brush cherry * hedge water berry

Origin and description: Introduced as ornamentals and street trees, these plants pose a threat as they are widely planted, are very adaptable and have on occasion been found growing wild. *S. cuminii* and *S. jambos*, both from Asia, have been found invading coastal bush in KwaZulu-Natal. *S. paniculatum* is an evergreen, perennial shrub or tree which, like the others, can reach a height of more than 6 m. It is planted as an ornamental and as a hedge throughout South Africa and is often promoted by nurseries as a good 'bird plant' on account of the succulent pinkish red berries which are indeed much favoured by birds such as hadedas. Unfortunately, this is how it is being spread into the wild where it can be found growing around Pretoria, in KwaZulu-Natal and in parts of the Western Cape such as Plettenberg Bay and Cape Town. Places it invades are urban open spaces, riverbanks, and coastal bush. It has also been found growing wild in the Kruger National Park.

Impact: The bright pinkish red berries of *S. paniculatum* are a familiar sight in late summer and the flowers are a source of nectar for honeybees. It is, however, like all syzygiums, a potential invader and can replace indigenous vegetation.

Control: Unwanted plants should be cut down and removed.

Syzygium paniculatum

PHYTOLACCACEAE

Phytolacca dioica
belhambra * bobbejaandruifboom

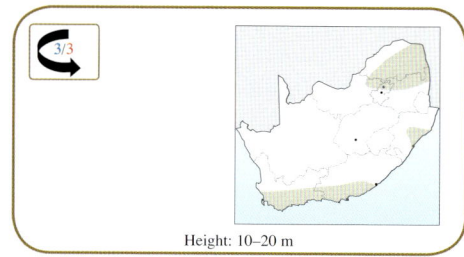

Height: 10–20 m

Origin and description: Introduced from South America as an ornamental evergreen shade tree, *P. dioica* is now found growing wild. It is characterised by the large, pendant clumps of green berries, which turn black as they mature, and the huge trunk buttress, which helps to store water. The plant is highly tolerant of dry conditions and stores water in the soft spongy wood, which can be as much as 80% water and very little wood

Other common names
monkey grape * ombu * poke berry * umbra tree * belambraboom * dikboom * uMzimuka (Z)

tissue. It is often found near old homesteads and trading stores.

Impact: These trees can be very competitive in dry areas and because of their size have a high visual impact. The roots and sometimes the fruits are said to be poisonous.

Control: These trees are best cut down and removed.

Phytolacca dioica

PINACEAE

Pinus halepensis
aleppo pine * aleppoden

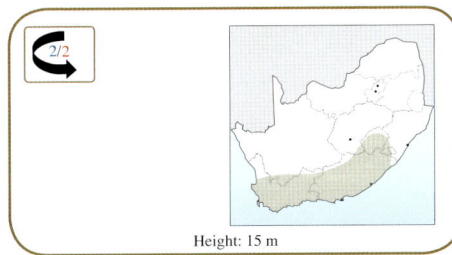

Height: 15 m

Origin and description: All *Pinus* species were introduced at one time or another for commercial use. *P. halepensis* is from the eastern Mediterranean region of Europe. The cones of *P. halepensis* are reddish brown and glossy; they do not have the distinct, raised ridges of *P. pinaster*.

Impact: The seeds of pine trees, which are dispersed by wind, squirrels, baboons and other animals, germinate easily. Aleppo pines invade veld, mountain and lowland fynbos, especially in the Eastern Cape. They transform the landscape, reduce carrying capacity and increase the risk of fire. Their problem status is increasing and they are severely depleting biodiversity.

Control: Large plants can be ring-barked or felled. Seedlings and saplings can be uprooted when the soil is moist or treated with a herbicide.

Other common names
Calabrian pine

Pinus halepensis

PINACEAE

Pinus patula
patula pine * treurden

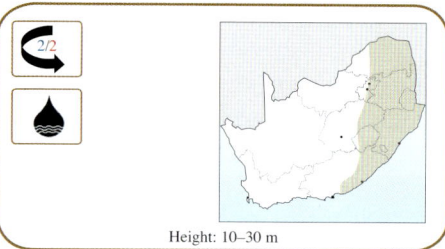

Height: 10–30 m

Origin and description: All *Pinus* species were introduced at one time or another for commercial use. *P. patula* is from Mexico and is still the most popular pine tree for commercial plantations. It was first planted in 1907 in an experimental plantation in Tokai in the Cape. It is a quick growing, drought-tolerant ornamental and as such is popular for golf courses. About 69 *Pinus* species have been cultivated in southern Africa at some time and eight have become naturalised.

Impact: The seeds of *P. patula*, which are dispersed by wind, squirrels, baboons and other animals, germinate easily and help to spread the tree into unintended areas such as moist grassland, forest margins and road cuttings. They can transform the landscape, reduce carrying capacity and, being very flammable, increase the risk of fire.

Other weedy *Pinus* species in South Africa
P. canariensis 2/3 (Canary pine * Kanariese den), a subtropical pine from the Canary Islands
P. elliottii 2/2 (slash pine * basden), from southeast North America
P. pinea 3 (stone pine * umbrella pine * sambreelden), from southern Europe
P. radiata 2/2 (Monterey pine * radiata pine * radiataden), from southern North America
P. roxburghii (=*P. longifolia*) 2/2 (chir pine * longifolia pine * tjirden), from the Himalayas
P. halepensis (see page 147),
P. pinaster (see page 149)
P. taeda 2/2 (loblolly pine * loblollyden), from southeastern North America

Control: Large plants can be ring-barked or felled and tebuthiuron is registered for application to the soil. Seedlings and saplings can be uprooted when the soil is moist or treated with a herbicide.

Pinus patula

PINACEAE

Pinus pinaster
cluster pine * sparden

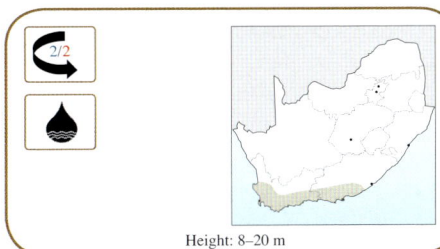
Height: 8–20 m

Other common names
maritime pine * mannetjiesden * trosden

Origin and description: All *Pinus* species were introduced at one time or another for commercial use. *P. pinaster* is from the western Mediterranean region of Europe and was first established in plantations by the French Huguenots in 1825, but compared to the other *Pinus* species it has little value as timber. The main infestations of *P. pinaster* are still around Franschhoek in the Western Cape where the Huguenots first settled. The scales on the cones of *P. pinaster* have distinct, raised ridges.

Impact: The seeds of pine trees, which are dispersed by wind, squirrels, baboons and other animals, germinate easily and establish themselves in cool, moist soil. Pine trees invade veld, mountain and lowland fynbos, transforming the landscape, reducing carrying capacity and increasing the risk of fire. Their problem status is increasing and they are severely depleting biodiversity. It is No. 70 on the list of the *World's Worst Invasive Species*.

Control: Large plants can be ring-barked or felled and tebuthiuron is registered for application to the soil. Seedlings and saplings can be uprooted when the soil is moist or can be treated with a herbicide.

Pinus pinaster

PITTOSPORACEAE

Pittosporum undulatum
Australian cheesewood * Australiese kasuur

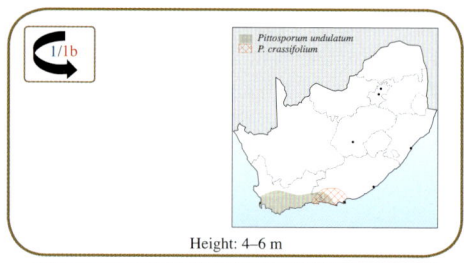

Height: 4–6 m

Origin and description: *P. undulatum* comes from eastern Australia but is being grown in South Africa as an ornament, for shade and for hedging. It is especially common as an ornament in the Cape provinces, but has been found to be invasive in certain areas; even in Australia it is considered a problem in areas where it does not belong. It has flourished in areas where the environment has been altered by humans, whether by habitat damage, fertiliser runoff or by the suppression of bush fires near suburbs. The fruit, which go orange when ripe, are very sticky when crushed and smell very distinctly of orangeade.

Impact: *P. undulatum* grows quickly and rapidly shades out many other plants. It seems to adapt to soils with higher nutrient levels much more readily than indigenous species and hence grows well in areas where the soil has been changed in this way. The seeds are much favoured by birds, which can neglect indigenous fruits as a consequence and thereby seriously impair their natural dispersal.

Control: There is no specific herbicide registration. Unwanted plants should be physically removed.

Other common names
mock orange * sweet pittosporum * Victorian box * soet pittosporum

Similar species
Pittosporum crassifolium X3/3 (karo * stiff-leaved cheesewood * styweblaarkasuur), from New Zealand is naturalised in the Cape

Pittosporum undulatum

POLYGONACEAE

Triplaris americana
ant tree * triplaris

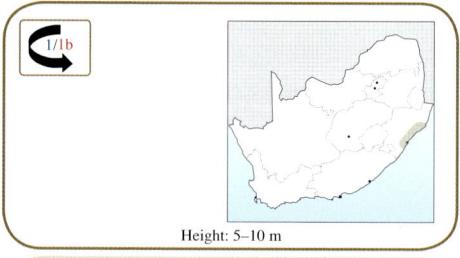

Height: 5–10 m

Other common names
Indian almond

Origin and description: Introduced from Central America only in about 1980, these plants rapidly became very popular as street and garden trees, especially in and around Durban. They are spectacular in full bloom, especially the female tree, which is smothered in clusters of showy red flowers as opposed to the male tree, which has less profuse, tawny blossoms. The stems are often hollow and are a haven for ants.

Impact: Unfortunately, these trees do so well in the subtropical regions of KwaZulu-Natal that they have been declared a Category 1 weed, which means they should be removed and that it is illegal to plant or trade them.

Control: These are now prohibited plants and should be removed. Physical removal is currently the best option.

Triplaris americana

PROTEACEAE

Grevillea robusta
Australian silky oak * Australiese silwereik

Origin and description: Originally from eastern Australia, the silky oak is now a common escapee. It is planted as an ornamental, but it produces a very good timber often used in making musical instruments and window joinery as it is resistant to rotting. Also planted as a windbreak in citrus plantations around Nelspruit in Mpumalanga. There is also a range of hybrids.

Impact: A prolific producer of nectar and seeds, *G. robusta* is now found in forest margins, riverbanks and bush land. It is tolerant of dry conditions but also grows well in moist areas. The pollen may

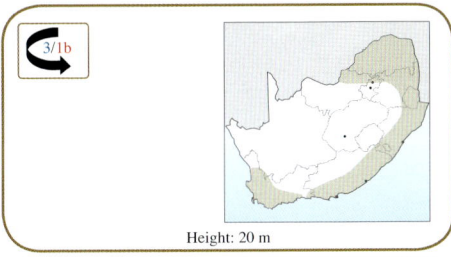

Height: 20 m

Other common names
silver oak * he-oka

Similar species
G. banksii ⤴ 1b (crimson oak * scarlet silky oak), naturalised in coastal KwaZulu-Natal
G. juniperina ⤴ (spider flower)
G. lanigera ⤴ (woolly grevillea)
G. rosmarinifolia ⤴ 3

Grevillea robusta

trigger hay fever, and contact with the foliage can cause allergies on the legs, arms and face, similar to those caused by poison ivy (*Toxicodendron* sp.). The intrusive roots can damage drainage systems.

Control: This tree is best controlled by physical removal but it is susceptible to herbicides used as cut-stump or basal-bark treatments.

Grevillea robusta

Grevillia banksii

Grevillea robusta

PROTEACEAE

Hakea sericea
silky hakea * syerige hakea

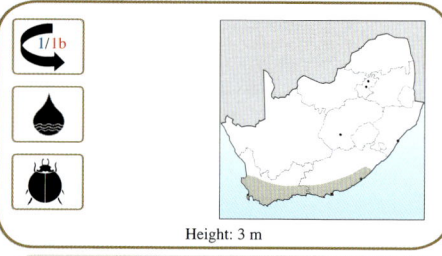

Height: 3 m

Origin and description: Several of the 149 species of Australian hakea are now seriously invasive weeds in South Africa. Introduced in the 1830s, probably for use as hedges and to control drift sand, they were well established in the Cape coastal region by 1900. At one time they covered 4 800 km² or 14% of the Cape Mountain Fynbos.

Impact: Old, dry *Hakea* stands are extremely flammable and cause such hot fires that all indigenous flora and fauna in the vicinity are destroyed. When a fire passes through an infested area, massed release of the seeds is triggered and the bare areas are quickly recolonised by the hakea. *H. gibbosa* and *H. sericea* are the major threats to Mountain Fynbos and form

Other common names
needle bush * naaldbos
Similar species
H. gibbosa ↻ 1/1b ● 🐞 (rock hakea * harige hakea * harige speldebos)
H. salicifolia ↻ X3 (hedge hakea * willow hakea * makspeldebos * wilgerhakea)
H. drupacea (=*H. suaveolens*) ↻ 1/1a (sweet hakea * soethakea * soetspeldebos)

huge monospecific stands, totally replacing indigenous species. They make mountain slopes unattractive and inaccessible to mountaineers, promote erosion and increase transpiration, with subsequent reduction of runoff to rivers and dams.

Control: Although biological control is having considerable success, it will still be necessary to use other control methods. Plants should be cut off below the lowest leafy parts and then stacked to dry for 8 to 12 months. Each stack should be weighted down or anchored so that it cannot blow about and should be left until all the seeds have been released. Rodents will eat many of these seeds, but those that germinate must be killed by fire or other methods. The soil-applied herbicide tebuthiuron has been registered but its use is risky and expensive. There is also a mycoherbicide, which is effective and much safer, but takes longer to work.

Hakea sericea

Hakea sericea *Hakea sericea* *H. salicifolia*

ROSACEAE

Eriobotrya japonica
loquat * lukwart

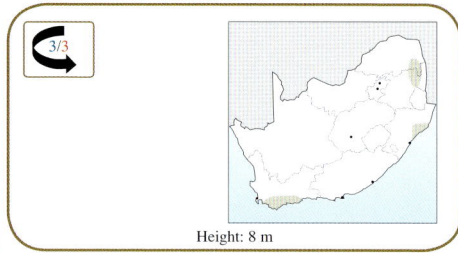
Height: 8 m

Other common names
Japanese medlar * Japanese plum * Japanese mispel

Origin and description: Planted as an ornamental, this plant has escaped into the wild. It is originally from China and was introduced into Japan 1 000 years ago. It is a medium-sized tree that produces quite large, edible fruit. The plant is spread by mammals and birds, which feed on the fruit and disseminate the seeds into the wild.

Impact: The fruits act as an alternative host for the fruit fly, that can exacerbate the fruit fly problem in areas such as the deciduous fruit-growing regions of the Western Cape. This fruit tree is unusual in that it flowers in autumn or early winter, with the fruit ripening in early spring.

Control: No herbicide registrations exist, but to control the loquat it should be cut down and the stump treated with a suitable herbicide. Unwanted plants should be removed.

Eriobotrya japonica

ROSACEAE

Prunus persica
peach tree * perskeboom

Origin and description: *P. persica* is a deciduous fruit tree from China and is cultivated throughout the world. There are many cultivars and subspecies.
Impact: In South Africa this small tree is now widely naturalised. It is found on roadsides, riverbanks and waste places, usually but not exclusively, in urban areas and in regions that experience frost in winter. It is competitive, replaces indigenous vegetation and acts as a host for fruit flies and other insect pests of commercially grown fruit.
Control: Unwanted plants should be cut down and the stump either removed or treated with a herbicide.

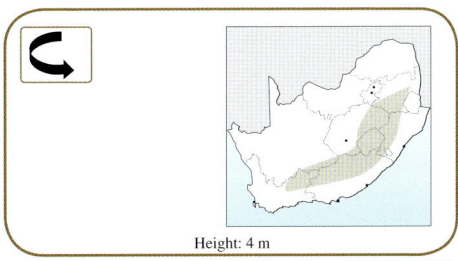
Height: 4 m

Prunus persica

SALICACEAE

Populus x canescens
grey poplar * gryspopulier

Populus alba **var.** *alba*
white poplar * abele

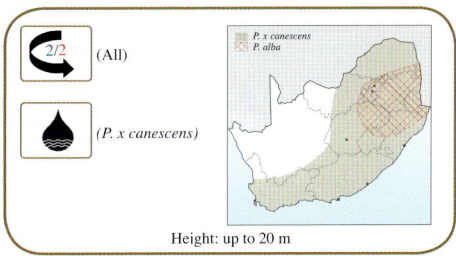
Height: up to 20 m

155

Origin and description: There are about 30 species of poplar, native to the northern hemisphere, and four of them have found a home in South Africa. *P. x canescens* (a hybrid between *P. alba* and *P. tremula*) was introduced from Eurasia, probably as an anti-erosion agent and for matchwood in the 1920s. Being a hybrid, it is sterile and does not produce seed, but it coppices when cut and regenerates vigorously from root suckers. Mature trees produce a good quality, light, pinkish timber but clumps of poplar must be carefully managed if suitable timber trees are to be produced. *P. alba* is less vigorous, but still quite widespread, invading riverbanks, vleis and dongas.

Impact: These poplars are found throughout the country on riverbanks and in vleis, where they can form dense and uniform stands. They can spread into surrounding veld.

Other common names
P. x canescens: populier * vaalpopulier * popoliri (S)

Similar species
P. deltoides 🍁 X3 (cottonwood * match poplar * vuurhoutjiepopulier), from southeastern North America
P. nigra 🍁 X2 (black poplar * Lombardy poplar * Italiaanse populier), from Eurasia

Control: *P. x canescens* is difficult to control mechanically. Large trees should be ring-barked or felled and the entire root system removed. Systemic herbicides can work but require repeated applications and follow-ups. There are several registrations for cut-stump treatments.

Populus x canescens

SALICACEAE

Salix babylonica
weeping willow * treurwilger

Salix fragilis
crack willow

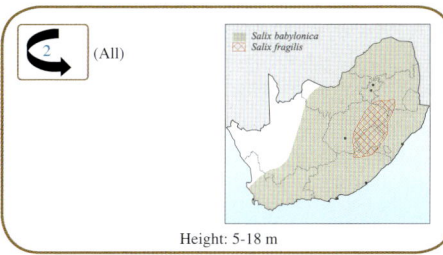
(All)

Height: 5-18 m

Origin and description: Originally from the dry areas of northern China, the weeping willow was introduced to South Africa from Europe in order to prevent erosion along riverbanks. There are several indigenous species of willow, but these alien species are the most widespread and best known. The indigenous *S. mucronata* is very similar but it does not droop or 'weep' as much; the branches of *S. babylonica* droop vertically towards the ground. *S. fragilis* is erect and can grow much taller. The twigs are brittle and old trees rot easily and break apart in storms.

Impact: These plants are a threat to the indigenous species, which they can replace. The large, spreading mass of roots can reduce the depth of waterways and thereby increase the risk of flooding. *S. babylonica* is one of the most prominent invaders of the Grassland Biome in South Africa (Henderson 2007). They were 'Declared Invaders Category 2' but do not feature on the NEMBA lists.

Control: After careful identification, unwanted plants should be cut down, physically removed and replaced with the indigenous species.

Other common names
S. babylonica: Babylon willow * Peking willow * willow tea * huilwilgeboom * treurboom * wilgerhout * moluoane (S)

Similar species
S. mucronata (Cape willow * Kaapse wilger), indigenous
S. subserrata (safsaf willow * safsafwilger), indigenous

Salix fragilis

Salix babylonica

SIMAROUBACEAE

Ailanthus altissima
tree of heaven * hemelboom

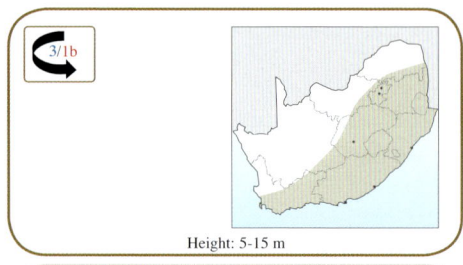
Height: 5-15 m

Origin and description: A deciduous tree from China, which was probably introduced as an ornamental and has been planted as such throughout the country. It is a very hardy and adaptable plant and can be found in very inhospitable conditions. In particular, it can withstand severe atmospheric pollution, which would kill other plants. It thrives in full sun and disturbed areas. All parts of the tree, especially the flowers, have an unpleasant odour. The fruits are papery, winged structures called samaras.

Similar species
Chinese sumac * copal tree * stinking cedar * stinking sumac * varnish tree

Impact: Its great adaptability can make it a threat to indigenous species. It is a prolific seed producer (up to 350 000 in a year), grows rapidly and has the potential to overrun indigenous vegetation. Established trees produce numerous root suckers and resprout vigorously from cut stumps and root fragments. It can suppress competition with an allelopathic chemical called ailanthone. A secretion from the young shoots causes an allergic skin reaction in some people.

Control: Plants should be cut down and physically removed. This should be followed up to ensure that root suckers are not formed.

Ailanthus altissima

SOLANACEAE

Cestrum aurantiacum
yellow cestrum * oranjesestrum

Cestrum laevigatum
inkberry * inkbessie

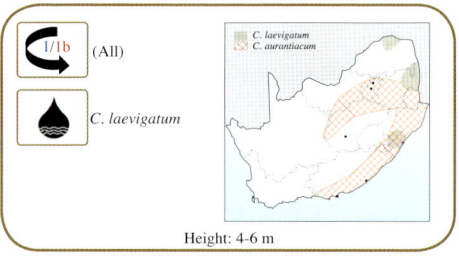
Height: 4-6 m

Origin and description: Introduced from Central and South America respectively, as ornamental shrubs, hedges and for windbreaks, *C. laevigatum* in particular has now invaded various areas in the central and eastern parts of the country. Hluhleka Nature Reserve near Port St Johns is particularly heavily infested, but it is found from the KwaZulu-Natal midlands, through Mpumalanga into Zimbabwe. The flowers have

Other common names
poison berry * uyinki (Z)
Similar species
C. parqui 1/1b (Chilean cestrum * green cestrum), from Chile
C. elegans 1/1b (crimson cestrum * karmosynsestrum), from Mexico

an unpleasant odour by day and are sweet-smelling by night.

Impact: These plants can form dense stands, eliminating indigenous flora and transforming the landscape. Many areas of coastal bush are seriously infested; the plants coppice vigorously. The name 'inkberry' is earned because of the juicy, black berries. The unripe berries, which are normally green in the months of June and July, and the young shoots are very poisonous. Inkberry poisoning in stock is also called 'Chase Valley disease' because it was first recognised from Chase Valley in Pietermaritzburg.

Control: Chemical control is the best method available at present. A suitable herbicide such as triclopyr or imazapyr is painted either onto the stems or cut stumps. Physical methods require the removal of the entire plant on account of the vigorous regrowth.

C. diurnum ⮌ (from the West Indies)
Should not be confused with *Phytolacca octandra,* also called 'inkberry', or *Nicotiana glauca*

Cestrum laevigatum

Cestrum elegans

Cestrum aurantiacum

SOLANACEAE

Nicotiana glauca
wild tobacco * tabakboom

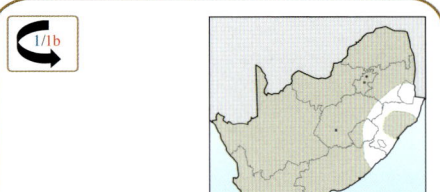

Height: 2-5 m

Origin and description: Indigenous to Argentina, this annual plant has been known as a weed in southern Africa since the 19th century. It is thought to have been introduced into Namibia in horse fodder during the German occupation. It is widespread, except in KwaZulu-Natal, and is usually found on roadsides, riverbanks, old lands and even as a cultivated garden plant. Closely related to commercial tobacco, it can become a large, woody shrub. The seed capsules contain hundreds of tiny seeds, which are easily transported by water. It cannot withstand flooding but is highly tolerant of arid conditions, as it is believed to be able to obtain moisture from fog. It is common in the beds of rivers that only flow occasionally.

Impact: *N. glauca* can cause poisoning of livestock, with symptoms similar to that of nicotine poisoning. Evidently, according to one Afrikaans name, it is well known for poisoning ostriches. The plant has been used as a rat poison in Italy and the dried flowers are said to kill cockroaches.

Control: *N. glauca* should be controlled when small. There are no specific herbicide registrations, but it should be susceptible to the usual ones.

Other common names
coneton * Mexican tobacco * mustard tree * tree tobacco * Jan-tak * volstruisgifboom * mohlafotha (S) * tabaka bume (Ss)

Similar species also recorded as naturalised
N. longiflora (longflower tobacco), also from South America
N. tabacum (tobacco), from South America, now widely cultivated

Nicotiana glauca

SOLANACEAE

Solanum mauritianum
bugweed * luisboom

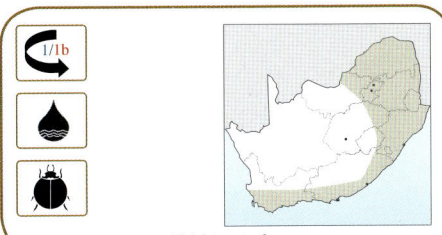

Height: up to 8 m

Origin and description: *S. mauritianum* was originally thought to be a native of tropical Asia, but it has since been discovered that it is indigenous to South America. It is now a perennial weed causing serious problems in Mpumalanga, Limpopo, KwaZulu-Natal and parts of the Cape provinces. It was first recorded in present-day KwaZulu-Natal in 1862.

Impact: *S. mauritianum* causes serious problems in plantations, sugarcane and on wasteland as it can very quickly reach a height of 3–5 m and shade out and replace other vegetation. It is the principal weed of South Africa's timber forests. Apart from being poisonous, the fruits act as a host for the fruit fly, which is a serious pest of orchards. The berries are attractive to birds, that eat them and transport the seeds elsewhere; peculiarly, the birds are not affected by the poison. In young forest plantations, feeding birds perch on young pine trees and break the growing tips. It is also suggested that birds, having developed a predilection for bugweed fruits, which are available in vast quantities, no longer bother to seek out and distribute fruits of indigenous plants. Bugweed can often be seen growing under other trees, which is where roosting birds drop the seeds.

Control: Fortunately *S. mauritianum* can be killed easily by cutting, stem painting or soil-applied or foliar herbicides if the foliage can be reached. There is evidence of resistance developing to some herbicides. Young plants can be hand-pulled. When mechanically cleared, the clouds of fine hairs that are dislodged contain toxins that have been blamed for respiratory problems in workers clearing these plants. Large numbers of seedlings often emerge under trees that have been killed by chemical means, from seeds that are unaffected by the herbicide. This makes follow-up treatments essential. There has been limited success recently with the introduction of the leaf-eating biocontrol agent *Gratiana spadicea*.

Other common names
bugtree * flannel weed * woolly nightshade * groot bitterappel * uB-hoqo (Z) * umbanga banga (Z)

Similar species
Solanum betaceum X3/3 (tree tomato * tamarillo), from South America

Solanum mauritianum

Solanum betaceum

TAMARICACEAE

Tamarix ramosissima
pink tamarisk * perstamarisk

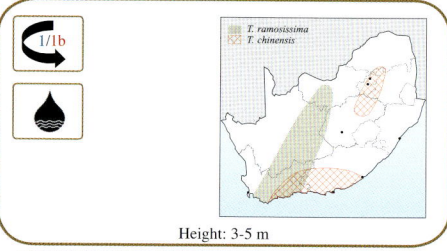

Height: 3-5 m

Origin and description: Believed to be originally from drier parts of Eurasia, the tamarisk is a popular garden plant. In some areas, like in Namibia, the naturalised populations are hybrids between *T. ramosissima* and the indigenous wild type. *Tamarix* species have twiggy, woody branches covered with small, scale-like leaves, which is an adaptation for arid conditions. The minute, hermaphroditic, pink flowers are borne in long, feathery, terminal spikes, often before the leaves appear.

Impact: The tamarisks replace indigenous vegetation and raise soil salinity. They invade watercourses, sandy riverbanks and drainage lines,

Other common names
saltcedar
Similar species
T. aphylla X3/1b (Athel tamarisk * Athel tree * desert tamarisk * woestyntamarisk)
T. chinensis 1/1b
T. gallica 1b (French tamarisk * Franse tamarisk)
Do not confuse with *T. usneoides* (indigenous) (wild tamarisk * abiekwasgeelhout)

removing precious moisture from an already arid environment. They are a serious threat to the water resources of the Karoo.
Control: These plants should be physically re-moved or cut down and the stumps treated with a herbicide. For extensive and dense infestations, there is an option for aerial application of a foliar herbicide.

Tamarix ramosissima

SHRUBS (WOODY OR HERBACEOUS), NOT DISTINCTLY TREE-LIKE OR SUCCULENT

Within the section, the plants are arranged alphabetically according to family and genus.

SPECIES	Needle leaves	Thorny stems	Seeds in multi-seeded capsules	Seed in a bean-like pod	Seed in a berry	Red flowers	Yellow flowers	Purple flowers	White flowers	Green or indistinct flowers
Cytisus scoparius, p. 183	●		●	●			●			
Spartium junceum, p. 183	●			●			●			
Ulex europaeus, p. 184	●	●	●				●			
Mimosa pigra, p. 178		●	●					●		
Alhagi maurorum, p. 175		●		●				●		
Rosa rubignosa, p. 191		●			●	●			●	
Lantana camara, p. 196		●			●	●		●		
Pyracantha spp., p. 190		●			●				●	
Duranta erecta, p. 195		●			●			●		
Caesalpinia decapetala, p. 175		●		●			●			
Rubus spp., p. 192		●			●				●	
Solanum sisymbriifolium, p. 193		●			●					
Pyracantha spp., p. 190		●			●				●	
Asparagus laricinus, p. 167		●		●						
Sesbania punicea, p. 182			●	●		●				
Ricinus communis, p. 174			●			●				
Leucaena leucophylla, p. 177			●	●			●			
Senna spp., p. 178			●	●			●			
Sesbania bispinosa, p. 181			●	●			●			
Ricinus communis, p. 174			●		●					
Tecoma stans, p. 172			●				●			
Hypericum perforatum, p. 184			●				●			
Nerium oleander, p. 165			●			●			●	
Buddleja davidii, p. 193			●					●		
Ardisia crenata, p. 187					●				●	
Ligustrum spp., p. 187					●				●	
Cotoneaster spp., p. 189					●				●	

	Needle leaves	Thorny stems	Seeds in multi-seeded capsules	Seed in a bean-like pod	Seed in a berry	Red flowers	Yellow flowers	Purple flowers	White flowers	Green or indistinct flowers
Chromolaena odorata, p. 168					■				■	
Tithonia rotundifolia, p. 171						■				
Montanoa hibiscifolia, p. 169							■		■	
Thevetia peruviana, p. 166							■			
Tithonia diversifolia, p. 171							■			
Seriphium plumosum, p. 170								■		
Plectranthus comosus, p. 186								■		
Atriplex nummularia, p. 173										■

APOCYNACEAE

Nerium oleander
oleander * selonsroos

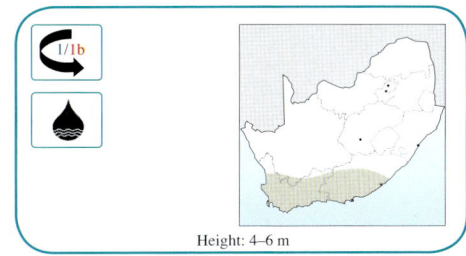

Height: 4–6 m

Origin and description: *N. oleander* was introduced from the Mediterranean region, including North Africa, as an evergreen ornamental shrub for hedges and as a screening plant. There are some (over 400) non-invasive, sterile and double-flowered cultivars; the flowers can be pink, red or white and there are even some variegated varieties. The seed capsules burst open and release little parachute seeds that blow away in the wind.

Impact: *N. oleander* is found as a weed in the veld, particularly in the cooler, moister areas, damp ravines and along rocky and gravelly watercourses in the southern and Western Cape. It is highly competitive and is very poisonous to both animals and humans; it has been reported that one leaf is sufficient to kill a sheep. Eating meat that has been skewered on the stems or simply inhaling the smoke from burning wood can produce symptoms of poisoning.

Other common names
Ceylon rose * dog-bane * double oleander * rose bay * rose laurel * South Sea rose

Control: Chemical control is the best method available at present. A suitable herbicide is painted either onto the stems or cut stumps. Physical methods require the removal of the entire plant on account of the vigorous regrowth. Unintended plants should be removed.

The sterility or non-invasiveness of cultivars has to be investigated.

Nerium oleander

APOCYNACEAE

Thevetia peruviana *(=T. yccotlii, T. neriifolia)*
yellow oleander * geeloleander

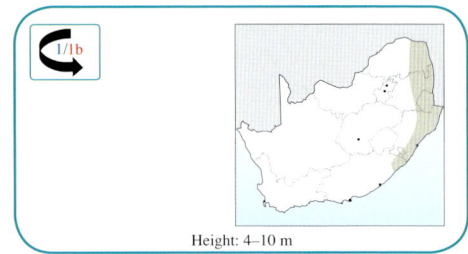

Height: 4–10 m

Origin and description: Native to the tropics of Central and South America, *T. peruviana* was introduced as an ornamental and has escaped into the wild, mainly in the northeastern regions.

Impact: The plant is now commonly found invading bush, watercourses and waste areas where it replaces indigenous vegetation. The whole plant, but especially the seeds, are highly toxic and potentially lethal. It is claimed that only about four seeds or two leaves can be fatal to a child.

Control: Chemical control is the best method available at present. A suitable herbicide is painted onto either the stems or cut stumps. Physical methods require the removal of the entire plant on account of the vigorous regrowth.

Other common names
lucky nut * Mexican oleander

Thevetia peruviana

ASPARAGACEAE

Asparagus laricinus (=*Protasparagus laricinus*)
wild asparagus * katbos

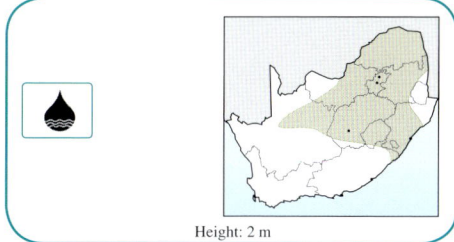

Height: 2 m

Other common names
bergkatdoring * fynkatbos * lang-beenkatdoring * ibutha (Z)

Origin and description: There are many indigenous species of *Asparagus* occurring in all parts of South Africa and they often become troublesome weeds. Some species have spines, while many have white flowers and red berries.

Impact: The plants are capable of forming impenetrable clumps along roads, fences and in waste places. Farmers in northern Free State in particular, find these plants a serious nuisance as they encroach on grazing land from established infestations on the perimeter of the land.

Control: These plants are difficult to remove manually on account of their strong tap roots and cannot efficiently absorb foliar-applied herbicides on account of their fine, feathery leaves. A herbicide has been registered however, but it does not cover all the various species and great care must be taken with application and follow-up.

Asparagus laricinus

ASTERACEAE

Chromolaena odorata *(=Eupatorium odoratum)*
paraffin weed * paraffienbos

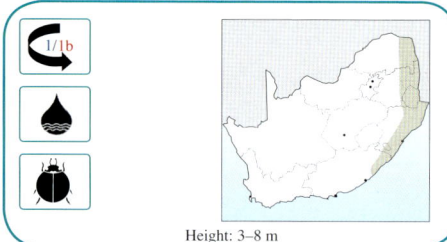

Height: 3–8 m

Origin and description: A native of South America, *C. odorata* was first recorded in the southwestern Cape in 1858 where it did not survive. It was probably reintroduced into present-day KwaZulu-Natal in about 1947 in packing materials contaminated with the seed that were off-loaded at Durban harbour. It soon established itself, and by 1962 was reported as spreading 'virulently' along the coastal areas. It is now a major perennial weed in the coastal region of KwaZulu-Natal and all eastern lowveld areas. It was first recorded in the Kruger National Park in 1997. It is one of the world's worst invaders.

It germinates easily and produces vast amounts of wind-blown seeds. One plant can produce up to a million seeds. It cannot survive frost, and favours moist areas.

Impact: *C. odorata* invades even undisturbed vegetation, totally swamping and replacing indigenous species, resulting in up to 100% loss in species diversity. It is highly inflammable, even when green, and is said to be poisonous to horses. In the iSimangaliso Wetland Park (formerly St Lucia), its presence is threatening the survival of crocodiles by shading and clogging nesting sites. The crushed leaves have a characteristic smell.

Control: Control of this weed is difficult and costly because it is capable of vigorous regrowth from stem coppice, root suckers and seed. Large

Other common names
Armstrong's weed * Eupatorium * Siam weed * triffid weed * usandanezwe (Z)

Similar species
C. odorata should not be confused with the indigenous vine *Mikania cordata* or with *Ageratina adenophora* (see page 210)

Chromolaena odorata

plants must be cut down and a suitable herbicide should be applied to the stump or regrowth. Small plants can be pulled out by hand. Follow-up inspections and treatment are essential to ensure that all traces have been eliminated. A beetle is raising hopes of biological control, but so far it has failed to establish itself where it was released in KwaZulu-Natal. The work continues with various other species of insect.

Chromolaena odorata

ASTERACEAE

Montanoa hibiscifolia
tree daisy * montanoa

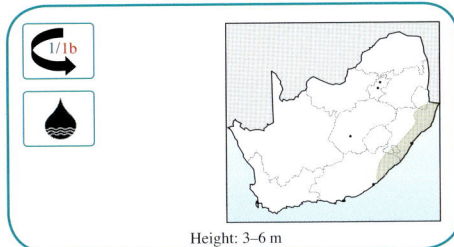

Height: 3–6 m

Origin and description: *M. hibiscifolia* is from Central America and was introduced as an ornamental. It is a semiwoody perennial shrub that can reach a height of 6 m.

Impact: It is not a particularly aggressive weed but forms unsightly infestations along roads and in disturbed areas, mainly in KwaZulu-Natal and the northern parts of the Eastern Cape. Because of its height, it shades out indigenous species and obstructs vision. There was a severe infestation of this plant in the Siteka Nature Reserve near Port St Johns where it had totally displaced the indigenous vegetation over large areas. This indicates the potential of the plant to become a very serious invader.

Control: Once established, the roots are difficult to remove by hand and there are no specific herbicide registrations. The mature plants should be cut down and the stumps treated if necessary. Normal follow-up procedures should be followed.

Other common names
ubhongobhongo (Z)

Montanoa hibiscifolia

ASTERACEAE

Seriphium plumosum *(=Stoebe vulgaris, Stoebe plumosa)*
bankrupt bush * slangbos

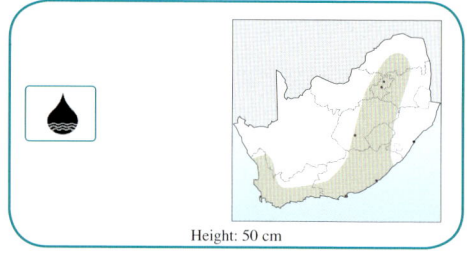

Height: 50 cm

Origin and description: An indigenous dwarf shrub that looks a bit like heather. It is much branched and usually grey, although the colour can vary according to the altitude. Pale brown bracts surround the small purple flowers, giving the flower spikes an attractive golden appearance. Several species in the genus *Stoebe* were recently combined with *Stoebe plumosa* as *Seriphium plumosum*, which is widespread in South Africa and is well known by hikers who use it to make a soft mattress when sleeping out of doors. It is also popular in the flower industry and often features in the South African exhibit at the Chelsea Flower Show.

Other common names
asbossie * bankrotbossie * Khoi-kooigoed * slanghoutjie * vaalbos

Impact: The form of *S. plumosum* that was previously known as *Stoebe vulgaris* is unpalatable to stock and is of relatively low nutritional quality. It invades productive grassland, thereby severely reducing the carrying capacity. The invasion of this plant into natural veld is a huge problem in arid and semi-arid grasslands, especially in the Free State and North West. It is also highly inflammable, thereby increasing the risk of uncontrollable veld fires.

Control: Tebuthiuron can be used as a soil treatment and metsulphuron-methyl as a selective

Seriphium plumosum

Seriphium plumosum

foliar spray, which should preferably be applied during the active growing period from October to April. Salt can also be used by simply sprinkling some at the base of each stem; it alters the salinity enough to inhibit growth. Thick infestations can be burned and the regrowth sprayed or slashed. Follow-up treatments and the adoption of proper land management practices are critical factors for long-term control.

ASTERACEAE

Tithonia diversifolia
Mexican sunflower * Mexikaanse sonneblom
Tithonia rotundifolia
red sunflower * rooisonneblom

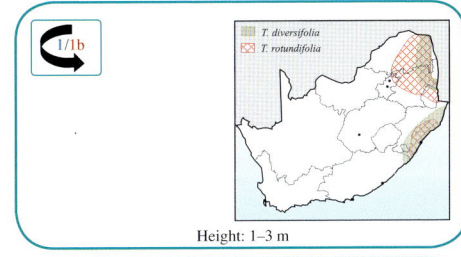
Height: 1–3 m

Origin and description: Natives of South America and introduced into South Africa as ornamentals, these sunflowers are now widespread as weeds in the warmer regions of the summer-rainfall area. *T. diversifolia* is especially common in Durban and Nelspruit, whereas *T. rotundifolia* becomes more frequent further north. Reproduction is only by seeds, but they can produce perennial clumps that resist fire and contact herbicides.

Impact: They can grow up to 3 m tall and form dense colonies on roadsides, railway embankments and waste places, particularly in urban areas. Infestations of these weeds not only cause serious and unsightly access problems, but they also totally swamp and replace the indigenous vegetation. On the positive side, Kenyan farmers have found great value in these plants as a green manure.

Control: Control is best achieved manually, with chemical follow-ups, if necessary, on seedlings that reappear in cleared areas.

Other common names
T. diversifolia: tree marigold * umbabane (Z) * umavayi (Z)

Tithonia diversifolia

Tithonia rotundifolia

Shrubs

BIGNONIACEAE

Tecoma stans
yellow bells * geelklokkies

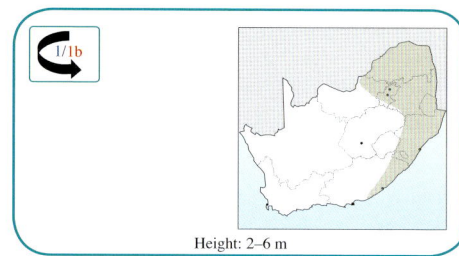

Height: 2–6 m

Origin and description: Probably originating in Central America, *T. stans* is now an invasive alien plant in many regions of the world, including the eastern parts of South Africa. It was introduced on account of its beautiful tubular, yellow flowers and its hardy, evergreen nature, thereby making it an ideal garden and landscaping plant. In fact, there are some garden varieties, which look only a little different, that have also escaped and become naturalised. It is the national flower of the Bahamas.

Impact: Yellow bells is frost tolerant but prefers full sunlight and invades disturbed areas such as roadsides, watercourses and waste areas. It can form dense stands that completely eliminate indigenous vegetation.

Control: Once established, *T. stans* is very difficult to remove with herbicides. The plant should be removed physically, taking care to destroy all seeds. The area should be inspected regularly to manage any regrowth.

Other common names
yellow elder * ginger-thomas * yellow trumpet bush * insimbephuzi (Z)

Tecoma stans

CHENOPODIACEAE

Atriplex nummularia* subsp. *nummularia
old man salt bush * oumansoutbos

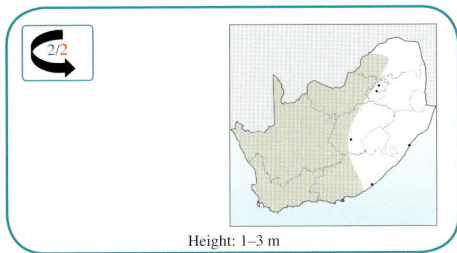
Height: 1–3 m

Origin and description: Introduced from Australia, this perennial, semideciduous shrub is now a common weed in the more arid parts of the Cape provinces, the central regions of the country and up into Namibia. It is edible and makes a good, nitrogen-rich fodder, being widely cultivated for this purpose in the arid regions of the Western Cape. It can survive drought years of only 50 mm rainfall and can withstand heavy frosts and saline soils. The seed of *A. nummularia* is in a papery or corky capsule called an utricle, that turns pink or straw colour as the plant matures.

Impact: This plant is commonly found in waste places and roadsides and can swamp indigenous fynbos once it becomes established. It is also known to invade coastal dunes, pans and other ecologically sensitive areas. It replaces indigenous vegetation.

Control: No specific herbicides have been registered for this species, but because of its perennial nature, it should be controlled physically before it becomes established. Being a Category 2 invader means that its cultivation should be carefully monitored and is subject to a permit.

Other common names
Alston's saltbush * Australian saltbush * Australiese brakbos * grootsoutbos

Atriplex nummularia

173

EUPHORBIACEAE

Ricinus communis var. *communis*
castor-oil plant * kasterolieboom

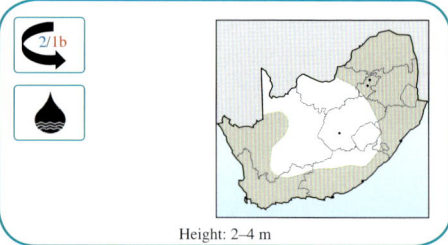
Height: 2–4 m

Origin and description: The origin of this species is not certain. It is probably from elsewhere in Africa and found its way here with Stone Age people, possibly 3 000 years ago. It is generally considered to be a perennial, although it is somewhat variable. It does not exude latex like most Euphorbiaceae. The three-lobed fruits are extremely toxic to humans and animals, one fruit being potentially fatal. Zulu people, however, use a paste for toothache and as a purgative. Although castor oil is extracted from the seeds of this plant, it has to undergo extensive purification before it is safe for consumption. There are cultivars with coloured leaves and fruits – they should be planted with great caution.

Impact: It is a common pioneering weed of roadsides, riverbanks, waste places and occasionally in perennial crops such as sugarcane. It can grow up to 4 m tall, hence the common names referring to a tree. It is very vigorous and competitive, but usually occurs only in disturbed habitats, limiting its threat to the indigenous biodiversity.

Control: Large plants can easily be controlled by chopping or uprooting them. The soil should be replaced and the site covered with leaves. The weed is generally sensitive to herbicides, with several registered as cut stump treatments.

Other common names
castorbean * wonder tree * bloubottelboom * bosluisboom * mohlafotha (S) * mokhura (P) * mufuta (Sh) * umfude (N) * umhlakuva (X; Z)

Ricinus communis

FABACEAE

Alhagi maurorum (=*A. camelorum*)
camel thorn bush * kameeldoringbos

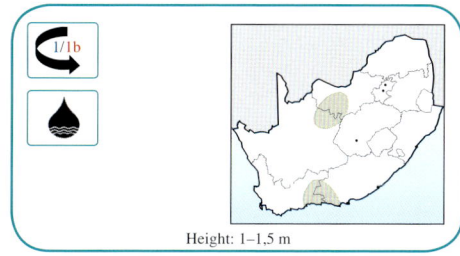

Height: 1–1,5 m

Other common names
caspian manna * volstruisdoring

Origin and description: *A. maurorum* is a native of Eurasia that has become naturalised in parts of the Cape provinces. It is thought to have first appeared at a horse station near Oudtshoorn during the Anglo-Boer War, which suggests it probably arrived as a contaminant of imported fodder. The plant established itself in the area and is now found throughout the Little Karoo and parts of the Northern Cape. It has a strong ramifying root system, making it difficult to remove by hand. New shoots can appear 5 m from the parent plant. The small pea-like flowers are borne directly on what appear to be green thorns, but are in fact spine-tipped branches.

Impact: The plant has been the subject of an intensive eradication campaign, which drastically reduced the infestation but did not eliminate it. If the pressure of the campaign is not maintained, it could once again multiply and spread, threatening the indigenous vegetation.

Control: A herbicide mixed with diesel is registered for the control of this weed but chemical control on any scale is laborious and expensive. Every effort should be made to keep this weed under control.

Alhagi maurorum

FABACEAE

Caesalpinia decapetala
Mauritius thorn * kraaldoring

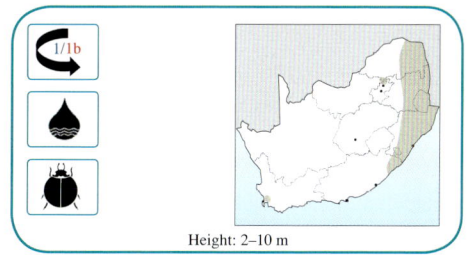

Height: 2–10 m

Origin and description: A native of India and Sri Lanka, *C. decapetala* has spread as a seriously invasive weed in the warm, high-rainfall areas of South Africa. Although, probably not the only reason for its introduction, this bush was once planted, along with *Agave sisalana*, to create an impenetrable barrier along the international border between Zululand and Mozambique – hence the Zulu name of 'ufenisi' or 'fence'. It is still planted as a hedge around kraals where it soon forms a barrier, keeping the cattle in and predators out. *C. decapetala*, which is evergreen, coppices when cut and trailing branches root where they touch the ground. Wider dispersal is generally achieved by means of the large seeds, that are readily transported by water and animals.

Impact: Mauritius thorn has spread along watercourses and has invaded natural forests. Although it favours riverine areas, it also invades arid hillsides, especially in areas where it has been used as a kraal hedge. Its dense, smothering habit can eliminate all other vegetation, transforming the habitat and the landscape. It will quickly form dense, monospecific, thorny thickets that are totally impenetrable to humans and livestock.

Control: This weed can be controlled by a combination of chemical (glyphosate or triclopyr) and mechanical means. Chemicals should be applied to small plants or to the regrowth of larger individuals that have been slashed down. Seedlings and saplings can be uprooted when the soil is moist. However, to prevent coppicing without the use of a herbicide, the entire rootstock must be dug out. Repeated follow-up inspections and treatments are usually necessary. Seed-feeding biological agents were released in 1999, but success to date is insignificant.

Other common names
cat's claw * mysore thorn * shoofly * ufenisi (Z) * ubobo-encane (Z)

Similar species
C. gilliesii 1b (bird-of-paradise flower * paradysvoëlblom), from tropical America
C. spinosa (algaroba), from South America

Caesalpinia decapetala

FABACEAE

Leucaena leucocephala* subsp. *leucocephala *(=L. leucocephala, L. glauca)*
leucaena * reusewattel

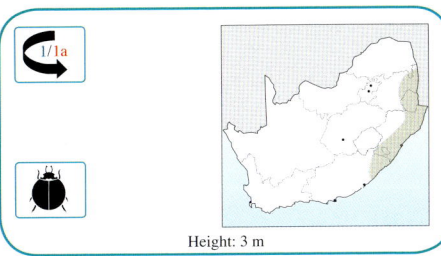
Height: 3 m

Origin and description: A perennial shrub from South America, *L. leucocephala* was introduced as a fodder crop and as a source of firewood. It is also efficient in nitrogen fixation. However, it is a conflict species since even though it is still promoted as a valuable contributor to reforestation, soil regeneration and the fight against global warming, it clearly has the ability to be seriously invasive.

Impact: If allowed to grow uncontrolled, it can escape into the wild and become a problem plant. It grows quickly and forms dense thickets, which crowd out any native vegetation. Even though it is considered a valuable fodder, if it constitutes more than 25% of an animal's diet, it is toxic. It is found as a weed throughout the subtropical areas, especially in KwaZulu-Natal. It is usually found in road and railway reserves, on riverbanks and in disturbed areas, but is becoming very common in

Other common names
lead tree * white popinac * stuipboom * ubobo (Z) * ulusina (Z)

all unattended areas on the KwaZulu-Natal coast, where it forms monospecific thickets. It is No. 46 on the list of the *World's Worst Invasive Species*.

Control: It can be controlled by cutting down the plants and digging out the roots. Chemical control is very difficult, but basal stem treatments and foliar sprays can work. Insects that attack the seeds were released as biocontrol agents in 1999. The success of this is not yet significant.

Leucaena leucocephala

FABACEAE

Mimosa pigra *(=M. pellita, M. asperata)*
giant sensitive plant * raak-my-nie

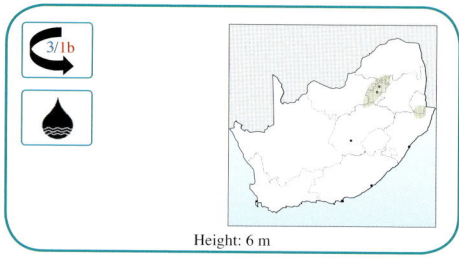

Height: 6 m

Other common names
amourette * catclaw mimosa * mimosa * kruidjie-roer-my-nie * umazifisa (Z)

Origin and description: From tropical America, *M. pigra* was introduced into South Africa as an ornamental. The leaves have the fascinating property of being sensitive to touch and will close up on contact.

Impact: This plant has been listed as one of the world's 100 worst invasive species. It forms dense, thorny, impenetrable and monospecific thickets, particularly in wet areas. It has relatively recently been found invading such places in the subtropical areas in northern KwaZulu-Natal and into Mozambique. It has also caused much concern north of Tzaneen, along the Letaba River. The seeds are dispersed by water and can float for an indefinite period, but require a dry period so that they can germinate in soil and grow. They are also dispersed by humans and animals, in mud adhering to fur, clothing or vehicles. *M. pigra* threatens the production, and cultural and conservation values of wetlands, blocking access and reducing biodiversity. It is regarded as one of the worst weeds in Australia and any plant that thrives in Australia can thrive in South Africa.

Mimosa pigra

Control: Small outbreaks can be removed by hand-pulling or grubbing, ensuring that as many roots as possible are removed. Glyphosate is registered as a foliar application. The spread of this weed must be prevented.

FABACEAE

Senna didymobotrya *(=Cassia didymobotrya)*
peanut butter cassia * grondboontjiebottercassia

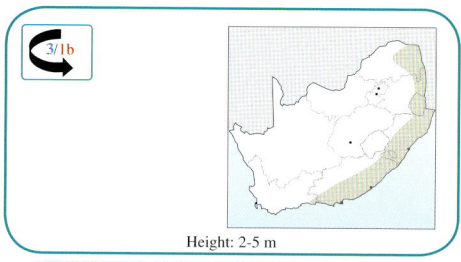

Height: 2-5 m

Other common names
oatmeal cassia * popcorn senna * wild senna * munwahuku (Sh)

Origin and description: Introduced from tropical eastern and central Africa as an ornamental plant, *S. didymobotrya* has become naturalised in South Africa and is now a common invasive weed. It has been used as a green manure and a cover crop. It is a semideciduous perennial shrub that only reproduces by seed and can tolerate only a light frost. In more temperate regions, it produces flowers throughout the year; these flowers smell of peanut butter. It is the most common and widespread of the several species of *Senna* that have found a home in South Africa.

Impact: *S. didymobotrya* is found throughout the summer-rainfall regions, especially in the eastern areas. It invades disturbed areas such as roadsides, waste areas, woodland and river banks, replacing indigenous vegetation. It is also obstructive and untidy.

Control: No specific herbicide has been registered for this weed and it is best controlled manually.

Senna didymobotrya

FABACEAE

Senna hirsuta (=*Cassia hirsuta*)
woolly senna

Origin and description: Introduced from South America and the West Indies for hedging and as an ornamental, this species has escaped into the wild and has been declared an invasive weed in South Africa. There are also some indigenous species of *Senna* that have been accused of being weedy, but they are not serious.

Impact: These plants invade disturbed areas such as roadsides and waste areas, replacing

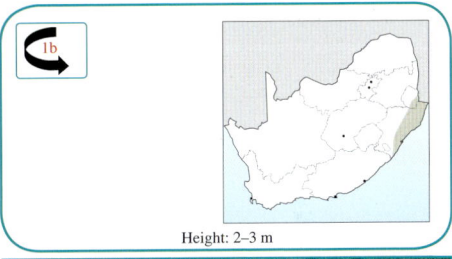

Height: 2–3 m

Senna hirsuta

Senna hirsuta

indigenous vegetation. They can also be obstructive and untidy. *S. hirsuta* is common on the edges of sugarcane fields in Zululand.

Control: No specific herbicides have been registered for this weed and it is best controlled manually.

179

FABACEAE

Senna pendula* var. *glabrata *(=Cassia coluteoides)*

Senna septemtrionalis *(=Cassia floribunda)*
smooth senna * arsenic bush

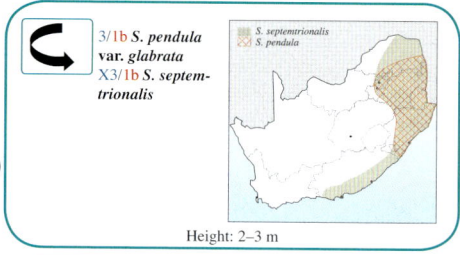

3/1b *S. pendula* var. *glabrata*
X3/1b *S. septemtrionalis*

Height: 2–3 m

Origin and description: Introduced from South America and the West Indies for hedging and as ornamentals, these species have escaped into the wild and have been declared invasive weeds in South Africa. There are also some indigenous species of *Senna* that have been accused of being weedy, but are not serious.

Impact: These plants invade disturbed areas such as roadsides and waste areas, replacing indigenous vegetation. They can also be obstructive and untidy.

Control: No specific herbicides have been registered for these weeds and they are best controlled manually.

Similar naturalised species from Central or South America
S. bicapsularis 3/1b (=*Cassia bicapsularis*) rambling cassia * winter senna
S. corymbosa (=*Cassia corymbosa*) buttercup bush * scrambled eggs
S. multiglandulosa (=*S. tomentosa*) wild senna * peulbos * wildesenna
S. occidentalis 1b(=*Cassia occidentalis*) stinking weed * wild coffee

Senna septemtrionalis

Senna pendula var. *glabrata*

FABACEAE

Sesbania bispinosa
spiny sesbania * stekelsesbania

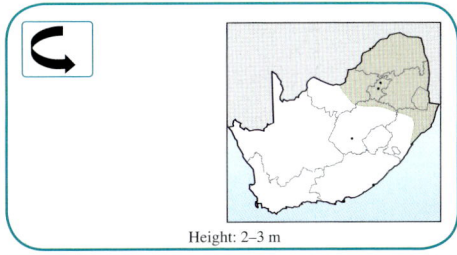

Height: 2–3 m

Origin and description: Introduced from Asia and North Africa, this plant is now widespread in the summer-rainfall regions of South Africa. From a distance *S. bispinosa* looks similar to *S. punicea* (see page 182), but it has cream and purple flowers and long, thin pods. It is a biennial, but the bare, dried plants are a familiar sight in winter.

Impact: In the Mpumalanga lowveld, at the Loskop Irrigation Scheme and in Swaziland it is a weed of cotton, rice and other croplands. It also occurs along roadsides and in disturbed areas and is capable of growing in saline soil.

Control: Control measures should be initiated before the plant becomes established. There are no herbicides registered for this weed.

Sesbania bispinosa

FABACEAE

Sesbania punicea
red sesbania * rooisesbania

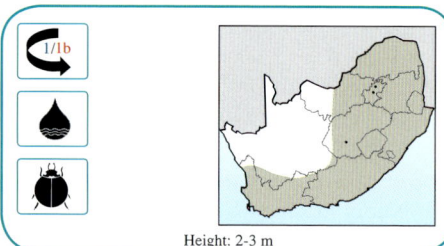

Height: 2-3 m

Origin and description: *S. punicea* is a perennial shrub and native of South America that was probably introduced as an ornamental early in the 20th century. However, it was not until 1960 to 1980 that it spread alarmingly in most parts of the country. It was declared a noxious weed in 1979. It is a deciduous shrub and is spread by seeds, which are contained in characteristic four-winged pods. All parts of the plant are poisonous but the seeds are particularly lethal to birds and sheep – as few as six seeds can kill a chicken. Birds and other animals are therefore unlikely to play a role in dispersal. In South Africa it is distributed from the Cape provinces through KwaZulu-Natal to Mpumalanga.

Impact: *S. punicea* is found mainly in permanently or seasonally wet places but is also able to establish itself in disturbed areas such as roadsides and refuse dumps. Many roadside infestations are probably a result of the use of soil containing sesbania seeds for road construction. It can occur in dense, monospecific stands, which can transform the landscape.

Control: Biological control of sesbania by means of a beetle from Argentina is proving very effective, a fortunate situation as chemical control is rather ineffective and costly. Several soil, foliar or cut-stump herbicides have been registered. Slashing induces vigorous regrowth, which must

Other common names
Brazilian glory pea * coffee weed * rattlepod * tango * Brasiliaanse glorie-ertjie

be sprayed or physically removed. The best time for this is after the spring flush, when the plants have exhausted their root reserves.

Sesbania punicea

FABACEAE

Spartium junceum
Spanish broom * Spaanse besem

Origin and description: *S. junceum* was introduced from Europe as an ornamental. In spring it has masses of bright yellow flowers that are very distinctive. The fruit is a pea-like pod that bursts open with an audible crack. Species of *Spartium*, *Genista* and *Cytisus* are all closely related.

Impact: *S. junceum* is now well established in parts of South Africa, especially in the Western Cape, invading stream banks, waste areas and fynbos. *Cytisus scoparius* is very similar and is commonly found in the high-lying areas of the eastern escarpment. It is particularly common around Harrismith and is found mainly along roadsides, but can be seen invading veld. Several other *Cytisus* species are cultivated as garden ornamentals.

Control: There are no particular recommendations for the control of these plants. They should be physically removed whenever they are encountered.

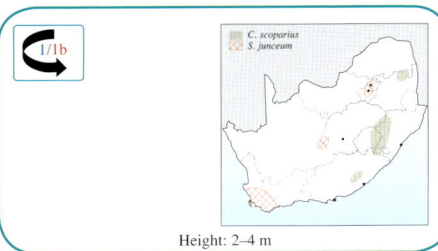

Height: 2–4 m

Similar species
Cytisus scoparius 1/1a (Scotch broom * Skotse brem), from Europe
Genista monspessulana 1 (=*Cytisus monspessulanus*) (Montpellier broom * Montpellier brem), has been found naturalised on Table Mountain

Cytisus scoparius

Spartium junceum

FABACEAE

Ulex europaeus
European gorse * gaspeldoring

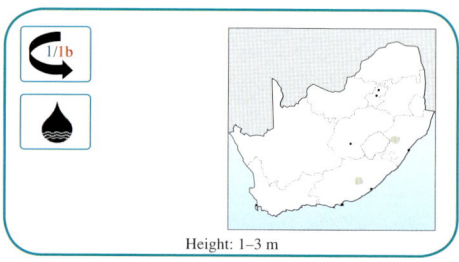

Height: 1–3 m

Origin and description: Introduced from Europe and Britain, this perennial plant was brought over as a hedging plant, an ornamental and as a source of honey. It was originally found invading mountain grassland, vleis and valleys in the moist highland soils of the Eastern Cape, but since then it has been discovered in some of the mountainous regions of KwaZulu-Natal. It is closely related to broom (species of *Cytisus* and *Spartium*) but differs in its extreme spininess, with the leaves being modified into 1–4 cm long spines.

Impact: Apart from its attractive flowers, gorse is very prickly and can quickly form a barrier, impenetrable to stock, people and wild animals. It is extremely flammable, has aggressive seed dispersal and is tolerant of fire, with the burnt stumps quickly re-sprouting. It fixes nitrogen and can thrive on dry, rocky soils where many other plants struggle and are quickly replaced by the gorse. Soil is often bare under the plants, increasing erosion where it has replaced grass. It has the potential to transform the landscape and is a serious threat to biodiversity.

Control: Great care should be taken to prevent the spread of this weed. Metsulphuron-methyl is registered in South Africa for control of gorse and should be applied to actively growing plants in late summer. Committed follow-up on regrowth and seedlings is essential since the seeds can lie dormant for up to 25 years.

Other common names
furze * whin

Ulex europaeus

HYPERICACEAE

Hypericum perforatum
St John's wort * Johanneskruid

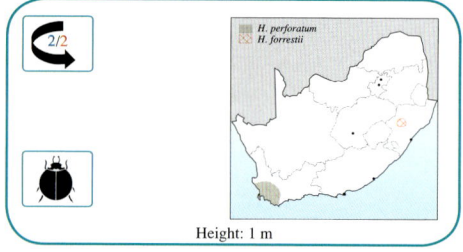

Height: 1 m

Origin and description: This species is a native of Europe, Asia and North Africa and a weed of most of the temperate regions of the world, being different from indigenous *Hypericum* species with its creeping underground stems. *H. perforatum* was originally introduced into South Africa in 1942 as an impurity in vetch seed that was sown at Helshoogte near Stellenbosch. Because of the ease with which it could spread, it soon covered large areas of the southwestern Cape. It is easily distributed by seed or by the rhizomes, which are often cut up and spread during cultivation. The plant has two distinct growth phases. In autumn and through the winter it is creeping and prostrate and in summer it produces several woody, upright flowering stems. The stems produce flowers from November to January and die off towards late summer, leaving characteristic brown stalks.

Impact: *H. perforatum* invades lowland and mountain fynbos, replacing indigenous vegetation. It is poisonous to livestock, causing photosensitisation.

Control: Biological control of *H. perforatum* with a beetle was introduced in 1961–1962. Periodic redistribution of the original beetles and further establishment of a gall midge has restricted the weed in most areas. At present no herbicides have been registered for the control of this weed.

Other common names
gammock * goatsbeard * herb-john * klamath weed * penny-john * rosin rose * Tipton's weed * touch-and-heal

Similar species
H. forrestii ⓒ (=*H. patulum*), from China and an emerging weed in the KwaZulu-Natal midlands

H. androsaemum ⓒ 1 (tutsan), also from Eurasia, has invaded parts of Australia and New Zealand. It has been proposed as a Category 1 in South Africa as a precaution

Several other *Hypericum* species and hybrids are commonly grown in southern Africa as garden plants and for the cut-flower market

Hypericum forrestii

Hypericum perforatum

LAMIACEAE

Plectranthus comosus *(=Coleus grandis)*
(*P. barbatus* misapplied in SA)
woolly plectranthus * Abessiniese coleus

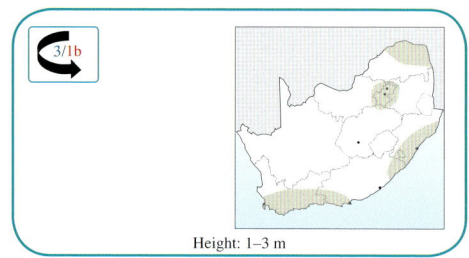

Height: 1–3 m

Origin and description: South Africa is home to over 30 species of *Plectranthus*, which are in turn, just some of the 350 species that are native mainly to regions of the southern hemisphere. Many have very attractive foliage and flowers, which is why they are used as garden plants throughout the world. *P. comosus* was introduced from India and Sri Lanka for this purpose and possibly because it is a vigorous grower. This property, of course, is what makes this shrub a risk and it can now be found growing wild and vigorously in hedges, roadsides and shady areas like forest margins. The same plant is also legitimately called *Coleus grandis* and is often referred to as *P. barbatus*, a confusing situation that has to be clarified as far as taxonomy is concerned.

Impact: *P. comosus* can smother and replace indigenous vegetation. It degrades valuable grazing land, for example, by invading grassy hillsides in the northern parts of the Eastern Cape where it is widely used as a kraal hedge. It is also an untidy and aggressive weed in unattended urban areas.

Control: Unwanted plants should be carefully identified and removed. Preference should always be given to known indigenous or non-invasive species.

Plectranthus comosus

MYRSINACEAE

Ardisia crenata
coralberry tree * koraalbessieboom

Ardisia crenata

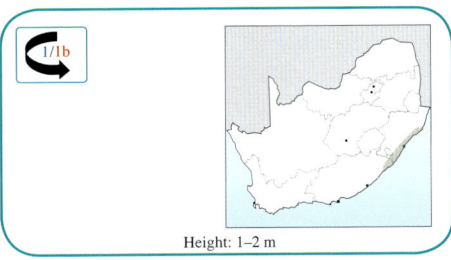
Height: 1–2 m

Other common names
Christmas berry * coral bush * hen's eyes * spiceberry

Origin and description: A native of Asia from India to Japan that was introduced into South Africa as a garden ornamental and as a pot plant. *A. crenata* has escaped into the wild in several eastern areas where it has been found invading forest margins, river banks and swamp forest. The bright, shiny red berries stay on the plant for many months, which make it an attractive garden plant. The berries, that are eaten and dispersed by birds, remain viable for many months.

Impact: *A. crenata* has been placed in Category 1 because of its invasive nature and ability to disrupt native plant communities, especially in forest understoreys.

Control: Unwanted plants should be uprooted and totally removed since they can re-sprout when cut down or when the foliage is removed by fire.

OLEACEAE

Ligustrum japonicum
Japanese wax-leaved privet * Japanese liguster

Ligustrum ovalifolium
California privet * Kaliforniese liguster

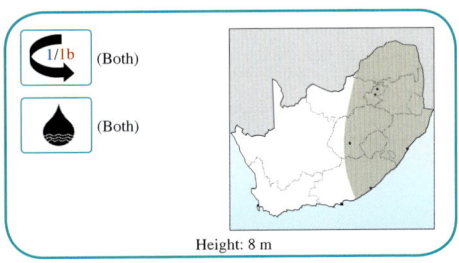
Height: 8 m

Origin and description: Planted as evergreen or semi-evergreen ornamentals and for hedges, privets have escaped into the wild. *L. japonicum* was originally from China and was introduced into Japan some 1 000 years ago. It is a medium-sized tree that produces masses of small berries, which persist into winter and are popular with birds, thereby spreading the plant far and wide into woodland, hedges, gardens and roadsides. Most of the other species (approximately 50) of *Ligustrum* are native to eastern and southeastern

Asia and Australia. There is also a wide range of cultivars for horticulture but their sterility or non-invasiveness has to be investigated.

Impact: Privets replace indigenous vegetation and their attractive berries are poisonous if eaten in large quantities.

Control: To control the privet it should be cut down and since the stumps can readily regrow, they should be treated with a suitable herbicide such as imazapyr. Follow-up is always required as the seeds, of which there are many, stay viable for many years.

Other common names
L. japonicum: glossy privet * Nepal privet * white wax tree
Similar species
Other species of privet that are recorded as naturalised in South Africa are:
L. lucidum ⊂ 3/1b ● (Chinese wax-leaved privet * Chinese liguster)
L. vulgare ⊂ 3/1b ● (common privet * gewone liguster)
L. sinense ⊂ 3/1b ● (Chinese privet * Chinese liguster)

Ligustrum ovalifolium

Ligustrum japonicum

ROSACEAE

Cotoneaster franchetii
orange cotoneaster * pronkbessiebossie

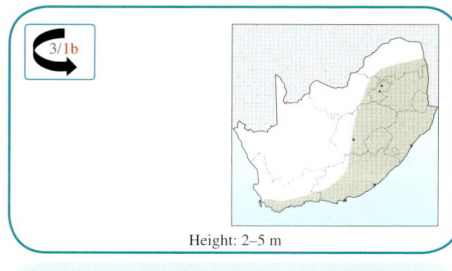
Height: 2–5 m

Origin and description: Cotoneasters come from temperate regions of Asia, Europe and even North Africa; *C. franchetii* comes from southwestern China. It is cultivated for hedging and as an evergreen ornamental but has escaped and is now growing wild in the more temperate areas of South Africa. They do not have thorns, unlike *Pyracantha* species.

Impact: Cotoneasters invade temperate grassland, forest margins, kloofs, riverbanks and rocky outcrops. The plentiful berries, which are toxic if consumed in any quantity, are eaten by birds, that spread the plant further afield and as with other fruit-bearing alien plants, distract these birds from disseminating the seeds of more threatened, indigenous species.

Control: Seedlings can be hand-pulled and plants up to 3 m tall can be sprayed with a herbicide. Above that it should receive a basal-bark treatment or be cut down and the stump chemically treated. Unintended plants should be removed.

Similar species
C. glaucophyllus 1b (glaucous cotoneaster * late cotoneaster), from Asia
C. pannosus 3/1b (silver-leaf cotoneaster), from China
C. salicifolius 1b (willow-leaved showberry), from western China
C. simonsii 1b (Himalayan cotoneaster * Simon's cotoneaster), from Asia
Over 30 species cultivated as garden ornamentals

Cotoneaster franchetii

ROSACEAE

Pyracantha angustifolia
yellow firethorn * geel branddoringbos
Pyracantha crenulata
Himalayan firethorn * rooivuurdoring

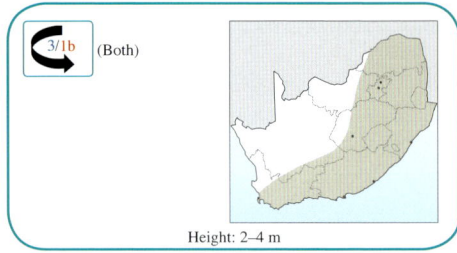

Height: 2–4 m

Origin and description: These very similar species of thorny, evergreen shrubs were introduced from China and the Himalayas respectively as ornamentals and as a security hedging. *P. angustifolia* is now widespread on the highveld and eastern escarpment and is commonly used as a screen on highways. It has masses of attractive red or orange berries (or pomes) that are favoured by birds. *P. crenulata*, which is different from the other firethorns only in that the leaves and calyx are hairless, also invades high-altitude grassland and is most invasive in the eastern Free State. The distribution of firethorns is restricted to such areas as freezing winter temperatures are probably needed to stimulate seed germination. There are many hybrids or cultivars, but most of them also have the potential to become invasive and have to be investigated.
Impact: Dense infestations seriously downgrade veld and transform the landscape. The rigid spines can hinder access. These plants have the potential to spread further and existing infestations should be monitored closely.
Control: There are no herbicides registered to control these plants. Plants should be chopped out by hand, a task that is the easiest to do when the plants are small and the soil is moist. Older plants should be handled with care on account of the unpleasant thorns.

Similar species (of seven cultivated in South Africa) on the proposed list
P. coccinea 1b (red firethorn * scarlet firethorn), also from Asia
P. crenatoserrata 1b (Chinese firethorn), from China
P. koidzumii, 1b from Taiwan

Pyracantha coccinea

Pyracantha angustifolia

ROSACEAE

Rosa rubiginosa *(=R. eglanteria)*
eglantine * wilderoos

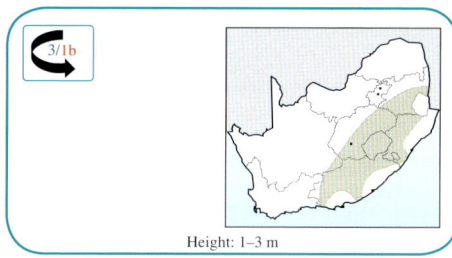

Height: 1–3 m

Origin and description: Introduced from Europe and Asia as an ornamental, this plant has found a home in the temperate, hilly areas of the eastern highveld in and around Lesotho. It invades high-altitude grassland, rocky ridges, roadsides and overgrazed land. It reproduces by seed and suckers freely from the roots. The bright red fruit (otherwise known as 'achenes' or 'hips'), are an attractive sight in the generally brown highveld autumn, and are sometimes collected for making juice. This juice is high in vitamin C.

Other common names
sweetbriar

Similar species
Rosa multiflora (=*Rosa polyantha*) (multiflora rose)
Various *Rubus* species (e.g. *R. cuneifolius*)

Impact: It is not as aggressive or as invasive as the bramble, but reduces the carrying capacity of the veld on account of its thorny stems. This plant has the potential to spread into other areas and should be monitored closely.

Control: There are no specific herbicide registrations for this plant, but it should be controlled in the same way as the various brambles.

Rosa rubiginosa

ROSACEAE

Rubus cuneifolius
American bramble * Amerikaanse braambos

Rubus fruticosus
European blackberry * braam

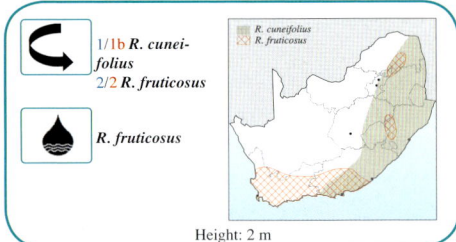

Height: 2 m

Origin and description: There are 17 species of 'bramble' in South Africa, some being indigenous, such as *R. rigidus*, and some being exotic. There are many forms, hybrids and subspecies of bramble and a specialist is required to identify them properly. *R. cuneifolius* was introduced from North America at the beginning of the 20th century by a farmer who wished to make a living by making bramble jam. *R. fruticosus* is from Europe and occurs mainly in the Cape.

The plants are spread by seeds in the fruit. The small seeds in the berries have a hard, resistant seedcoat, which ensures their survival through the digestive tract of birds, other animals and humans. The bramble also spreads vegetatively by means of tip-rooting and sucker formation on the roots. Sucker development is stimulated when the aerial part of the plant has been destroyed by fire, mowing or inadequate herbicide application.

Impact: *R. cuneifolius* is just one of the species that is now becoming a serious problem, mainly in parts of KwaZulu-Natal, where it forms dense stands in the veld, on roadsides and in forests. The thorny bushes create impenetrable barriers, restricting the movement of workers, equipment and animals.

Control: The underground runners make the *Rubus* species very difficult to eradicate and specialised herbicides are normally used. These herbicides are most effective in autumn when downward sap movement can transport the chemical to the roots. The plants can be controlled by cultivation if the rhizome is removed. It is important to make follow-up inspections and treatments, for several years if necessary, in order to ensure complete eradication of this weed from a specific area.

Other common names
R. cuneifolius: blackberry * Gozard's curse * sand bramble * sandbraam * ijingijoye (Z)
R. fruticosus: blackberry * bramble * bosbraam
R. rigidus: ijikijolo (Z) * monokotswai-wa-banna (S)

Some similar species recorded as naturalised
R. flagellaris X1/1b (=*R. baileyanus*, *R. canadensis*, *R. procumbens*): bramble * braam
R. niveus 1b (=*R. albescens*, *R. lasiocarpus*): Ceylon raspberry * hill raspberry * java bramble
R. x proteus: 1/1b American bramble

Rubus fruticosus

SCROPHULARIACEAE

Buddleja davidii
Chinese sagewood * Chinese saliehout

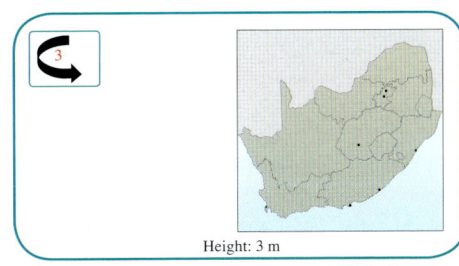

Height: 3 m

Origin and description: Native to northwestern China and Japan, this shrub is planted throughout the world as an ornamental on account on its large, colourful and fragrant flowers. In South Africa it has been proposed as a Category 3 plant but in Britain it is in the top five alien invasive plants and it is also a noxious weed in the USA. This indicates its toughness and ability to adapt, so it should be cultivated with caution. This is also

Other common names
butterfly bush * summer lilac * somerlila
Similar species
B. madagascariensis (Madagascar sagewood), from Madagascar, similar, but with orange flowers
B. salviifolia (indigenous)

Buddleja davidii

the parent plant of many colourful cultivars.
Impact: Chinese sagewood can escape from gardens and has the potential to flourish and replace indigenous vegetation. It is very attractive to butterflies.
Control: Unintended plants should be removed and destroyed.

SOLANACEAE

Solanum sisymbriifolium
dense-thorned bitter apple * doringtamatie

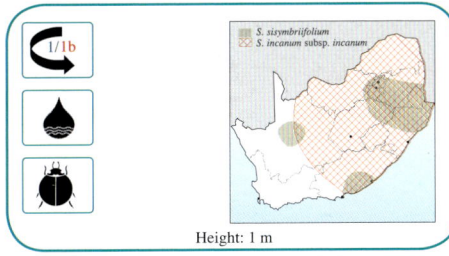

Height: 1 m

Origin and description: *S. sisymbriifolium* was introduced from South America during the Anglo-Boer War. It is a spiny, woody shrub, with a very extensive root system, that is highly resistant to nematodes and as such is used as a trap crop for potato cyst nematode in the UK. Often found growing along fences in open veld, as this is where the birds that have eaten the fruit will sit and deposit the seeds. There are many other species of *Solanum* often referred to as 'bitter apple' or 'wild tomato'. Many of them have thorns on the stems and the leaves. Some of them are toxic, with the unripe fruit being more toxic than the ripe fruit. The ripe fruit does not fall off easily and often remains on the plant in winter. The fruit are then spread around in hay or by birds and other animals that eat them.

Other common names
wild tomato * sticky nightshade * fire-and-ice plant * tamatiedissel * digdoringbitterappel
Similar species
S. incanum subsp. *incanum* (grey bitter apple * thorn apple * gifappel * dinjinsa (Sh) * intfuma (Si) * morola (P) * umdulukwa (N)), indigenous
S. chrysotrichum (giant devil's fig), naturalised in the Eastern Cape

Impact: It is found in such places as roadsides, orchards and tramped-out veld. This plant is a very resilient and aggressive invader. Once established, it is very difficult to remove and can replace large areas of indigenous vegetation.

Control: *S. sisymbriifolium* can be controlled with a foliar application of triclopyr. Unfortunately this is an expensive operation. Biocontrol investigations are under way, but so far with minimal success.

Solanum sisymbriifolium

Solanum chrysotrichum

Solanum incanum

VERBENACEAE

Duranta erecta *(=D. repens, D. plumieri)*
forget-me-not-tree * vergeet-my-nie-boom

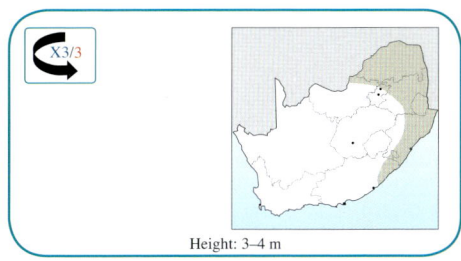

Height: 3–4 m

Origin and description: A native of tropical America from Florida to Brazil and including the West Indies, that was introduced into South Africa as an ornamental and as a hedge plant. The wild-type has pale mauve-blue flowers but there is a white-flowered cultivar that is often planted as a companion. There is also a form with variegated leaves, but it is as yet uncertain as to how invasive these cultivars are. Sometimes there are spines at the leaf axils and the berries are very decorative, but can be toxic.

Impact: This plant can invade moist sites along roads, in forests and rivers and is capable of forming impenetrable hedges and thickets. The toxic berries are said to be able to cause severe discomfort if eaten by humans and are even potentially fatal. Birds eat the berries, thereby aiding the spread of unwanted plants and potentially ignoring the more threatened indigenous plants.

Other common names
golden dewdrop * pigeon berry * sky flower * geelbessie * kraaldoring * wolwedoring

Control: Unintended plants should be removed by hand. The sterility or non-invasiveness of cultivars has to be investigated.

Duranta erecta

VERBENACEAE

Lantana camara
common lantana * gewone lantana

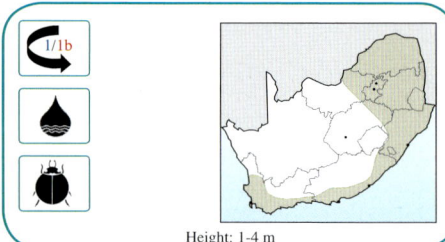

Height: 1-4 m

Origin and description: *L. camara* is a native of tropical America and is considered one of the world's 10 worst weeds, with some 650 varieties being major weeds in over 60 countries. It is one of the most serious invader species in South Africa as well as in some other countries such as Sri Lanka and India. First recorded in South Africa in 1858, it spread rapidly into present-day KwaZulu-Natal and other eastern areas. It is also invasive in parts of Namibia and into Botswana.

The colour of the flowers and the sharpness of the spines vary, but all these forms, and most of the cultivated ones, are considered to be the same species. Over 50 different variants are recognised in South Africa alone and it has been discovered that these variants differ in their susceptibility to herbicides and even to some introduced biocontrol agents. There is doubt about the sterility of some so-called 'sterile' cultivars such as *L. montevidensis* (see page 40). The fruit of the weed is a small, black berry, which is easily spread by birds. Sometimes sudden invasions are experienced when seed is washed down from an infested area in a flash flood and deposited on a floodplain downstream.

Impact: Lantana forms dense, impenetrable thickets, replacing indigenous plants, increasing erosion and seriously interfering with farming and forestry activities. The plant can be toxic to cattle.

Control: Eradication is laborious and expensive. Chopping the dense bushes and then painting the stumps or spraying the regrowth with herbicide is the usual and most effective method. Foliar sprays on large, uncut bushes are expensive and not very successful. Best results are achieved if the plants are sprayed in summer to autumn. Small plants can be pulled out by hand when the soil is moist. A comprehensive follow-up and maintenance programme is essential.

Other common names
bird's brandy * cherry pie * tickberry * gomdagga * sumba (Sh) * ubukhwebezane (Z) * ubutywala bentaka (X)

Similar species
Four indigenous species in southern Africa, e.g. *L. rugosa*

Lantana camara

HERBS

Within the section, the plants are arranged alphabetically according to family and genus.

SPECIES	Exudes white sap	Thorny, prickly or spiny	Seeds in spiny or hairy capsules	Seeds in multiseeded capsules	Fluffy seeds that blow away easily	Seeds in a berry	Pungent when crushed	Red/pink flowers	Yellow flowers	Purple flowers	White flowers	Blue flowers	Green or insignificant flowers
Argemone spp., p. 286	✓	✓	✓	✓			✓		✓		✓		
Sonchus spp., p. 239	✓	✓			✓				✓				
Euphorbia spp., p. 266, 268	✓			✓									✓
Tragopogon spp., p. 242	✓				✓				✓	✓			
Lactuca serriola, p. 232	✓				✓				✓				
Cichorium intybus, p. 222	✓											✓	
Xanthium spinosum, p. 243		✓	✓	✓									✓
Berkheya spp., p. 216		✓			✓				✓				
Picris echioides, p. 234		✓			✓				✓				
Cirsium vulgare, p. 223		✓			✓					✓			
Centaurea melitensis, p. 220		✓							✓				
Salsola kali, p. 264		✓									✓		✓
Amaranthus spinosus, p. 206		✓											✓
Datura spp., p. 299			✓	✓			✓			✓	✓		
Papaver spp., p. 286			✓	✓				✓					
Crotalaria sphaerocarpa, p. 269				✓					✓				
Canna spp., p. 255				✓					✓				
Bidens spp., p. 217				✓				✓	✓		✓		
Agrimonia procera, p. 297				✓									
Achyranthes aspera, p. 202				✓						✓	✓		✓
Xanthium strumarium, p. 244				✓									✓
Acanthospermum spp., p. 316				✓									✓
Physalis spp., p. 300				✓		✓			✓				

SPECIES	Exudes white sap	Thorny, prickly or spiny	Seeds in spiny or hairy capsules	Seeds in multiseeded capsules	Fluffy seeds that blow away easily	Seeds in a berry	Pungent when crushed	Red/pink flowers	Yellow flowers	Purple flowers	White flowers	Blue flowers	Green or insignificant flowers
Nicandra physalodes, p. 300				■				■			■		
Sesamum triphyllum, p. 289				■						■			
Oenothera rosea, p. 284				■				■					
Striga asiatica, p. 285				■				■					
Hedychium spp., p. 307				■				■	■	■	■	■	■
Raphanus raphanistrum, p. 252				■					■	■			
Hibiscus spp., p. 276				■					■	■			
Corchorus trilocularis, p. 275				■					■				
Oenothera spp., p. 283				■					■				
Sisymbrium spp., p. 254				■					■				
Sida spp., p. 279				■					■				
Cleome spp., p. 258				■						■			
Lavatera arborea, p. 277				■						■			
Oenothera tetraptera, p. 284				■							■		
Cleome gynandra, p. 258				■							■		
Capsella bursa-pastoris, p. 250				■							■		
Lilium formosanum, p. 274				■							■		
Silene gallica, p. 260				■							■		
Lepidium spp., p. 251				■									■
Polycarpon tetraphyllum, p. 259				■									■
Lupinus angustifolius, p. 269				■					■				
Lupinus luteus, p. 269				■								■	
Dittrichia graveolens, p. 227					■		■		■				
Hypochaeris radicata, p. 231					■				■				
Schkuhria pinnata, p. 235					■				■				

Herbs

SPECIES	Exudes white sap	Thorny, prickly or spiny	Seeds in spiny or hairy capsules	Seeds in multiseeded capsules	Fluffy seeds that blow away easily	Seeds in a berry	Pungent when crushed	Red/pink flowers	Yellow flowers	Purple flowers	White flowers	Blue flowers	Green or insignificant flowers
Senecio spp., p. 236					●				●				
Conyza spp., p. 224					●				●				
Pseudognaphalium luteo-album, p. 235					●				●				
Solanum spp., p. 302,303								●		●	●		
Rivina humilis, p. 291													
Phytolacca octandra, p. 289													●
Anthemis spp., p. 214							●		●		●		
Foeniculum vulgare, p. 209							●		●				
Tagetes minuta, p. 241							●		●				
Melilotus indicus, p. 271							●		●				
Ageratum spp., p. 212							●			●		●	
Ceratotheca triloba, p. 289							●			●			
Erodium moschatum, p. 272							●			●			
Melilotus albus, p. 271							●				●		
Ciclospermum leptophyllum, p. 208							●				●		
Campuloclinium macrocephalum, p. 219								●		●			
Mirabilis jalapa, p. 281								●	●				
Persicaria spp., p. 294								●					
Zinnia peruviana, p. 245								●					
Galinsoga parviflora, p. 229									●		●		
Rapistrum rugosum, p. 253									●				
Amsinckia menziesii, p. 245									●				
Arctotheca calendula, p. 214									●				

SPECIES	Exudes white sap	Thorny, prickly or spiny	Seeds in spiny or hairy capsules	Seeds in multiseeded capsules	Fluffy seeds that blow away easily	Seeds in a berry	Pungent when crushed	Red/pink flowers	Yellow flowers	Purple flowers	White flowers	Blue flowers	Green or insignificant flowers
Coreopsis lanceolata, p. 225									■				
Flaveria bidentis, p. 228									■				
Malvastrum coromandelianum, p. 278									■				
Reseda lutea, p. 296									■				
Sigesbeckia orientalis, p. 238									■				
Verbesina encelioides, p. 242									■				
Chrysanthemum segetum, p. 221									■				
Arctotis venusta, p. 215										■	■		
Cosmos bipinnatus, p. 226										■	■		
Echium plantagineum, p. 246										■			
Fumaria muralis, p. 272										■			
Lamium amplexicaule, p. 274										■			
Mantisalca salmantica, p. 220										■			
Verbena bonariensis, p. 306										■			
Centaurea cyanus, p. 220										■			
Achillea millefolium, p. 210											■		
Ammi majus, p. 207											■		
Ageratina adenophora, p. 210											■		
Heliotropiium europaeum, p. 248											■		
Parthenium hysterophorus, p. 232											■		
Nothoscordum gracile, p. 202											■		

SPECIES	Exudes white sap	Thorny, prickly or spiny	Seeds in spiny or hairy capsules	Seeds in multiseeded capsules	Fluffy seeds that blow away easily	Seeds in a berry	Pungent when crushed	Red/pink flowers	Yellow flowers	Purple flowers	White flowers	Blue flowers	Green or insignificant flowers
Limonium sinuatum, p. 293											☒	☒	
Echium vulgare, p. 246												☒	
Stachytarpheta urticifolia, p. 305												☒	
Trichodesma zeylanicum, p. 249												☒	
Ambrosia artemisiifolia, p. 213													☒
Atriplex lindleyi, p. 260													☒
Cannabis sativa, p. 257													☒
Galium spurium, p. 298													☒
Urtica spp., p. 304													☒
Amaranthus spp., p. 204–206													☒
Chenopodium spp., p. 261–263													☒
Gamochaeta pensylvanica, p. 230													☒
Plantago spp., p. 291													☒
Rumex spp., p. 295–296													☒
Scleranthus annuus *, p. 259													☒
Nephrolepis exaltata, p. 280													☒
Polypodium aureum, p. 280													☒

ALLIACEAE

Nothoscordum gracile *(=Allium gracile)*
fragrant false-garlic * basterknoffel

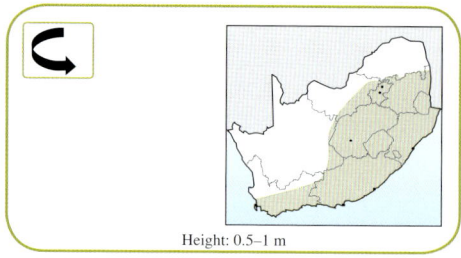
Height: 0.5–1 m

Origin and description: A native of South America that is now widespread. *N. gracile* resembles an onion or garlic, but without the smell. The plant regenerates from underground bulbs as well as seeds. It is sometimes confused with garlic chives and also with *N. borbonicum* (see below).
Impact: In South Africa it is a common weed of gardens and nurseries.
Control: Mainly because of its waxy leaves and underground bulbs, it is not susceptible to herbicides and none have been registered. It is best removed by hand, but care should be taken that no bulblets remain that can only exacerbate the problem.

Other common names
onion weed
Similar species
N. borbonicum: similar, also called onion weed but is a hybrid between *N. gracile* and *N. entrerianum* from Argentina

Nothoscordum gracile

Nothoscordum gracile

AMARANTHACEAE

Achyranthes aspera **var.** *aspera*
(=*A. repens*)
burweed * haak-en-steek-bossie

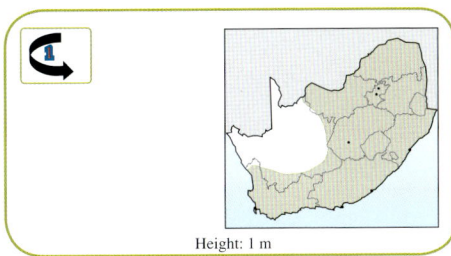
Height: 1 m

Origin and description: Of uncertain origin. It is probably exotic, but is now pantropical in distribution and occurs in many parts of the world. Before production of the white flowers, this plant is rather plain and insignificant. As the flowers mature, the fruits turn downwards until they are lying along the stem, pointing to the ground. Being armed with sharp and unpleasant barbs, these fruits can penetrate the skin if the stalk is held carelessly. This perennial weed is found throughout the summer-rainfall regions. It occurs mainly under shady conditions such as in hedges, along streams, at the edge of patches of bush and in overgrown areas of gardens.

Impact: It is in gardens where it is best known as an unwanted plant. It grows profusely in shady areas and is straggly and untidy.

Control: Burweed is easily controlled by cultivation, but mature plants should be pulled with care.

Other common names
chaff flower * grootklits * bohomane (S) * isinama (Z) * lenamo (Si) * moxato (T) * udombo (N) * usibambangubo (Z)

Achyranthes aspera

AMARANTHACEAE

Amaranthus deflexus
perennial pigweed * meerjarige misbredie

Amaranthus thunbergii
red pigweed * rooimisbredie

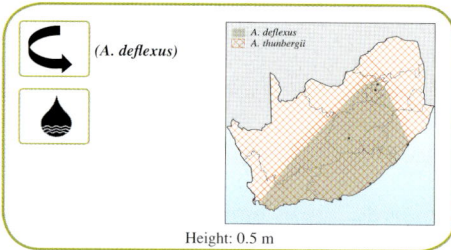

Height: 0.5 m

Origin and description: *A. thunbergii* is indigenous, whereas *A. deflexus* is from the Americas. They are both found throughout southern Africa, but are most abundant in northern and western Mpumalanga and the Free State. They tend to be more common on roadsides and in gardens than in crops.

A. thunbergii has a much flatter growth habit than its relatives, often forming a mat, with only the one central shoot growing erect. It has stout taproots, bears flowers in the axils and will also often turn red as it matures. *A. deflexus* is perennial, sprawling and with a stout, fleshy taproot.

Impact: These amaranths, especially *A. thunbergii*, can become very dense and competitive in any crop.

Control: With the notable exception of bendioxide, all species of *Amaranthus* are susceptible to the normal broadleaf herbicides and are easy to remove when small.

Other common names
A. thunbergii: Cape pigweed * poor man's spinach * red devil * hanekam * hondebosch * kalkoenslurp * kraanvoëlbossie * sprinkaanbossie * imbuyu (Z) (Si) * mohwa (Sh) * theepe (P) (S) * uqupose (X)

Amaranthus deflexus

Amaranthus thunbergii

AMARANTHACEAE

Amaranthus hybridus **subsp.** *hybridus*
pigweed * misbredie

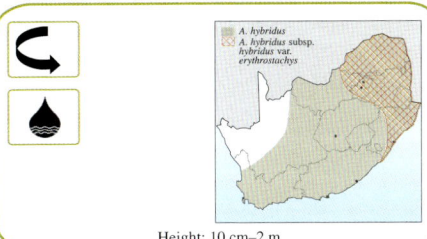

Height: 10 cm–2 m

Origin and description: There are approximately 150 species of *Amaranthus* in South Africa, most of which are indigenous but *A. hybridus* subsp. *hybridus* is one that is exotic, coming from various parts of the Americas. *Amaranthus* species are weeds worldwide.

A. hybridus is very variable in colour and size but is usually much taller than the other species of *Amaranthus*. However, the genus has recently been revised and there is still confusion with other species and the various subspecies. These weedy species of *Amaranthus* are closely related to the ornamental varieties as well as those grown as grain crops in parts of Africa. *Amaranthus hybridus* subsp. *hybridus* var. *erythrostachys* illustrates how complicated this can become. Pigweeds are much favoured as a spinach by some people and are sold at some open-air markets.

Other common names
Cape pigweed * redshank * hanekam * kalkoenslurp * rooibossie * sprinkaanbossie * isheke (Si) * poea (S) * umbhido (Z) * umbuya (Z) * umfino (X) * umtyutu (X) * yhepe (Ss) (T)

Impact: *A. hybridus* (now *A. hybridus* subsp. *hybridus*) is said to be the most abundant and widely distributed broadleaf weed in southern Africa and is found as a weed of crops throughout the region. It competes strongly for available moisture, light and nutrients. These weeds have also been implicated as alternative hosts for verticillium, a fungus disease that attacks several crop plants, including potatoes, tomatoes and cotton.

Control: With the notable exception of bendioxide, all species of *Amaranthus* are susceptible to the normal broadleaf herbicides used in agriculture and are easy to remove when small.

Amaranthus hybridus

205

AMARANTHACEAE

Amaranthus spinosus
thorny pigweed * doringmisbredie

Amaranthus viridis
slender amaranth * skraal misbredie

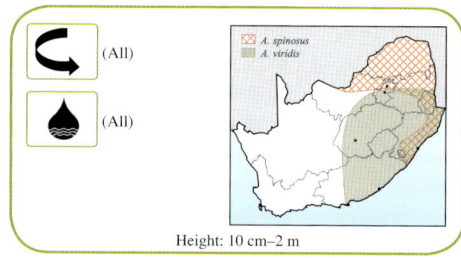

Height: 10 cm–2 m

Origin and description: These two species are exotic, *A. spinosus* coming from Central America and *A. viridis* probably from Eurasia.

A. spinosus has a pair of spines at the base of the leaves and is more troublesome in subtropical areas. It is well known in cotton and tobacco, but it is less common in maize. *A. viridis* is similar to *A. deflexus* (see page 204) but its central stem

Other common names
A. spinosus: needle burr * prickly careless weed * soldier weed * mohwa-guru (Sh) * sere pelêlê (P)
Similar species
*A. deflexus (*see page 204)

grows vertically and there are no flowers in the axils. They are frequently confused with each other. *A. viridis* is less frequently observed in croplands.

Impact: All species of *Amaranthus*, but *A. spinosus* in particular, are often found in rich, disturbed soils such as around kraals or cattle pens. Such plants contain high levels of nitrates, which in ruminants are converted to highly toxic nitrites by the micro-organisms present in the rumen. If eaten in excessive quantities by livestock, this weed can cause severe poisoning and even death.

Control: With the notable exception of bendioxide, all species of *Amaranthus* are susceptible to the normal broadleaf herbicides and are easy to remove when small.

Amaranthus spinosus

Amaranthus viridis

APIACEAE

Ammi majus
lace flower * kantblom

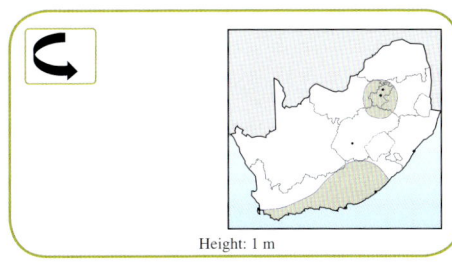
Height: 1 m

Origin and description: Introduced from Eurasia, this weed was and still is planted as an ornamental. *A. majus* has established itself in the wild and although found throughout South Africa it is only properly established in the Cape provinces and Gauteng.

Impact: Although recorded from most provinces, probably as simple garden escapes, it is most common in the southern Cape between Knysna and Port Elizabeth where it is well established and a nuisance in vineyards and orchards. The flowers are much valued for use in flower arrangements.

Control: No herbicides have been registered for its control but the plants should be removed while still small as they become more resistant as they mature.

Other common names
bishop's weed * Queen Anne's lace

Ammi majus

APIACEAE

Ciclospermum leptophyllum *(=Apium leptophyllum)*
wild celery * wildeseldery

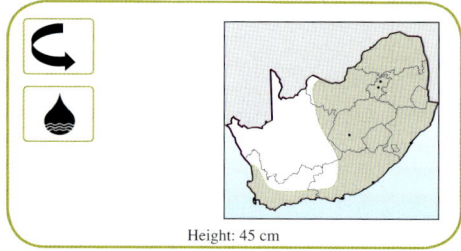
Height: 45 cm

Origin and description: A native of Central America, *C. leptophyllum* is now widely distributed throughout the world. It is a widespread annual weed in South Africa.

Impact: It is usually found in damp places in gardens, cultivated lands and on riverbanks. In the eastern lowveld in particular, it can become dense and competitive and is occasionally a serious weed in sugarcane.

Control: Chemical control is not effective on mature plants as once the above-ground foliage has been destroyed, it will readily regrow from parts remaining in the soil. Systemic herbicides should therefore be used where this weed is a problem. It is easily controlled by shallow cultivation in the seedling stage.

Other common names
fir-leafed celery * lawn celery * marsh parsley * slender celery

Ciclospermum leptophyllum

Ciclospermum leptophyllum

APIACEAE

Foeniculum vulgare
wild fennel * vinkel

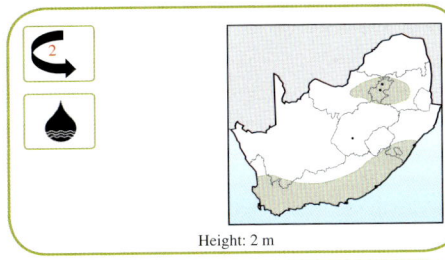
Height: 2 m

Origin and description: Introduced from Europe as a horticultural crop during the early settler days, *F. vulgare* has escaped into the wild and is now found throughout South Africa. It is especially common in the southern and Western Cape where it has become seriously invasive. It is found on roadsides, in waste places and in croplands. *F. vulgare* is a perennial plant, reproducing only by seed. It acts as a perennial in most instances, but in crops such as wheat it must seed down every season. Different varieties of this species exist.

Impact: When crushed, it has a very strong, characteristic smell and is therefore a nuisance in wheat, especially in the Swartland. When infested fields are harvested, the fennel can be smelt from a considerable distance. The grain from these fields, being contaminated with the smell of fennel, is rejected at the silo. On roadsides it is unsightly and obstructive.

Control: A herbicide is registered for this weed in cereals. The sterility or non-invasiveness of cultivars has to be investigated.

Other common names
bobbejaanvinkel * wilde anyswortel * imbozisa (Z)

Herbs

Foeniculum vulgare

209

ASTERACEAE

Achillea millefolium
milfoil * duisendblaar-achillea

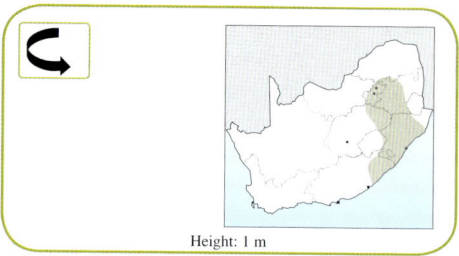
Height: 1 m

Origin and description: Milfoil was introduced from the northern hemisphere into South Africa as an ornamental and has now become naturalised in many parts of the country.
Impact: It is commonly found on roadsides and in waste places. Milfoil is not usually a problem in cultivated lands, except when it occurs in perennial pastures. It has been reported as a contaminant of various kinds of commercial seed, especially grass seed. It is used in herbal medicine and Chinese proverbs claim that yarrow brightens the eyes and promotes intelligence!
Control: No specific herbicide has been registered for this weed, but as it is a perennial, it will probably require the use of a systemic chemical to eradicate it successfully.

Other common names
common yarrow * dog-daisy * sneezeweed * thousand-leaf

Achillea millefolium

ASTERACEAE

Ageratina adenophora (=Eupatorium adenophorum)
crofton weed

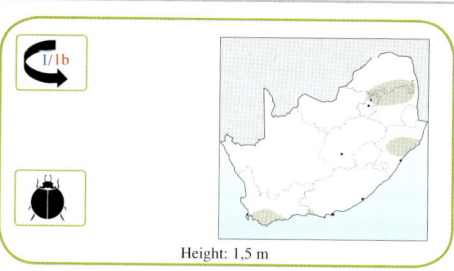
Height: 1,5 m

Origin and description: Introduced from Central America as an ornamental, *A. adenophora* has established itself in various parts of the country, invading roadsides, plantations and waste areas. It is a perennial, evergreen shrub, growing from seed and regenerating strongly from cut stems.

Other common names
Mexican devil * snake root

It should not be confused with *C. odorata* (see page 168), which it resembles strongly. When crushed, *A. adenophora* does not have the distinctive paraffin smell of *C. odorata*.

Impact: Although found in various provinces, it is most prevalent around Pietermaritzburg in KwaZulu-Natal but this limited distribution is showing signs of expanding. It has already been found invading places as far apart as Magoebaskloof in Limpopo and Stellenbosch in the Western Cape. In some areas the infestations are very dense and competitive. It is unpalatable to cattle and said to be toxic to horses.

Control: There are no herbicide registrations for this weed. Since it is closely related to *C. odorata* (see page 168) it can probably be controlled in a similar fashion. The tips of the stems are often galled by an introduced gallfly, *Procecidochares utilis*, which was introduced in 1984 but is not yet having a significant effect.

Other common names
Mexican devil * snake root
Similar species
A. riparia C (creeping crofton weed, mistflower) is also a 1/1b species but is not yet well established

Ageratina adenophora

ASTERACEAE

Ageratum conyzoides
invading ageratum * bokkruid

Ageratum houstonianum
garden ageratum * tuinageratum

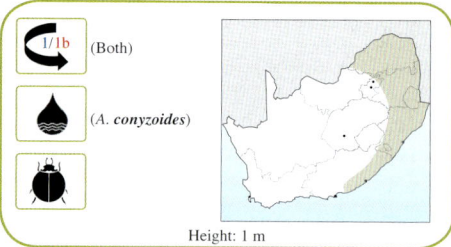

Height: 1 m

Origin and description: These plants were introduced from South America as ornamentals. *A. houstonianum* is called 'Todd's curse' after a nurseryman who lived in Pietermaritzburg during the 19th century. It was first recorded as a weed in 1883. These plants are often cultivated in gardens but have become very widespread weeds. *A. conyzoides* is common in most of the warmer areas of the eastern part of the country, whereas *A. houstonianum* is dominant in the KwaZulu-Natal midlands. It is distinguished by its relatively large flower heads. In various parts of the world *A. conyzoides* is used in medicines and folk remedies. The leaves are often used as a wound dressing and in the West Indies, for example, a tea is made that is used to treat a variety of ailments.

Other common names
A. conyzoides: billy-goat weed * blue weed * tropical whiteweed * indringer-ageratum
A. houstonianum: floss flower * Mexican ageratum * Todd's curse

Impact: Two of the 'Big Six' of invasive alien Asteraceae, these plants are common weeds of annual crops and can invade perennial crops such as sugarcane. They are also found on roadsides, riverbanks and in waste areas. On occasion they can be a serious problem in cotton, where the wind-blown seeds contaminate the lint and irritate the pickers' eyes. The seeds also reportedly stick in the throat of livestock and can choke and kill them.

Control: They are relatively easy to control by cultivation and pre-emergence herbicides. However, once the plants have matured they are difficult to control with many of the post-emergence chemicals. The sterility or non-invasiveness of cultivars has to be investigated.

Ageratum conyzoides

ASTERACEAE

Ambrosia artemisiifolia
common ragweed

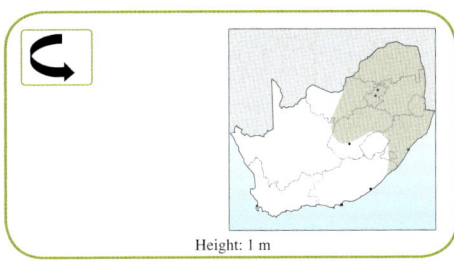
Height: 1 m

Origin and description: Originally from North America, ragweed is now a common weed in subtropical KwaZulu-Natal as well as on the highveld. Do not confuse with parthenium (*Parthenium hysterophorus*), which has similar leaves but white flowers as opposed to the yellow / green ones of ragweed.

Impact: Ragweed is commonly found on roadsides and in waste places, but is also a significant and competitive weed of sugarcane, especially on the sandy coastal soils. It is also implicated as being a major cause of hay fever. *A. artemisiifolia* is spread by seeds that can remain viable for many years and is browsed by stock.

Control: No specific herbicide has been registered for this weed, but it is susceptible to cultivation and should respond to the usual broadleaf-weed herbicides used in sugar cane.

Other common names
annual ragweed * bitterweed * hay fever weed

Ambrosia artemisiifolia

ASTERACEAE

Anthemis arvensis
corn chamomile * wildekamille

Anthemis cotula
dog-fennel * stinkkamille

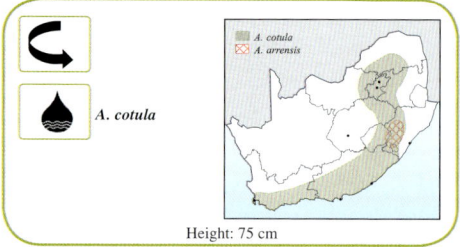

Origin and description: These are two species of chamomile from Europe and Asia that are now established weeds in South Africa. *A. cotula* in particular was probably introduced as an ornamental. The seeds are easily dispersed as a contaminant of pasture seeds.
Impact: Both are now weeds of pastures, cereals and, to a lesser extent, other crops. *A. cotula* in particular is smelly when crushed and is unpalatable.
Control: A herbicide registration exists in groundnuts, so these plants should be susceptible to chemical control. Otherwise they can be removed by cultivation, especially when young.

Other common names
A. arvensis: wild camomile
A. cotula: chamomile * mayweed * pig-sty-daisy * poison daisy * stinking mayweed

Anthemis cotula

ASTERACEAE

Arctotheca calendula (=*Cryptostemma calendulaceum*)
Cape marigold * soetgousblom

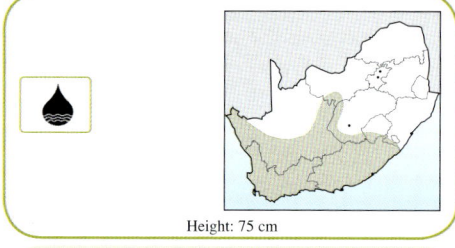

Origin and description: This is an indigenous plant. As the common names suggest, this plant is widespread in the Western Cape, and is especially prevalent along the southwestern coast and in vineyards. It has also been recorded in a few areas of KwaZulu-Natal, the Free State and North West Province. A characteristic feature is the light grey wooliness on the underside of the leaf.
Impact: *A. calendula* is well known in most of the wheat areas of the Western Cape and is found as a widespread weed in lawns and golf courses. It is thought to taint the milk of animals that eat it.
Control: *A. calendula* is easily controlled in wheat by most post-emergence herbicides, especially MCPA. Diquat is usually better than paraquat in minimum tillage situations. Glypho-

Other common names
Cape dandelion * Cape weed * botterblom * Kaapse madeliefie * tonteldoek

Arctotheca calendula

sate and glyphosate mixtures are very effective in vineyards, especially before flowering. It is also susceptible to various pre-emergence chemicals, but unfortunately there are no herbicides registered for pastures.

Arctotheca calendula

ASTERACEAE

Arctotis venusta
Free State daisy * witgousblom

Origin and description: This is an indigenous species that has become a troublesome weed in the summer-rainfall regions, especially the Free State and parts of the Karoo. It is planted in gardens as an ornamental.
Impact: Although unpalatable, it is thought to be mildly poisonous and can taint milk. It is a common weed of wheat and along roadsides.
Control: When it occurs in crops it responds to conventional herbicides and can be removed by cultivation.

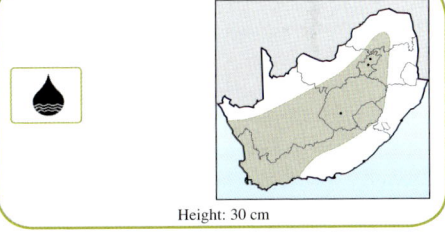
Height: 30 cm

Other common names
gousblom * lehaha-la-tsela (S)
Similar species
Arctotis leiocarpa: Karoo daisy * witgousblom

Arctotis venusta

215

ASTERACEAE

Berkheya erysithales

Berkheya rigida
disseldoring

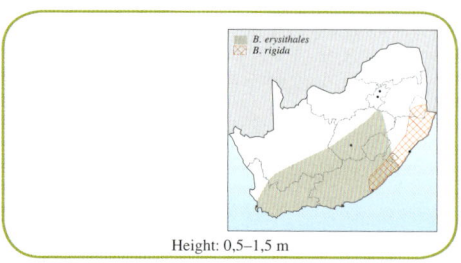

Height: 0,5–1,5 m

Origin and description: There are many indigenous species of *Berkheya*. Some, like these two, are thistle-like and can become a nuisance because of their unpleasant spines. *B. erysithales* can be found in KwaZulu-Natal and the Eastern Cape, whereas *B. rigida* is fairly widespread but especially common and troublesome in parts of the southern Cape.

Other common names
B. rigida: isihlungu (Z)

Impact: These weeds become a problem in overgrazed veld and mismanaged pastures, severely reducing the carrying capacity. They can be hazardous to livestock, especially if incorporated into baled hay. They are also common on roadsides, waste areas and fallow lands.

Control: There are no herbicides registered for these weeds and they are best controlled by physical means. Repeated slashing is necessary in order to deplete the deep tap roots.

Berkheya erysithales

Berkheya rigida

ASTERACEAE

Bidens bipinnata
Spanish blackjack * Spaanse knapsekêrel

Bidens pilosa
common blackjack * gewone knapsekêrel

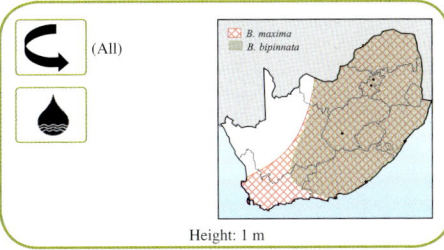

Height: 1 m

Origin and description: Both these weeds were introduced during the last century, *B. bipinnata* from Eurasia and *B. pilosa* from South America. These two closely related weeds, which often hybridise, are well known to South Africans. Their fruit are the unpleasant 'blackjacks' that stick to clothing and hair, being able to burrow rapidly through several layers of clothing. Each little blackjack has tiny barbs and rough edges allowing it to burrow like this. This is an extremely efficient method of dispersal. The leaves are eaten by some people as a spinach.

Impact: These weeds are common, widespread and extremely troublesome, being found in most crops and disturbed areas. They have been implicated as alternative hosts for verticillium, a fungus disease that attacks several crop plants, including potatoes, tomatoes and cotton.

Control: Blackjacks often germinate in dense mats. This uniform, shallow germination fortunately means that they are relatively easy to control, especially with post-emergence herbicides. Because of the large seeds, pre-emergence control can be erratic.

Other common names
B. bipinnata: beggar tick * black fellows * black jack * bur marigold * cobbler's pegs * pitchfork * baster kakiebos * monyane (P)
B. pilosa: beggarsticks * blanket-stabbers * cobbler's pegs * wedevrouens * amalenjane (Z) * mokolonyane (Ss) * muchize (P) * mushiji (V) * ucucuza (Z) * ugamfe * umesisi (Z) * umhlabangubo (X) * uqadolo (Z) (Si)

Similar species
Another exotic species of blackjack, *B. biternata* (five-leaved black jack), is occasionally found. It is similar to *B. pilosa* with its broader leaves, but the leaves are usually divided into five leaflets, with the lowermost pair re-divided into two or three segments

Bidens pilosa

Bidens pilosa

Bidens bipinnata

ASTERACEAE

Campuloclinium macrocephalum *(=Eupatorium macrocephalum)*
pompom weed * pompom-bossie

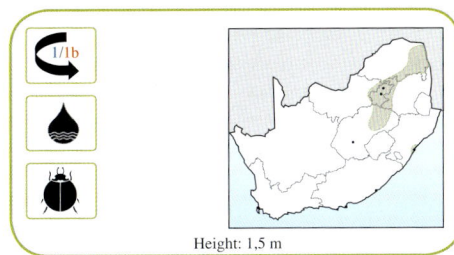

Height: 1,5 m

Origin and description: *C. macrocephalum* is a native of Argentina and Brazil. It was introduced as an attractive ornamental flower, but has escaped into the wild, with the first South African records being from Durban in 1972. Over the past few years it has rapidly established itself in many areas. It is most common around Pretoria, but has been recorded from as far as Tzaneen, Barberton, Bloemfontein and KwaZulu-Natal. A very large infestation appeared in the Rietvlei Nature Reserve near Pretoria and it has spread aggressively into the Waterberg region in Limpopo Province. All known infestations in KwaZulu-Natal have been treated and killed. The presence of purple top (*Verbena bonariensis*) seems to indicate a suitable habitat for pompom weed and the two are often seen growing together.

Impact: The plant has attractive pink flowers, and although it spreads easily by seed, it can also regenerate from underground rhizomes. Once established, it causes serious degradation of the veld, lowering the biodiversity and reducing the grazing capacity by being unpalatable to large herbivores. It can colonise a wide range of habitats and has the potential to be a very serious invader.

Control: Three herbicides have now been registered for this weed but it appears as if control by this means will not be achieved easily. Destroying the above-ground parts of the plant by means of herbicides or fire can actually make the problem worse. This is because it stimulates the rhizomes to shoot and thereby produce more flowers. However, if it is done repeatedly, it is possible that the plant may eventually be exhausted. The registered herbicides should be applied early, before the plant produces the tall upright stalks. It can be controlled by mechanical means, as long as the entire plant is removed. Landowners are encouraged to eliminate this plant wherever possible. Three insect species are being tested for biological control and a fungus has been found growing on pompom weed, which could possibly weaken and kill the plant.

Campuloclinium macrocephalum

ASTERACEAE

Centaurea cyanus
cornflower * koringblom

Centaurea melitensis
Malta centaurea * Maltadissel

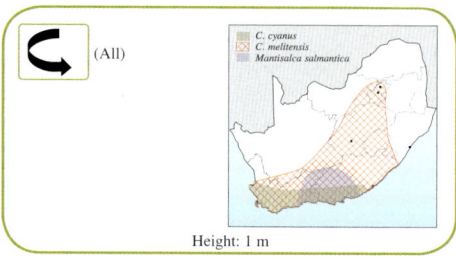

Height: 1 m

Origin and description: These species of *Centaurea* from Eurasia were introduced as ornamentals or as seed impurities up to a hundred years ago, and are now weeds in South Africa. The blue-flowered *C. cyanus* occurs in wheat fields of the Western Cape. The yellow-flowered *C. melitensis* from Malta is better known in lucerne fields and along roadsides. Both these plants have spread as impurities in cultivated seed. *C. melitensis* was first collected from the Cape Flats in 1865 but it is thought that it was probably introduced into present-day KwaZulu-Natal and Mpumalanga with contaminated oat seed 10 years previously. The closely related, purple-flowered *Mantisalca salmantica* is naturalised mainly around Aberdeen in the Eastern Cape.

Impact: These plants are untidy, unpalatable and they replace indigenous vegetation. They are toxic to horses.

Control: No herbicides have been registered for the control of these weeds. When small they should succumb to normal broadleaf weed herbicides and shallow cultivation.

Other common names
C. cyanus: bachelor's button * blue bottle * bloukoringblom
C. melitensis: cockspur * napa thistle * saucy jack * koringdissel * vaaljacob

Similar species
Mantisalca salmantica (=*C. salmantica*), from Eurasia
C. calcitrapa (purple star thistle * sterdissel), from Europe
C. solstitialis (burweed * yellow cockspur * yellow star thistle * geeldissel * skaapdissel * vaaljacob), from the Mediterranean and first appeared in South Africa in the 1920s

Mantisalca salmantica

Centaurea cyanus

Centaurea melitensis

ASTERACEAE

Chrysanthemum segetum
corn chrysanthemum * koringkrisant

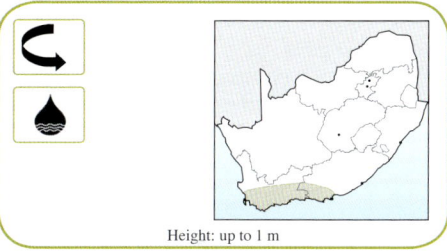

Height: up to 1 m

Other common names
corn marigold * goudsblom

Origin and description: An annual plant from western Asia and southern Europe that is now a cosmopolitan weed. It was first recorded in South Africa in 1904 in the Belfast district of present-day Mpumalanga where it was possibly cultivated as an ornamental and then escaped. It has now spread to the southern and Eastern Cape and is no longer found in Mpumalanga.
Impact: *C. segetum* is a weed of wheat and vegetables, especially in the areas around George. When dense, it can smother vegetable crops and is a strong competitor for water and nutrients.
Control: Susceptible to the usual broadleaf weed herbicides used in wheat. Chemical control in vegetables is more difficult and where a problem exists, a rotation with a cereal will offer the best prospects of long-term control.

Chrysanthemum segetum

221

ASTERACEAE

Cichorium intybus subsp. *intybus*
chicory * sigorei

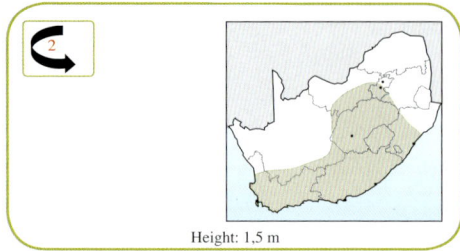

Height: 1,5 m

Origin and description: Originally from Eurasia, *C. intybus* was introduced into South Africa for cultivation, where it is still grown commercially, especially in parts of the Eastern Cape near Grahamstown. The roasted roots are used as a supplement or substitute for coffee mainly because of the absence of caffeine. This is often a desirable quality and in the US, for instance, chicory is used as a total replacement for coffee in prisons. The plant exudes a milky juice when broken and has stout taproots and large basal leaves.

Impact: *C. intybus* has escaped into the wild and is now a widespread, perennial weed occurring in most situations, especially fallow land, waste areas and roadsides. It is not often a serious weed in cultivated lands, but is an untidy and obstructive weed on roadsides.

Control: No specific herbicides have been registered, but it should be susceptible to the usual herbicides as long as the plant is not allowed to get too big.

Other common names
blue sailors * coffee weed * succory * tjiekoriebos * witloof

Cichorium intybus

ASTERACEAE

Cirsium vulgare
spear thistle * speerdissel

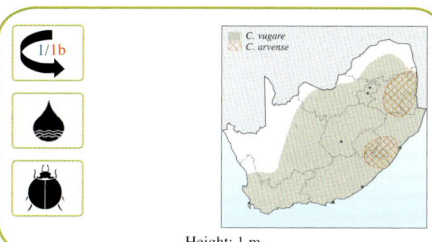

Height: 1 m

Origin and description: *C. vulgare* is a native of Europe and Asia that has spread to virtually all temperate zones of the world. It was first recorded at Van Reenen in KwaZulu-Natal and is thought to have been introduced with imported fodder during the Anglo-Boer War. It is now widespread in South Africa except in lowveld areas.

The species is biennial, taking two years to produce a flower. It flowers late in spring or early summer, producing an abundance of seeds, each with a silky plume for dispersal. Some seeds wash away during floods or cling to mud on vehicles, animals and implements. Birds eat the seeds and also collect the silky plumes for their nests. The seeds are also spread around farms in bales of contaminated stock feed.

Impact: It is common in pastures, waste places and along roadsides, preferring moist, rich soil. It does not thrive in regularly cultivated lands. An infestation of *C. vulgare* is an indication of poor veld management. A heavy infestation severely reduces the carrying capacity of the veld and can cause injury to man and animals.

Control: *C. vulgare* is easily controlled with regular cultivation and is susceptible to hormone and contact type herbicides. If possible, it should be controlled before flowering. Biocontrol agents have been introduced, but to date success is minimal.

Other common names
bull thistle * plume thistle * Scotch thistle * disseldoring * karmedik * skaapdissel * Skotse dissel * hlaba (S) * ntsoa-ntsane (Ss)

Similar species
C. arvense (creeping thistle * rankdissel) was probably introduced into South Africa as an impurity of seed. Although a national symbol of Canada, it is in fact a native of Eurasia and northern Africa. The plant has underground rhizomes and is more difficult to remove

Cirsium vulgare

ASTERACEAE

Conyza sumatrensis* var. *sumatrensis
(=Conyza albida)
tall fleabane * vaalskraalhans

Conyza bonariensis *(=Erigeron bonariensis)*
flax-leaf fleabane * kleinskraalhans

Conyza canadensis *(=Erigeron canadensis)*
horseweed fleabane * Kanadese skraalhans

Origin and description: Of the 15 species of *Conyza* that occur in South Africa three, including *C. sumatrensis*, were introduced from the Americas and they are major weeds. They are closely related species, similar in habit and distribution, which is widespread. *C. sumatrensis* grows to over 2 m and, unlike *C. bonariensis*, never has side stems taller than the main stem. *C. bonariensis* grows to about 1,2 m and has lateral branches taller than the main stem. *C. canadensis* is also relatively short but has smaller flowers and is unbranched. These three similar species can often be found growing together.

Impact: *Conyza* spp. are common annual weeds of gardens, roadsides, fallow land, forests and, to a lesser extent, annual crops and even golf greens. Perennial crops often become infested. These plants can become a serious problem in sugarcane, particularly ratoon cane or cane that has been neglected for a while. They are an important weed in crops under minimum tillage and must not be allowed to become too tall before treating it with herbicides. They are becoming

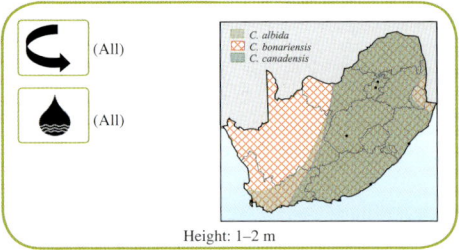
Height: 1–2 m

Other common names
C. sumatrensis var. *sumatrensis*: radiatorbossie
C. bonariensis: horseweed fleabane * armoedskruid

Similar species
C. chilensis (primroseleaf fleabane) from Central America is naturalised but less common

Conyza canadensis

Conyza sumatrensis

well known as winter weeds of maize lands and are also known to harbour the 'kromnek' or tomato spotted wilt virus, which is a problem in tomatoes, potatoes, tobacco and peas.
Control: *Conyza* spp. are susceptible to pre-emergence herbicides but shallow cultivation and the application of post-emergence herbicides must be completed before plants form a rosette, otherwise it will not be effective. The addition of 2,4-D or MCPA to glyphosate can improve the control of mature plants.

Conyza bonariensis

ASTERACEAE

Coreopsis lanceolata
tickseed

Origin and description: *C. lanceolata* is a native of the eastern United States that was introduced as an ornamental. It has now become naturalised in South Africa and is commonly found on roadsides and railway embankments in the summer-rainfall areas. In KwaZulu-Natal it was first recorded as having escaped from cultivation on the roadside at Kloof in 1955. It is therefore a relatively recent introduction. It can on occasion be observed near neglected graveyards where seed has escaped from cut plants put on graves.
Impact: It is a perennial with underground rhizomes and, once it becomes established, it is difficult to eradicate. It replaces indigenous vegetation.
Control: Physical removal or systemic herbicides should be used for effective eradication of this weed. The sterility or non-invasiveness of cultivars has to be investigated.

Height: 60 cm

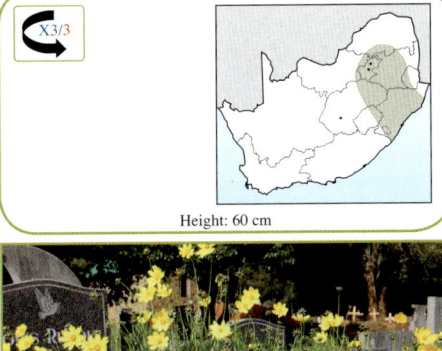

Coreopsis lanceolata

ASTERACEAE

Cosmos bipinnatus *(=Bidens formosa)*
cosmos * kosmos

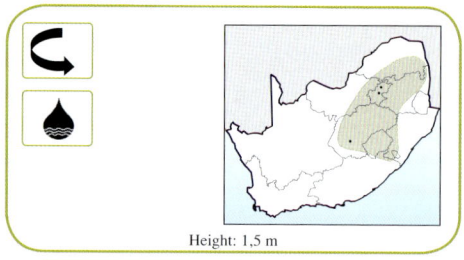

Height: 1,5 m

Origin and description: A native of Central America and the West Indies, *C. bipinnatus* is now widespread in the highveld areas of South Africa. It was introduced in fodder during the Anglo-Boer War and was originally recorded as being naturalised in Pretoria in 1904. This weed did not spread to KwaZulu-Natal until 1945.

The flowers are white, pink and occasionally dark purple, the ratio varying from place to place. Sometimes the flowers in a certain area are nearly all white and sometimes they are predominantly pink. Very occasionally, banks of the dark purple variety can be seen. This is a genetic characteristic and is not due to the nature of the soil or climate. Several varieties with larger flowers have been bred for the garden and can flower at other times of the year.

Impact: *C. bipinnatus* is not always an unwanted weed, as the dazzling autumnal displays of this flower along roadsides are spectacular. However, they spread into nearby fields and become a nuisance in cropland. These displays are most memorable in the higher-lying areas of KwaZulu-Natal, the Free State and the Mpumalanga highveld.

Control: *C. bipinnatus* is susceptible to cultivation and many broadleaf-weed herbicides.

Other common names
Mexican aster * mieliepes

Cosmos bipinnatus

ASTERACEAE

Dittrichia graveolens (=*Inula graveolens*)
Cape khakiweed * Kaapse kakiebos

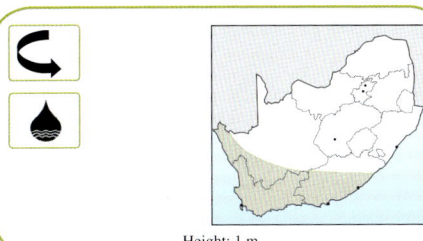
Height: 1 m

Origin and description: Originally from the Mediterranean region of Europe and having been introduced at the time of the Anglo-Boer War, this plant is now a major perennial weed in parts of the Cape. It is not found north of the Karoo. The plant is covered with glandular hairs that make it sticky to the touch. An oil extracted from the plant is highly valued in aromatherapy.

Impact: *D. graveolens* is a major problem in wheat lands as it has a very strong and distinctive smell, which can contaminate the grain. It will continue to grow after the field has been cut and is a common sight in the stubbled wheat fields of the Western Cape. It is also common in other crops, waste places and on roadsides. It is unpalatable to cattle, but if they do eat it, it can taint the milk and also cause oxalate poisoning.

Control: Once it has become established, *D. graveolens* is very difficult to control on account of its strong root system. It should be controlled before it reaches the five-leaf stage, either physically or chemically.

Other common names
camphor inula * Bolandse-kakiebos * Caledonbos * kanfer-inula

Similar species
Note that the common name 'khakiweed' is also applied to *Alternanthera pungens* and *Tagetes minuta*

Dittrichia graveolens

ASTERACEAE

Flaveria bidentis
smelter's bush * smelterbossie

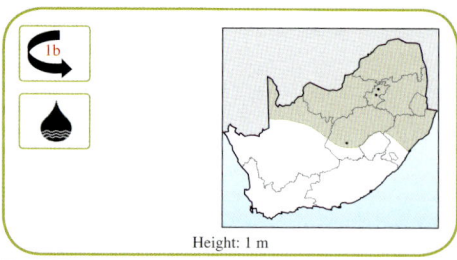

Height: 1 m

Origin and description: A native of tropical America, *F. bidentis* was probably introduced in imported fodder during the Anglo-Boer War. It has spread rapidly in South Africa and is most common in northern and eastern Mpumalanga, the Northern Cape and Namibia, but is found throughout the country with the exception of the southern and Eastern Cape.

Impact: It is a common annual weed of crops, gardens and waste places, occasionally becoming dense and competitive.

Control: *F. bidentis* is very easy to control with shallow cultivation and conventional herbicides.

Flaveria bidentis

ASTERACEAE

Galinsoga parviflora
small-flowered quickweed * knopkruid

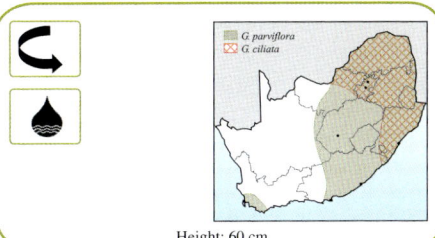

Height: 60 cm

Origin and description: *G. parviflora* is originally from South America but is now a cosmopolitan annual weed. It occurs throughout South Africa.
Impact: It is troublesome in a wide range of crops throughout South Africa, except the south-western Cape. When dense, it can become very competitive. It is on record as an alternate host for some nematode species as well as for the tobacco and cucumber mosaic virus.
Control: It is easy to control in most situations as it is a shallow germinator. In wheat, however, it is relatively tolerant to some of the non-hormonal herbicides. For this reason herbicide mixtures may have to be considered.

Other common names
galinsoga weed * gallant soldier * yellow weed

Similar species
G. ciliata, also from South America, can easily be confused with *G. parviflora*, but is more hairy and only occurs in KwaZulu-Natal and eastern Mpumalanga. The two species can often be found growing together

Galinsoga parviflora

ASTERACEAE

Gamochaeta pensylvanica *(=Gnaphalium pensylvanicum)*
roerkruid

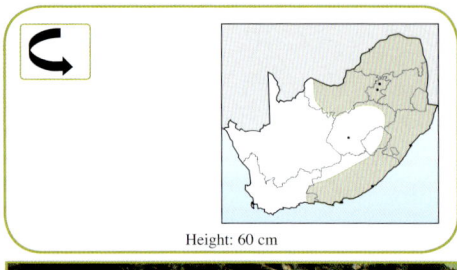

Height: 60 cm

Origin and description: Originally from North America, *G. pensylvanica* is now a widespread annual weed in the warmer parts of the world, including Australia, the Middle East and China. In South Africa it was first recorded in KwaZulu-Natal in 1865 and is now common in most of the eastern parts of the country, from the Eastern Cape through KwaZulu-Natal to Swaziland and the Mpumalanga lowveld.

Impact: It occurs as a weed of gardens, lands, orchards and other damp places up to an altitude of about 1 200 m. It is well known as a weed in conservation tillage in maize and in sugarcane. It is also common in lawns and golf courses.

Control: It can be controlled by shallow cultivation and is susceptible to conventional herbicides.

Gamochaeta pensylvanica

ASTERACEAE

Hypochaeris radicata
hairy wild lettuce * harige skaapslaai

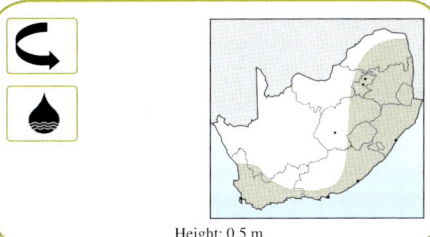
Height: 0,5 m

Origin and description: Introduced from Europe, this plant is now a cosmopolitan weed and is widespread in South Africa. The leaves of young plants are eaten by some people as a spinach.

Impact: *H. radicata* is mostly a weed of pastures, gardens, roadsides and waste places and is seldom a weed of annual crops, except in crops like maize when under conservation tillage. In the Cape provinces it infests orchards, vineyards and other perennial crops.

Control: It has fleshy rhizomes and numerous fibrous roots, making chemical control difficult once it has become established. It will require systemic broadleaf herbicides.

Other common names
false dandelion * spotted cat's ear * katoor * umkhothane (X)

Similar species
Hypochaeris glabra (smooth cat's ear)
Taraxacum officinale (common dandelion) (see page 320)

Hypochaeris radicata

ASTERACEAE

Lactuca serriola
wild lettuce * wildeslaai

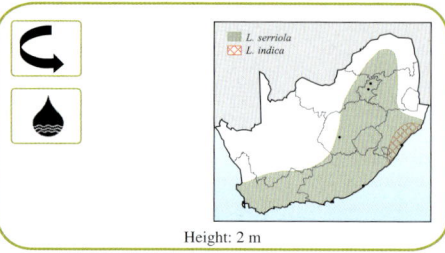

Height: 2 m

Origin and description: *L. serriola* is a native of Europe and is now a widespread weed in South Africa. It is the closest wild relative of the cultivated lettuce (*L. sativa*). The flowers vary in their shade of yellow and the plant exudes a milky sap when broken. Young leaves are used as a pot herb.
Impact: This weed and several other similar species occur on roadsides, in crops and gardens. They are untidy and replace indigenous vegetation.
Control: Although *L. serriola* can be perennial, it often behaves as an annual, especially in annual crops. It is therefore easy to control with shallow cultivation and is susceptible to pre- and post-emergence herbicides.

Other common names
compass plant * prickly lettuce * melkdissel
Similar species
L. indica, also from Europe and Asia

Lactuca serriola

ASTERACEAE

Parthenium hysterophorus
parthenium * Demoina bossie

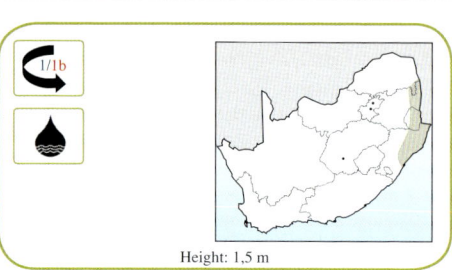

Height: 1,5 m

Origin and description: *P. hysterophorus* is from North America and the Caribbean and has been known in KwaZulu-Natal since 1880. It is found throughout the warmer areas of South Africa, Zimbabwe and Mozambique and is particularly prevalent in Swaziland. In the eastern Mpumalanga lowveld, where they call this plant Demoina weed, it suddenly appeared after cyclone Demoina in 1986. It is quite possible that seeds were picked up by the strong cyclonic winds in Swaziland and transported to the lowveld where it is now a major nuisance in sugarcane and banana orchards, and has recently found its way into the Kruger Park. Apart from wind, long-distance dispersal is by animals, vehicles and in mud. In Australia, it spread rapidly after introduction and is now considered to be one of their worst weeds.

Impact: *P. hysterophorus* not only exhibits a high degree of allelopathy (the chemical inhibition of one plant by another), but it also causes asthma and dermatitis in humans. It is an aggressive coloniser of crops, roadsides and waste places. It has also invaded the natural vegetation of game reserves in Swaziland to a considerable degree, especially where the veld has been overgrazed.

Control: *P. hysterophorus* seems to be tolerant to many chemicals. Systemic, nonselective chemicals should be used and, where feasible, individual plants should be pulled out by hand before they set seed. It is important to ensure a good cover of grass after any control programme, in order to prevent re-invasion. Further spread of this weed can be prevented by cleaning all vehicles, machinery and animals that come from infested areas. Meanwhile researchers are searching for biological control agents.

Parthenium hysterophorus

ASTERACEAE

Picris echioides
bristly oxtongue * stekel-picris

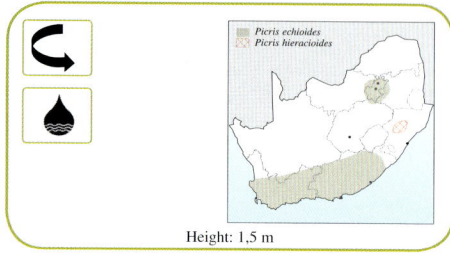

Height: 1,5 m

Origin and description: An annual or biennial weed, which is a native of central and southern Europe, *P. echioides* is now fairly widespread, especially in the southern Cape. The leaves, branches and stem have small bristles, hence the common name.
Impact: It is usually found in waste places and on roadsides, but can become a problem in orchards and vineyards. It is untidy and replaces indigenous vegetation.
Control: It is susceptible to nonselective herbicides such as paraquat and glyphosate, especially when young.

Other common names
ostong * stekelrige beestong
Similar species
P. hieracioides (hawkweed oxtongue) is very similar and occurs in KwaZulu-Natal.
Echium plantagineum (purple echium); the leaves of this plant do not have bristles (see page 246)

Picris echioides

ASTERACEAE

Pseudognaphalium luteo-album
(=Gnaphalium luteo-album)
jersey cudweed * roerkruid

Origin and description: *P. luteo-album* is from Europe and was introduced by early settlers. It is now common and widespread.
Impact: It is well known as a winter weed of maize lands. It will grow densely the following summer if the field is not cultivated, as it occurs in reduced tillage systems. This weed has readily become a problem in wheatfields in parts of the Free State.
Control: When mature, it is rather tolerant to herbicides but is susceptible to shallow cultivation as a seedling. It does not seem to respond well to the sulphonyl urea group of herbicides.

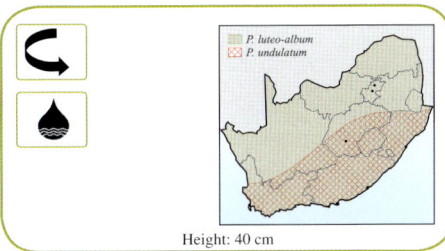
Height: 40 cm

Other common names
manku (S) * mgilane (Z) * musuwane (S)
Similar species
P. undulatum (cudweed * groenbossie * mothepetelle (S)), an indigenous species

Pseudognaphalium luteo-album

ASTERACEAE

Schkuhria pinnata
dwarf marigold * kleinkakiebos

Origin and description: *S. pinnata* is a common annual weed from South America. It was first recorded in South Africa in 1898. The Afrikaans common name derives from the fact that British soldiers ('Kakies') introduced this weed with imported fodder for their horses during the Anglo-Boer War.

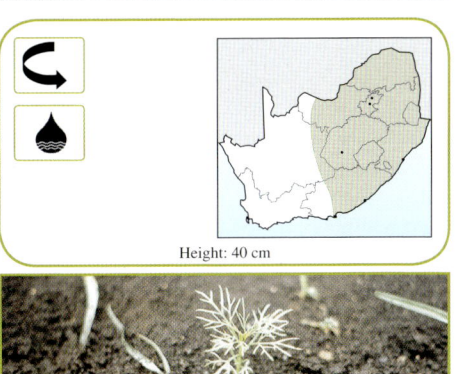
Height: 40 cm

Impact: *S. pinnata* is now a pest of many crops, especially in the summer-rainfall region. It occasionally becomes competitive. It is said that this plant will taint the milk of cattle that have eaten it.

Control: *S. pinnata* is generally well controlled by the normal pre- and post-emergence herbicides.

Other common names
yellow tumbleweed * bitterbossie * bokrambossie * hardebossie * waaibossie * letapiso (S)

Similar species
This plant should not be confused with those called 'khakieweed' (*Alternanthera pungens*) or 'kakiebos' (*Tagetes minuta, Dittrichia graveolens*)

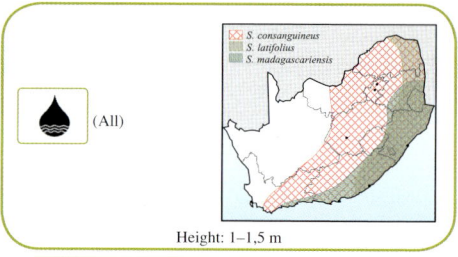

Schkuhria pinnata

ASTERACEAE

Senecio consanguineus
starvation senecio * hongerbos-senecio

Senecio latifolius
ragwort * krakerbossie

Senecio madagascariensis
canary weed * geelopslag

Height: 1–1,5 m

Other common names
S. consanguineus: ragwort * bankrotbos * radiatorbossie * bankrupt bush
S. latifolius: iyeza (X) * uqedizwe (Z)

Origin and description: All species of *Senecio* in South Africa are indigenous. Over 20 have been recorded as being weedy. *S. consanguineus* is found in the Northern Cape, Free State, Botswana and the central highveld. It is an annual that can survive more than one season. *S. madagascariensis* is very common in the eastern parts of the country.

Impact: These are some of the *Senecio* species that are common weeds of crops, gardens, roadsides and waste places. *S. consanguineus* is an annual, germinating in late summer and remaining green during winter. During soil preparation the following season, the light, fluffy seeds

Senecio latifolius

tend to block the radiators of tractors, hence the common name 'radiator weed', particularly in the western regions. *S. latifolius* is known to be toxic and occurs mainly in KwaZulu-Natal. *S. madagascariensis* is also found mainly in KwaZulu-Natal and is a common winter weed of harvested wheat and sugarcane lands.

Control: These weeds can be controlled by shallow cultivation when young or by post-emergence herbicides.

Senecio madagascariensis   *Senecio consanguineus*

Senecio madagascariensis

ASTERACEAE

Senecio ilicifolius
sprinkaan-senecio * sprinkaanbossie

Senecio inaequidens (=S. burchellii)
Molteno disease senecio * geelopslag

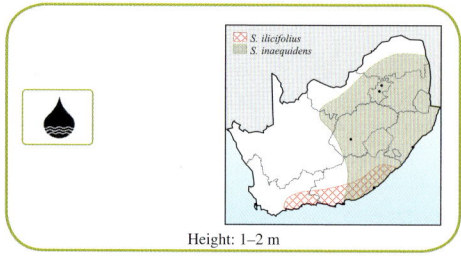

Height: 1–2 m

Origin and description: All *Senecio* species in South Africa are indigenous. The genus *Senecio* is the largest of all the flowering plant genera, with over 2 000 known species worldwide and up to 350 occurring in South Africa of which over 20 are known to be weedy.

Impact: These two species were proclaimed noxious weeds because they were implicated in cases of 'bread poisoning' in the Riversdale region of the Western Cape. Parts of the plant contaminated harvested wheat and poisoning occurred when the bread made from this wheat was consumed. Many species of *Senecio* contain toxic substances and some cause animal or even human poisoning, although most are not known to be toxic. Fortunately, grazing animals find *Senecio* unpalatable and will usually only eat the plants by accident. However, if sufficient quantities are consumed, fatal poisoning can occur. In horses in particular, the poison can be accumulative. Apart from being common in wheatlands, these weeds are also found in other crops, gardens, on roadsides and in waste places.

Control: These plants can be controlled by shallow cultivation when young, and are susceptible to many of the usual broadleaf weed herbicides.

Other common names
S. inaequidens: canary weed * Burchell-senecio * khotolia (S)

Senecio ilicifolius

ASTERACEAE

Sigesbeckia orientalis
St Paul's wort * Pauluskruid

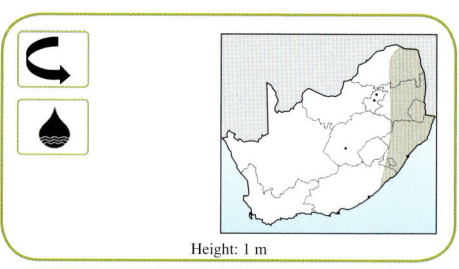

Height: 1 m

Origin and description: *S. orientalis* is of uncertain origin, but is probably exotic, being a widespread weed in the tropics of the Old World. In South Africa it occurs from the Soutpansberg to Port St Johns. It has not been recorded in the Free State.

Impact: St Paul's wort is commonly found on for-

Other common names
Indian weed * kleefgras

est margins, in waste areas and gardens, only occasionally becoming a weed of economic importance and then only in high-rainfall areas.
Control: It is not usually aggressive, but where control measures are necessary, it is susceptible to chemicals and cultivation.

Sigesbeckia orientalis

ASTERACEAE

Sonchus asper **subsp.** *asper*
spiny sowthistle * doringsydissel

Sonchus oleraceus
sowthistle * sydissel

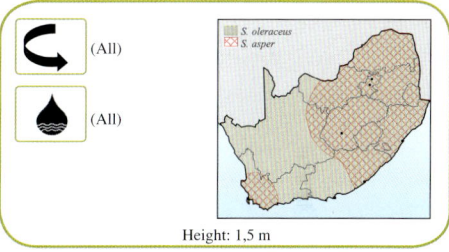

Height: 1,5 m

Origin and description: These are two of the earliest weeds introduced into South Africa from Europe, *S. oleraceus* having been recorded at the Cape as early as 1685. They are both anuals and very similar in appearance but the leaf margins of *S. asper* are spiny, especially when mature. The leaves are eaten as a salad and the milky roots are said to be good for bread-making. There are several indigenous species of *Sonchus*, but they are not considered weedy.
Impact: These weeds are widespread in cultivated fields, gardens, roadsides and waste areas.
Control: They are most effectively controlled by clean cultivation, followed by hand-weeding of scattered plants. The plants should be destroyed before the seeds are set. They are well

Other common names
S. oleraceus: milkthistle * tuindissel * bono-sa-lekhoaba (S) * ihahabe (Z) * ihlaba (X) * ingabe (Si) * lesabe (Ss) * lesese (P) * shashe (V)

Sonchus oleraceus

controlled by pre- and post-emergence broadleaf herbicides.

Sonchus oleraceus

Sonchus asper

ASTERACEAE

Tagetes minuta
tall khaki weed * langkakiebos

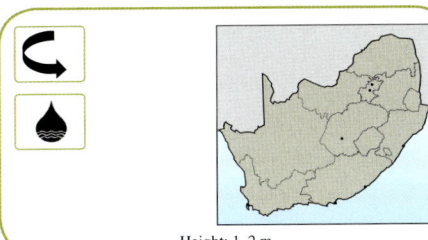

Height: 1–2 m

Origin and description: Introduced from South America, this plant is a very common, widespread and serious weed in many crops.

Impact: Anyone who has smelled khaki weed or the closely related garden marigold will instantly recognise the distinctive and clinging aroma. When maize, for instance, is infested with this weed and harvested, the grain may become tainted with the smell. This can lead to the crop being downgraded, with considerable financial consequences for the farmer. The smell has its benefits, however, as it is said to drive away nematodes. It has been suggested that the weed can be encouraged on fallow land and then ploughed in before a nematode-sensitive crop such as potatoes is planted. An extract from the leaves of this plant is used in the perfume industry.

Control: It is a weed that needs sunlight for germination, therefore germinating on or near the soil surface. This makes it susceptible to most pre-emergence herbicides as long as they do not leach into the soil. Even without soil disturbance, the seed of khaki weed is able to germinate throughout the summer season, provided it is not shaded by the developing crop. Best long-season control is therefore often achieved with a herbicide programme that incorporates a post-emergence element. Cultivation will control the seedlings but will bring fresh seed to the surface where they will germinate.

Other common names
Master John-Henry * Mexican marigold * stinking Roger * kakiebos * kleinafrikaner * jeremane (S) * insangwana (Z) * mbanje (N) (Sh)

Tagetes minuta

ASTERACEAE

Tragopogon dubius
yellow goat's beard * geelbokbaard

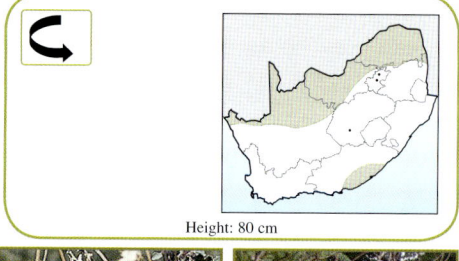

Height: 1 m

Origin and description: These semiperennial plants are natives of Europe. The plants are now widespread in South Africa, although they usually occur in the cooler areas higher than 1 500 m above sea level. These plants have a thick, fleshy taproot and the purple-flowered *T. porrifolius*, in particular, has been cultivated as a vegetable. The roots

Similar species
T. porrifolius: purple goat's beard * salsify * persbokbaard * wildeskorsenier

are cooked in the same manner as a parsnip, which they somewhat resemble. They are closely related to the dandelions (*Taraxacum* spp.) and, like the dandelion, the whole plant exudes a white juice when broken.

Impact: They are commonly found on roadsides, waste places, old fields and occasionally in vineyards.

Control: No herbicides have been registered for these weeds and they are very tolerant to most herbicides on account of the strong roots. They should respond to industrial chemicals used on roadsides but must be removed mechanically in croplands.

Tragopogon dubius

ASTERACEAE

Verbesina encelioides
wild sunflower * wildesonneblom

Height: 80 cm

Origin and description: A native of South America, this annual weed was first recorded in 1934 where it was found growing on an old course of the Orange River near Prieska. It has since spread rapidly on sandy soil, especially on riverbanks. Another species also called the wild sunflower, and possibly the wild type of the cultivated sunflower, *Helianthus annuus* (from North America) is a plant that can be found growing as a weed in certain parts such as around the Durban docks. It is similar in appearance to *V. enceliodes*, except the leaves are not serrated.

Impact: Occasionally troublesome in croplands such as those in the Vaalharts Scheme.

Control: There are no specific recommendations for its control, but it can be removed by cultivation whilst it is still small.

Verbesina encelioides   *Helianthus annuus*

ASTERACEAE

Xanthium spinosum
spiny cocklebur * boetebossie

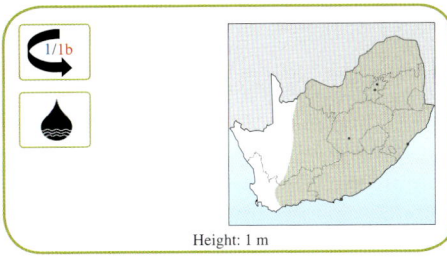

Height: 1 m

Origin and description: *X. spinosum* was introduced from South America and is now found throughout South Africa. It was the first declared noxious weed, when in 1860 the Cape government promulgated a law whereby the extermination of this weed was made obligatory under penalty of a fine, hence the well-known Afrikaans name 'boetebossie' – bush for which one is fined. It is still a Category 1 invader plant and should be removed by hand and burnt wherever it occurs.

Impact: It is usually found invading pastures and waste areas, less often in annual crops. The young plants are toxic and the burs cause serious damage to sheep's wool.

Control: Effective control is usually achieved with post-emergence, industrial herbicides or by physical removal.

Other common names
Bathurstbush * burweed * clotbur * dagger cocklebur * pinotiebossie * hlaba-hlabane (S) * iligcume (Z) * lepero (T)

Xanthium spinosum

ASTERACEAE

Xanthium strumarium
large cocklebur * kankerroos

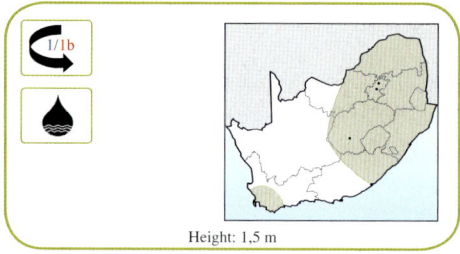

Height: 1,5 m

Origin and description: *X. strumarium* was introduced from South America and is now found throughout South Africa. All subspecies of *X. strumarium* are now considered to have originated from the same species, but there will be some variation from place to place.

Impact: It is a common, poisonous and serious arable weed. It is competitive, difficult to control and the burs contaminate sheep's wool. It is a major pest of maize, cotton and other annual crops.

Control: Pre-emergence herbicides do not work well because it has extended germination from various depths and large seeds with large food reserves. It can be controlled by shallow cultivation during the seedling stage or with post-emergence herbicides. There are nearly always escapes or late germinators, but these are usually shaded out by the crop. These escapes, however, eventually form seeds, which will maintain high seed levels in the soil once a field has become infested.

Other common names
hlaba-hlabane (S)

Xanthium strumarium

ASTERACEAE

Zinnia peruviana
redstar zinnia * wildejakobregop

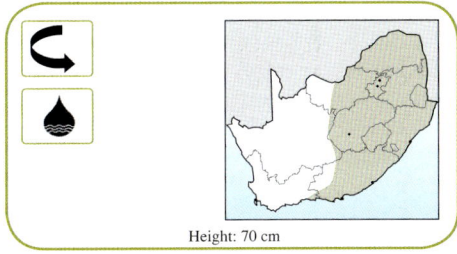
Height: 70 cm

Origin and description: A native of South America, *Z. peruviana* was introduced as a garden plant. It has been regarded as a weed in South Africa for over a century.
Impact: *Z. peruviana* is now a widespread annual weed of waste places, roadsides, other disturbed areas and dry river valleys.
Control: It is rarely a serious problem, but if control measures are necessary, it should be susceptible to conventional herbicides and cultivation during the seedling stage.

Other common names
Engelsmannetjies * fluitjiesbossie * lipii (S)

Zinnia peruviana

BORAGINACEAE

Amsinckia menziesii
fiddleneck * vioolnek

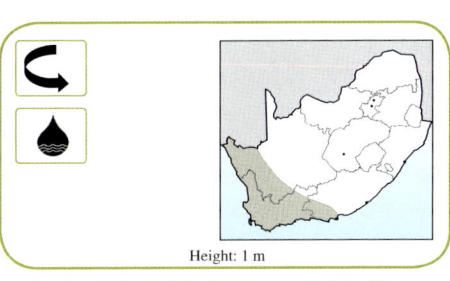
Height: 1 m

Origin and description: This plant was introduced into South Africa during the 1950s, probably from South and North America. It is now a weed of cereals in the Western and southern Cape. It is believed to be spreading into the Free State and western Mpumalanga. It gets its common name from the flower stems, bearing many small flowers, which curl over at the top, rather like the head of a fiddle.
Impact: It is a serious and competitive weed of wheat in the Sandveld and on the sandy soils of

Herbs

the Swartland. The fine, thorny hairs that cover the plant cause severe discomfort at harvesting.
Control: *A. menziesii* is not adequately controlled by the hormone-type chemicals, but is susceptible to post-emergence contact herbicides and the sulphonyl ureas.

Amsinckia menziesii

BORAGINACEAE

Echium plantagineum (=E. lycopsis)
purple echium * pers-echium

Echium vulgare
blue echium * blou-echium

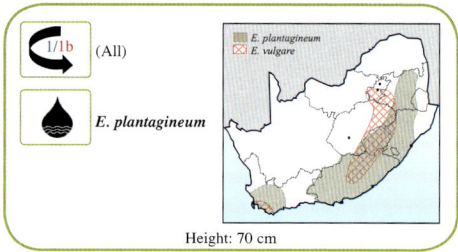

Height: 70 cm

Origin and description: These two species of *Echium* are natives of Europe and Asia and were originally introduced as ornamentals. They are now common weeds in South Africa as well as worldwide. *E. plantagineum* is the most common and prolific, especially in southern KwaZulu-Natal through to the northern parts of the Eastern Cape (e.g. Franklin) where fields frequently turn completely purple in spring. It is also a common weed of roadsides, orchards and vineyards in the Western Cape, where it appears in winter and produces a significantly larger plant.
Impact: *E. plantagineum* has a strong taproot and large, smothering leaves, which can strongly compete with pasture crops for space and moisture. In a badly infested field the pasture will die from moisture deprivation whereas the *Echium* can flourish. It is well known as a weed

Other common names
E. plantagineum: Franklin weed * Patterson's curse * salvation Jane * bloudisseldoring * natterkop
E. vulgare: blue devil * blueweed * viper's bugloss * bloubossie

Echium plantagineum

of pastures and lucerne in particular, being easily spread with contaminated seed or in hay used for animal bedding, which is then spread in the lands. It can be toxic if eaten by stock, but the plant is unpalatable.

Control: These weeds can be difficult to control. Herbicides must be used early in the spring before the plants are established or before commencing vertical growth. This is when the plants are small and susceptible to the chemicals, but unfortunately it is also when they are relatively small and insignificant that farmers tend to delay control measures until it is too late. In Australia, where *E. plantagineum* is also a major weed, work is under way to study the feasibility of biological control. (A man named Patterson was blamed for introducing *E. plantagineum* into Australia in 1880.)

Echium plantagineum

Echium vulgare

BORAGINACEAE

Heliotropium europaeum
European heliotrope

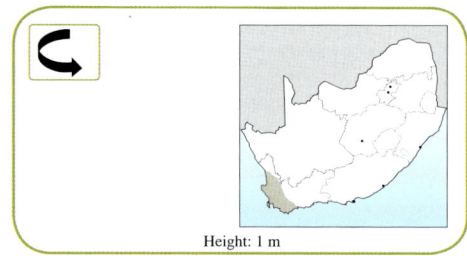
Height: 1 m

Origin and description: One of up to 300 species of *Heliotropium* that occur worldwide, and are usually referred to as 'borages'. This particular species is native to Europe, Asia and North Africa but has been distributed throughout the world, becoming weedy in many places such as in Australia and North America. The exact route of introduction of this particular species into South Africa in general or the Western Cape in particular, is uncertain, but it is easily spread because of seeds that attach to fur and clothing. Some heliotropes are popular garden plants.

Impact: European heliotrope has only recently invaded the Western Cape and Swartland and is now found covering large areas of harvested wheat land. It is known as a summer weed, meaning that it will grow after the first rains in the usually dry summer months, which is before the preparation of lands for the planting of wheat. The plant will remove precious moisture and the large volume of plant material, whether dead or alive, will clog cultivation and planting equipment. Heliotropes are toxic. It was reported that in 1974 an estimated 35 000 people were affected when they ate grain contaminated with heliotrope seed and 1 900 eventually died (Mohabbat *et al*. 1976, *Lancet* 2: 269–271). *H. europaeum* is also reported to be toxic and contains alkaloids that will damage the liver of livestock, which can lead to fatalities.

Control: There are no registered control methods, but this plant is easily killed by cultivation.

Other common names
caterpillar weed * common heliotrope * European turnsole

Heliotropium europaeum

BORAGINACEAE

Trichodesma zeylanicum
late weed

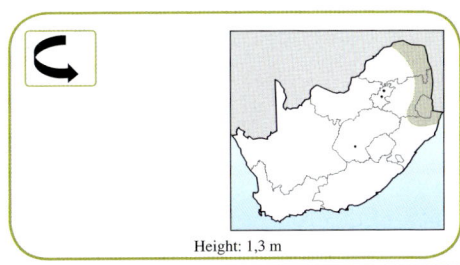
Height: 1,3 m

Origin and description: Although there are conflicting accounts of its origin, it is thought that this plant was introduced from Europe and Asia. It is now common in the northern regions, especially the northeastern lowveld of South Africa and northwards into central Africa where it is a problem in cultivated lands. It is called 'late weed' because it germinates and becomes noticeable relatively late in the summer. It reproduces only by seed, but plants often survive for more than a year.

Impact: It is commonly found on roadsides, waste places and orchards, becoming dense and competitive at times.

Control: No herbicides have been registered for this weed. It should be removed when young since the mature plants have sharp hairs that can irritate the skin when pulled out by hand.

Trichodesma zeylanicum

BRASSICACEAE

Capsella bursa-pastoris
shepherd's purse * herderstassie

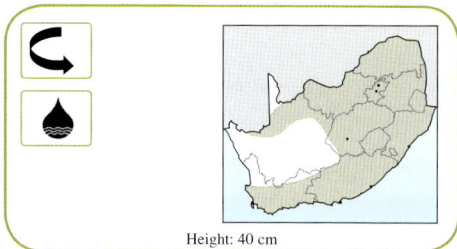

Height: 40 cm

Origin and description: *C. bursa-pastoris* is a native of Europe and was introduced to South Africa at the beginning of the 19th century, probably as a contaminant of crop seeds. Although occurring throughout the year, it is most noticeable in winter and of economic importance in vegetable crops grown in the winter-rainfall area as well as in irrigated crops elsewhere. It is a serious weed in wheat and other crops along the Orange River near Upington.

Impact: *C. bursa-pastoris* is now a common and competitive weed, particularly of vegetables, gardens and waste places. It is known to be a secondary host to various *Brassica* diseases.

Control: It is usually controlled effectively by herbicides and shallow cultivation.

Other common names
case weed * pepper plant * geld-beursie * wagter-se-sakkie* sebitsa (S)

Capsella bursa-pastoris

BRASSICACEAE

Lepidium bonariense
pepperweed * peperbossie

Lepidium africanum
pepperweed * peperbossie

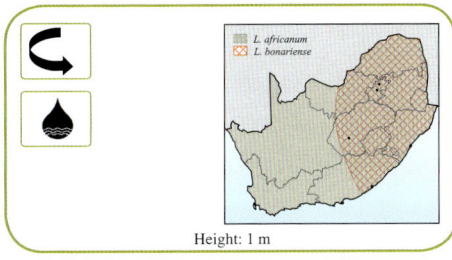
Height: 1 m

Origin and description: *L. bonariense* was probably introduced from South America. *L. africanum* and all its subspecies are indigenous. These two species are very similar in appearance and often occur together, the main difference being the shape of the leaves. Those of *L. bonariense* have deeper marginal grooves and it also tends to have larger seeds. They both have a sharp taste, hence the common name 'pepperweed'.

Impact: Both species are common throughout South Africa. Being frost resistant, often cause problems in winter crops, especially vegetables. They are also common in many other situations such as waste places and margins of fields.

Control: Although usually annual plants, they may persist for two years. They have strong stems, which makes them difficult to remove by cultivation once they have become established. They are susceptible to most of the conventional herbicides.

Other common names
L. bonariense: birdseed * pepper cress
L. africanum: birdseed * Cape pepper cress * peppergrass * pepperwort * tonguegrass * kanariesaadgras * sterkkos * sterkgras

Similar species
L. draba 1/1b (hoary cardaria * peperbos), an ornamental from the Mediterranean region and Eurasia, has been found in several central locations. It is a serious invader in the US and has been declared a Category 1 species in South Africa

Lepidium africanum

Lepidium bonariense

BRASSICACEAE

Raphanus raphanistrum
wild radish * ramenas

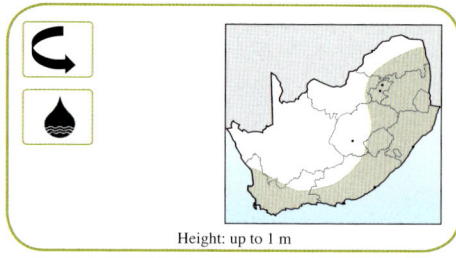
Height: up to 1 m

Origin and description: *R. raphanistrum* originated in Europe and is now widespread in South Africa. It can be distinguished from similar weeds such as wild mustard by its deeply divided leaves and by the deep constrictions between the seeds. The seedpods (not true pods but technically 'siliquae') eventually break into single-seed portions. The flowers are variable in colour, often being white and even purple as well as the predominant yellow.

Impact: It is the most important broadleaf weed in winter cereals in the Cape provinces. It is also a major weed of many crops in most other areas of the country. For example, in the Eastern Cape it is a major problem in potatoes, in many areas of the country it is a weed of pastures and it is a winter weed in maize lands under conservation tillage. Not only is *R. raphanistrum* a competitive and unsightly weed, but it also harbours insect pests and diseases of crops such as cabbages and Japanese radish. Such crops close to fields full of these weeds always have problems with diamond-back moth and aphids.

Other common names
charlock * field wall-flower * wild mustard * knopherik * wildemosterd * wilde radys

Control: *R. raphanistrum* is susceptible to certain sulphonyl-ureas and to hormone-type herbicides, but in sensitive crops pre-emergence control is the best option. It is well controlled in vineyards with non-selective herbicides. *R. raphanistrum* is not controlled by the acetanelide herbicides such as alachlor and acetochlor.

Raphanus raphanistrum

BRASSICACEAE

Rapistrum rugosum
wild mustard * wildemosterd

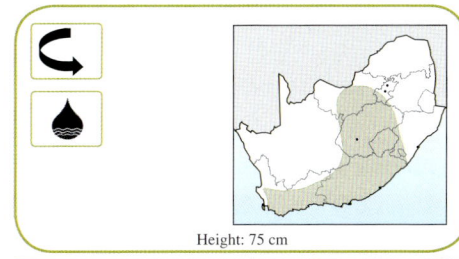

Origin and description: Introduced from Europe and Asia, this plant is now a widespread weed in South Africa. It can easily be confused with the various *Sisymbrium* species, *Erucastrum strigosum* and *Raphanus raphanistrum*, but the 'pods' of *R. rugosum* have two segments: one making up the stalk, while the other is a terminal, globular seed. (These are not the true pods of legumes but are called 'siliquae'.) The terminal leaf lobe is much larger than that of *Raphanus raphanistrum*.

Height: 75 cm

Impact: It is found in crops, orchards, waste places and roadsides, and can become dense and competitive if not controlled. It is especially common in the Cape and is most noticeable in winter.

Control: As for most members of the Brassicaceae family, *R. rugosum* is not susceptible to the acetanelide group of herbicides but is controlled effectively by the hormone-type and certain sulphonyl-urea herbicides. Non-selective, post-emergence sprays are very effective in vineyards, especially when the plants are young.

Rapistrum rugosum

BRASSICACEAE

Sisymbrium capense
Cape wild mustard * strandwildemosterd

Sisymbrium orientale
Indian hedge mustard

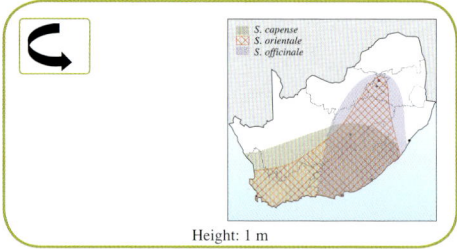

Height: 1 m

Origin and description: There are several weedy species of *Sisymbrium* in South Africa. (*Erucastrum austroafricanum*, that is indigenous, occurs throughout the country and is very similar but with relatively thick leaves and large 'pods' (siliquae). It has recently been removed from the genus *Sisymbrium*.) *S. orientale*, from Europe and Asia, is fairly widespread and has slender, rigid pods held away from the stem. *S. capense*, which is also indigenous, is not found in Mpumalanga but it occurs elsewhere throughout the summer- and winter-rainfall areas.

Impact: All these weeds are found in crops, especially cereals, in waste areas, orchards and on roadsides. The plants are edible when young.

Control: Although these weeds are susceptible to the hormone-type herbicides, they are not susceptible to the acetanelide group, which is often used for broadleaf weed control in vegetable crops. Non-selective, post-emergence sprays are very effective in vineyards, especially when the plants are young.

Similar species
Erucastrum austroafricanum (=*S. thellungii*) (wild mustard * wildemosterd), indigenous
S. officinale (hedge mustard * heiningwildemosterd), from Europe and North Africa
All these species are similar in behaviour and appearance to wild radish, *Raphanus raphanistrum*, and *Rapistrum rugosum*, which is also called wild mustard

Sisymbrium capense

Erucastrum austroafricanum  *Sisymbrium orientale*

CANNACEAE

Canna indica
garden canna * tuinkanna

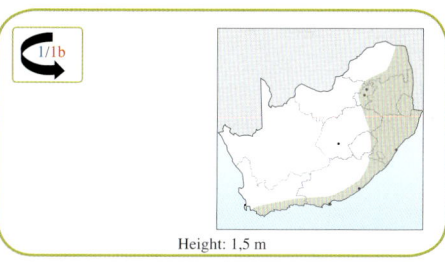

Height: 1,5 m

Origin and description: Canna is a well-known garden plant that has escaped into the wild and has become widely naturalised. Originally from Central America and the Caribbean, it is now found in the warm, moist areas of the eastern parts of the country. Many cultivars have been bred as ornamentals and there are about 30 different species names that are now considered to be simply synonyms for *C. indica*; others are *Canna x generalis*, which often has coloured foliage. The seeds are hard, black and perfectly round, hence the alternative common name 'Indian shot' as they could once have been used in

Other common names
Indian shot * wild canna * udumbe-dumbe (Z)
Similar species
C. glauca ⮌ and *C. x generalis* ⮌ and hybrid cultivars are also occasionally found in the wild but have not been declared invasive

old flintlock pistols when lead pellets were unavailable.

Impact: It is usually found on stream banks and on plantation edges, where it competes with and replaces indigenous species. It is becoming a serious problem in many areas.

Control: The plant has strong rhizomatous roots and is difficult to eradicate with herbicides. It should be removed physically, taking care to dig up and destroy the rhizomes.

Canna x generalis *Canna* sp.

Canna indica

CANNABACEAE

Cannabis sativa
dagga * hennep

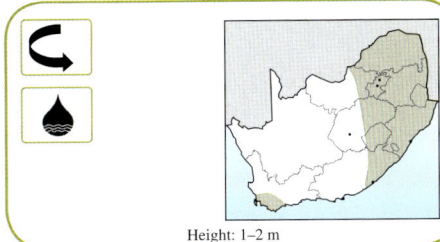
Height: 1–2 m

Origin and description: A native of Asia, this is now a cosmopolitan weed. It is an annual and if unattended can grow to a considerable height (up to 4 m). The seedlings are very distinctive, with one of the first two leaves being smaller than the other. A narcotic resinous substance is secreted by glands that are found mainly in the female flowers. Although the female flowers form the basis of the drug trade, the leaves can also be used. The male and female flowers occur on different plants. The presence or absence of pollen is the easiest way of differentiating between them.

Impact: In terms of the Medical and Dental Pharmacy Act, 1928 (Act No. 13 of 1928) it is an offence to cultivate *C. sativa* in South Africa. Nevertheless it is grown in most areas and illicit trade in the drug continues.

Control: It takes approximately 80 man-days to pull up and destroy 1 ha of these plants after they have grown to a reasonable size. Manual

Other common names
fragrant weed * gallow grass * grass * Indian hemp * marijuana * dwaalbos * isangu (Si) * matokwane (S) * nsangu (Z) * umya (X)

eradication programmes are therefore laborious and expensive. *C. sativa* is susceptible to the non-selective herbicides but must be treated at a young stage. If the plant is not killed effectively or if it is sprayed too late, then tillering can occur, thus creating a denser and more prolific bush.

Cannabis sativa

CAPPARACEAE

Cleome monophylla
single-leaved cleome * enkelblaar-cleome

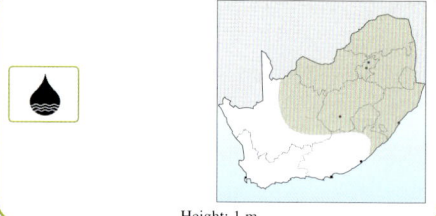

Height: 1 m

Origin and description: All these species of *Cleome* are indigenous.

Impact: *C. monophylla* is the most widespread and troublesome of the *Cleome* spp. and is a problem in field crops in most areas of the country. It can sometimes be quite a serious problem, as in maize in the northern and eastern Free State and Mpumalanga, for example. One of the Afrikaans names 'rusperbossie' indicates that it is a host plant of the lucerne caterpillar. The other related indigenous *Cleome* species are only occasional weeds, mainly in the northern areas. *C. gynandra* has been implicated as an alternative host for verticillium, a fungus disease that attacks several crop plants, including potatoes, tomatoes and cotton. All these species are subjects of at least one herbicide registration.

Control: *C. monophylla* can be controlled by pre- and post-emergence herbicides, as well as shallow cultivation when in the seedling stage.

Other common names
spider flower * spindlepod * rusperbossie * isiwisa esiluhlaza (Z) * matlepelo (S) * munyenyae (S) * mushangishangi (Sh)

Similar species
C. gynandra: bastard mustard * spider-wisp * oorpynpeultjie * snotterbelletjie * vingerblaartee * lerotho (P) * lube (N) * tsuna (Sh) * umzonde (Z)
C. rubella: pretty lady * mooi-nooientjie
C. monophylla seedlings can be confused with *Datura* spp. and with *Crotalaria*

Cleome monophylla

CARYOPHYLLACEAE

Polycarpon tetraphyllum
four-leaved allseed * naaldvrug

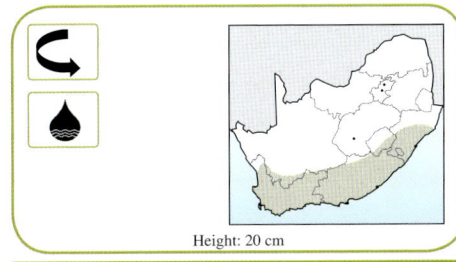

Height: 20 cm

Origin and description: Of uncertain exotic origin, *P. tetraphyllum* possibly originated from Eurasia. It is now common in parts of the Cape provinces, the Free State and KwaZulu-Natal. It was first recorded in South Africa at the beginning of the 19th century. It favours shaded places in waste areas, gardens and roadsides and has characteristic small white- and green-striped flowers.

Impact: It is an untidy plant that grows easily as a garden weed and is a common weed of grain crops in the Cape.

Control: It is easily controlled by post-emergence herbicides commonly used in grain crops.

Polycarpon tetraphyllum

CARYOPHYLLACEAE

Scleranthus annuus
annual scleranthus * knawel

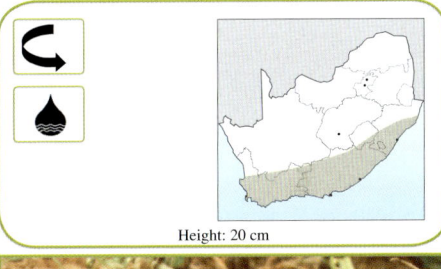

Height: 20 cm

Origin and description: *S. annuus* is a native of Europe that was introduced into South Africa in the 19th century and is now a widespread annual weed. It occurs throughout the country.

Impact: *S. annuus* is most troublesome in lucerne, wheatlands and vineyards in the Western Cape. It is also occasionally a problem in vegetable crops, particularly onions.

Control: The species is a weak competitor and rarely requires specific control measures. It can be controlled by shallow cultivation during the seedling stage and is susceptible to conventional herbicides.

Scleranthus annuus

CARYOPHYLLACEAE

Silene gallica
French silene * Franse silene

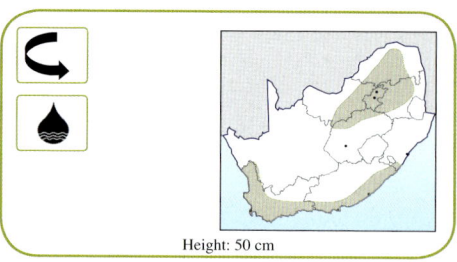
Height: 50 cm

Origin and description: A native of Europe, *S. gallica* has now become a cosmopolitan annual weed. It was introduced into South Africa at the beginning of the 19th century. The plant is fairly widespread but is most common in the Cape

Silene gallica

Other common names
French catchfly * gunpowder weed * windmill pink * eierbossie * jobskrale * kruitbossie

provinces, usually occurring in orchards and vineyards. It is an arable weed in East Africa.
Impact: It can become quite dense and competitive if left uncontrolled.
Control: It responds well to both chemical control and cultivation when in the seedling stage.

CHENOPODIACEAE

Atriplex lindleyi **subsp.** *inflata* (=*Blackiella inflata*)
sponge-fruit salt bush * blasie-soutbos

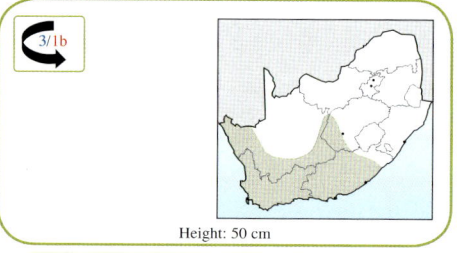
Height: 50 cm

Origin and description: Introduced, probably from Australia, this perennial, semideciduous shrub is now a common weed in the more arid parts of the Cape provinces and in Namibia. *A. lindleyi* subsp. *inflata* is a low-growing bush with its seeds contained in a characteristic balloon-like structure.
Impact: It is commonly found in waste places and roadsides and can swamp indigenous fynbos once it has become established. It is also known to invade coastal dunes, pans and other ecologically sensitive areas.
Control: No specific herbicides have been registered for this species, but because of its perennial nature it should be controlled physically before it becomes established.

Other common names
Lindley's saltbush * kleinbrakbossie

Atriplex lindleyi

CHENOPODIACEAE

Chenopodium album
white goosefoot * withondebossie

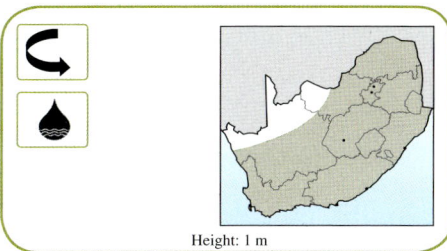
Height: 1 m

Origin and description: *C. album* is of European origin and is now found throughout South Africa. The seedlings appear furry on the top and are often tinged with purple underneath. Under certain growing conditions the young plants turn bright purple in the centre. This contrasting green and red/purple colouring can be quite striking. Seeds excavated from 2 000-year old settlements in Denmark were still able to germinate. Young plants are edible and the seeds are used to make flour.

Impact: It is a common and important weed of many crops throughout southern Africa. It is frost-tolerant and therefore also well known in winter crops, especially in wheat in the Free State. It has also been implicated as an alternative host for tomato spotted wilt virus as well as verticillium, a fungus disease that attacks several crop plants, including potatoes, tomatoes and cotton.

Control: *C. album* is easily controlled by cultivation and most of the pre- and post-emergence broadleaf weed herbicides. However, the waxy surface of the seedlings means that special attention should be given to post-emergence sprays, with the probable addition of a wetting agent.

Other common names
common pigweed * fat hen * lamb's quarters * wild spinach * bloubossie * hondepisbossie * misbredie * seepbossie * varkbossie * serue (S) * umbikicane (Si) (Z)

Similar species
C. album should not be confused with *Amaranthus* spp., which are also referred to by the common names 'pigweed' and 'misbredie'

Chenopodium album

Chenopodium album

CHENOPODIACEAE

Chenopodium ambrosioides
wormseed goosefoot * kruiehondebossie

Chenopodium murale **var.** *murale*
nettle-leaved goosefoot * muurhondebossie

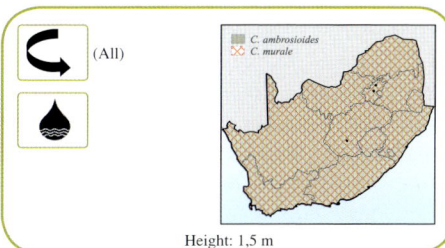

Height: 1,5 m

Origin and description: *C. ambrosioides* is from South America and a common weed of many warm countries. *C. murale* is from Europe. There are several other exotic species of *Chenopodium* that are considered weeds.

Impact: In South Africa they are widespread weeds but are usually found only on roadsides and in waste areas. *C. ambrosioides* is common in Namibia and has a characteristic smell when crushed. *C. murale* is similar to *C. album* (see above) but tends to have a darker colour. It is of economic importance in irrigated vegetables.

Control: All these species can easily be controlled by cultivation while they are still in the seedling stage. They are susceptible to many pre- and post-emergence herbicides.

Other common names
C. ambrosioides: Mexican tea * sandworm plant * galsiektebos * vlooibossie * insukumbili (X)
C. murale: chuana soap * wheat bush * gansevoet * khola-bosiu (S) * puniyi (X)

Chenopodium murale

Chenopodium ambrosioides

CHENOPODIACEAE

Salsola kali *(=S. australis)*
Russian tumbleweed * Russiese rolbossie

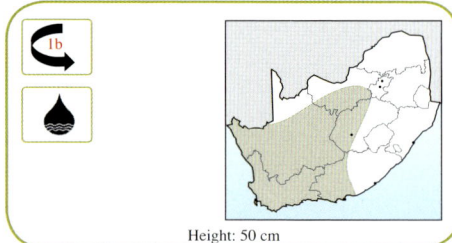

Height: 50 cm

Origin and description: Introduced from Asia, probably during the Anglo-Boer War, this is now a serious annual weed in parts of South Africa. It is very common in parts of the Cape provinces, the Free State and Namibia, being found mainly in waste places and on roadsides but also in orchards and gardens. The plant spreads easily by being broken off and rolled along the ground by a high wind, often for considerable distances.
Impact: *S. kali* is a very aggressive invader, being able to rapidly colonise new areas. As it is such a tough, unpalatable and un-pleasant plant, it is a highly undesirable weed. Its only benefit is that it can act as an anti-erosion agent in bare areas.
Control: This weed must be controlled while it is in the seedling stage.

Other common names
glasswort * rolypoly * saltwort * kakiebos * taaibos * tolbos

Salsola kali

DENNSTAEDTIACEAE

Pteridium aquilinum
bracken * adelaarsvaring

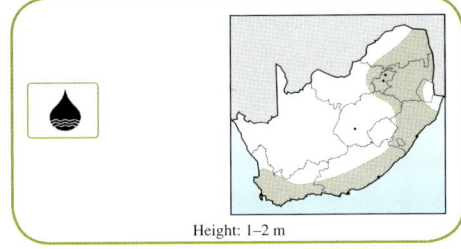
Height: 1–2 m

Origin and description: *P. aquilinum* is an indigenous perennial plant. It is widespread in South Africa and especially common in the high-rainfall areas. It is also a weed of the highlands of East Africa. It spreads within patches by means of hairy underground rhizomes. It is spread further afield by the copious amounts of spores produced in specialised structures on the leaves. The method of spore production is peculiar to the ferns (Pteridophyta).

Impact: *P. aquilinum* is a pest of damp places. In areas of high rainfall it also occurs in the open veld and is often accused of being an encroacher in grass-dominated ecosystems, smothering grasses and paving the way for woody encroachers and alien pioneer plants. Although the plant is poisonous to animals, particularly horses, instances of poisoning are rare. Poisoning may occur when the animals eat the young, curled, tender leaves in spring or after a fire. Instances

Other common names
eagle fern * pasture brake * fonteinbultvaring

of bracken poisoning have been recorded at Estcourt and Ixopo in KwaZulu-Natal and Knysna in the southern Cape. This weed is also highly flammable when dry.

Control: Control of this weed is difficult and therefore systemic herbicides such as imazapyr and metsulfuron-methyl are required and are best sprayed towards the end of the growing season. Otherwise, it can be removed mechanically by cutting stands continuously over six-week cycles during active growing periods.

Pteridium aquilinum

Pteridium aquilinum

EUPHORBIACEAE

Euphorbia helioscopia
umbrella milkweed * sambreelmelkkruid
Euphorbia peplus
stinging milkweed * brandmelkkruid

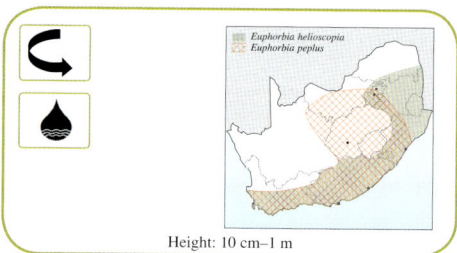

Height: 10 cm–1 m

Origin and description: Seventeen species of *Euphorbia* are safely cultivated in South Africa but 11 species have become naturalised. These two species both originate from Europe. *E. helioscopia* has been known in South Africa for over a century and is now widespread, except in the Free State. Like all Euphorbiaceae, they secrete a white fluid when broken, which can cause skin irritations.

Other common names
E. helioscopia: cat's milk * spurge * sun euphorbia * wartwort
E. peplus: petty spurge * purple spurge * sun euphorbia * hondekruid * melkbossiegras * son-euphorbia * wolfsmelk * bolila (S)

Impact: These weeds do not grow very tall, but can be moderately competitive if they become dense. They are a nuisance in gardens and are common weeds of vineyards in the Western Cape.

Control: As they are well controlled by post-emergence contact herbicides, they are seldom problems in situations such as orchards. However, in broadleaf crops these weeds are less sensitive to selective herbicides, thus posing a threat. They should be removed by shallow cultivation when in the seedling stage.

Similar species
Anagallis arvensis (pimpernel) (see page 343)
Stellaria media (chickweed) (see page 323)

Euphorbia peplus

Euphorbia helioscopia

EUPHORBIACEAE

Euphorbia heterophylla (=*E. geniculata*)
wild poinsettia * wildepoinsettia

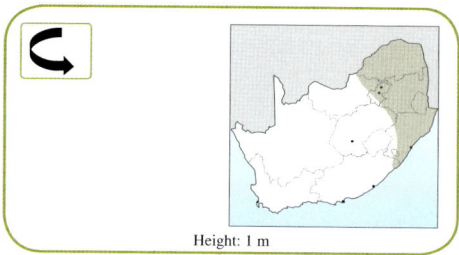
Height: 1 m

Origin and description: Originating from North America and Mexico, *E. heterophylla* is now a widespread weed, especially in the subtropical parts of the summer-rainfall regions. It is still planted as an ornamental in gardens and is a serious weed of cotton fields in India and Thailand. Milky sap oozes from damaged stems, and leaves can lose their red colour when growing as a weed. It is spread by seeds that are released explosively from ripe pods.

Impact: *E. heterophylla* forms generally isolated but resilient populations in crops such as sugarcane in the subtropical regions of the country. It is also a common invader of wasteland and railway embankments. It has been reported as toxic to stock.

Control: This weed is much more difficult to control once established and infestations should be removed by hand or spot-sprayed with systemic herbicides. In Brazil, this weed has developed multiple herbicide resistance (two modes of action).

Other common names
cruel plant * Japanese poinsettia * Mexican fireplant * milkweed * mole plant * paint leaf * painted euphorbia * gekleurde euphorbia

Similar species
E. pulcherrima (poinsettia * karlieboom): a common ornamental with cultivars, from Mexico but nevertheless found growing wild in KwaZulu-Natal

Euphorbia heterophylla

Euphorbia pulcherima

FABACEAE

Crotalaria sphaerocarpa subsp. *sphaerocarpa*
wild lucerne * wildelusern

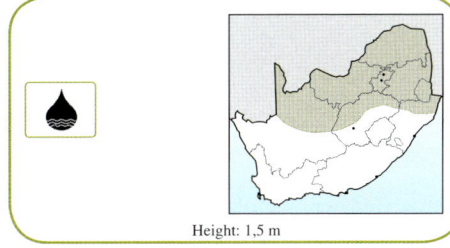

Height: 1,5 m

Origin and description: *C. sphaerocarpa* is a very common indigenous weed in the western maize-growing areas. About 80 indigenous species of *Crotalaria* have been recorded in South Africa. Several of them are minor weeds and only *C. sphaerocarpa* poses a serious threat, especially in maize and especially on sandy soils.

Impact: Wild lucerne can become dense and competitive and since seeds of this weed are poisonous, contaminated maize grain is downgraded and even totally rejected at the silo.

Control: *C. sphaerocarpa* is difficult to control pre-emergence as it germinates over the whole season and is large-seeded. It is both a shallow and deep germinator. The deep-germinating individuals can escape conventional herbicide programmes and become a nuisance later in the season. It is not well controlled by pre-emergence herbicides and is tolerant to most post-emergence herbicides once it is past the seedling stage. Care should therefore be taken with both the choice and application of herbicides if this weed is to be controlled effectively. It is important to add a wetting agent to post-emergence herbicides in order to enhance absorption.

Other common names
mielie-crotalaria * osipundula
Similar species
Crotalaria agatiflora (canarybird bush * voëltjiebos) is a popular ornamental that is growing wild around Pretoria (Category X3/1b)

Crotalaria sphaerocarpa

FABACEAE

Lupinus angustifolius
blue lupin * blou lupine
Lupinus luteus
yellow lupin * geel lupine

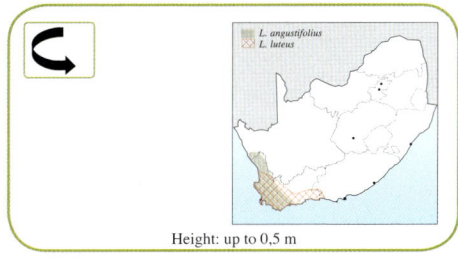

Height: up to 0,5 m

269

Origin and description: About 10 species of lupin (*Lupinus* spp.) are cultivated in South Africa for livestock and poultry feed and four have become naturalised in the wild. The yellow lupin and blue lupin are the most common and both are from the Mediterranean region, but the others are from the Americas.

Impact: These plants are now widely naturalised in the Western and Northern Cape, where they invade areas such as pastures, roadsides, forest margins and waste places. Being members of the legume family, they are capable of fixing nitrogen and turning it into ammonia, which can be of benefit in pastures, but it also enables the weed to establish itself in infertile soil and replace the indigenous plants, not only by direct competition, but also by changing the characteristics of the soil.

Control: Unwanted plants should be removed and naturalised infestations eradicated where possible. No specific herbicide registrations.

Lupinus angustifolius

Lupinus luteus

FABACEAE

Melilotus albus *(=M. alba)*
white sweet clover * witstinkklawer

Melilotus indicus *(=M. indica)*
annual yellow sweet clover * eenjarige geelstinkklawer

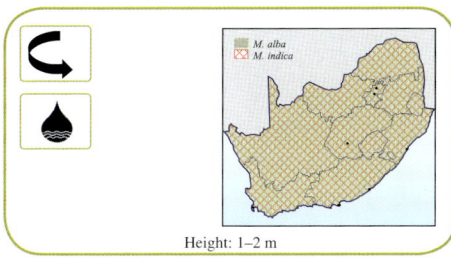

Height: 1–2 m

Origin and description: These weeds are natives of Europe and Asia and have been known in South Africa for over a century. They are now widespread annual weeds, having escaped from cultivation. They are found throughout the country in any disturbed ground, especially the roadsides of Gauteng. These species are said to be an excellent green manure, a good fodder crop and are favoured by bees. (The recent change in the spelling of these names has been brought about by taxonomists realising after 250 years, that Linnaeus had intended the Latin genus name to be masculine and it has erroneously been considered feminine until now.)

Impact: The plants contain coumarin, which has a very characteristic odour and can taint the meat, milk and eggs of animals that eat them. Flour made from wheat contaminated with seeds can also become tainted. Otherwise, they are untidy and obstructive.

Control: These weeds are susceptible to many of the pre-emergence herbicides and shallow cultivation.

Other common names
bokhara clover * hexham scent * bokhaarklawer * melilot

Similar species
M. officinalis (yellow sweet clover), from Eurasia. Also recorded as naturalised in South Africa

Herbs

Melilotus albus

Melilotus indicus

FUMARIACEAE

Fumaria muralis *(=F. officinalis)*
fumitory * duiwekerwel

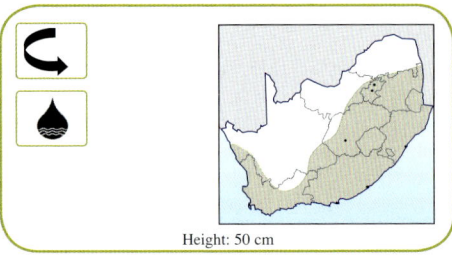

Height: 50 cm

Origin and description: A native of Europe, *F. muralis* is now a widespread annual weed in South Africa.

Impact: *F. muralis* is troublesome in wheat, particularly in Mpumalanga and the southern Cape. It is also common in lucerne and many vegetable crops in the winter-rainfall region. It can become dense and competitive.

Control: It is relatively tolerant to some herbicides, particularly in cereals, and care must be taken to cultivate or apply herbicides before the plants become well established. Seedlings are more susceptible to the various herbicides than mature plants.

Other common names
drug fumitory * pink weed * wall fumitory * duiwelskerwel * kerwel

Fumaria muralis

GERANIACEAE

Erodium moschatum
musk heron's bill * turknael

Height: 20–30 cm

Origin and description: *E. moschatum* was introduced from Europe by the early settlers. It has been a common weed in the southern and southwestern coastal areas of the Cape for more than a century. It gives off a musk-like odour when crushed but is still eaten by sheep, goats and ostriches. However, where these plants are found adjacent to the coast, it is thought that the odour becomes too strong with the result that livestock will not eat them.

Other common names
musk clover * white-stemmed filaree * muskuskruid * oorlosie * reiersbek

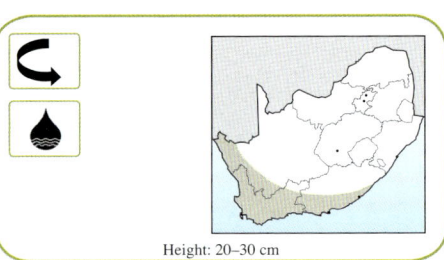

Impact: It is found in orchards, vineyards, cereals, gardens and all disturbed areas. Because the young plants have a flat growth habit, they compete with crops from a relatively early stage. This means that early control is essential.

Control: These weeds are controlled effectively by shallow cultivation during the seedling stage and by most broadleaf herbicides. Well controlled in vineyards by non-selective, post-emergence sprays.

Erodium moschatum

LAMIACEAE

Lamium amplexicaule
henbit deadnettle * turksenael

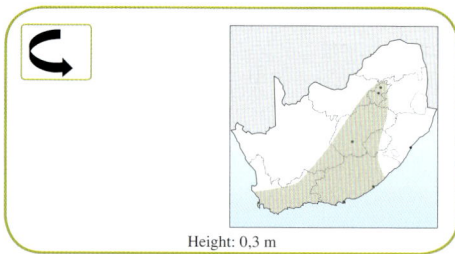
Height: 0,3 m

Origin and description: *L. amplexicaule* is a native of Europe, western Asia and even into northern Africa. It is now a widespread and common weed in South Africa, except in KwaZulu-Natal. When it occurs in significant numbers, the flowers are an important source of nectar and pollen for honey bees.
Impact: This plant is a frequent weed of disturbed areas, cultivated fields, orchards and gardens. The seeds can germinate throughout the season but it is most noticeable in late winter. It can be competitive, untidy and can replace indigenous vegetation.
Control: No herbicides are registered for this weed, but it should respond to the usual chemical and physical control programmes.

Lamium amplexicaule

LILIACEAE

Lilium formosanum
Formosa lily * trompetlelie

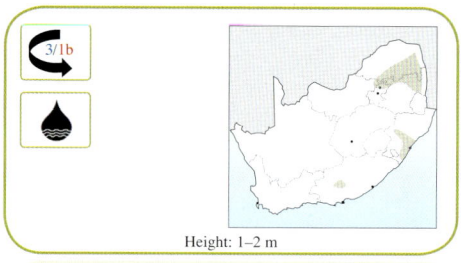
Height: 1–2 m

Origin and description: *L. formosanum* was introduced as an ornamental from Taiwan and has established itself in the wild. It is still a popular garden plant.
Impact: It is often found on forest edges and roadsides, especially on the N3 to Durban and

Other common names
St Joseph's trumpet * Sintjosefslelie

the N4. It has been suggested that it was deliberately spread along such tourist routes in an attempt to beautify the scenery. It can tolerate infertile, dry and sandy soils and can out-compete indigenous vegetation. It is a major and aggressive weed in Hawaii, so it has the potential to be a real problem in South Africa, and is seen as a serious contributor to the general degradation of the Grassland Biome. There are many indigenous alternatives for gardeners who are very fond of the Formosa lily.

Control: Unwanted plants should be dug out, but the bulb and any small bulbils forming around the old bulb must also be carefully removed and disposed of safely (by burning or deep burial). There are no herbicides registered for these plants and none appear particularly effective.

Lilium formosanum

MALVACEAE

Corchorus trilocularis
wild jute * wildejute

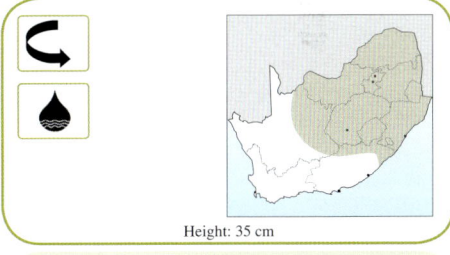
Height: 35 cm

Origin and description: Although this particular species of *Corchorus* originates from Eurasia, there are also some indigenous species that are generally not as weedy. They are usually annuals, reproducing only by means of seeds, but they can survive for more than one season. Some people eat the leaves of young plants as a spinach.

Impact: *C. trilocularis* occurs in the warmer parts of South Africa such as the lowveld areas and coastal regions, becoming a nuisance in crops, gardens and orchards in these areas.

Control: *C. trilocularis* will succumb to the conventional broadleaf-weed herbicides and shallow cultivation during the seedling stage. However, a mature plant is difficult to remove by hand on account of its fibrous stem and strong root system.

Similar species
C. olitorius (Indian jute) – probably indigenous
C. tridens (wild jute * wildejute) – also from Eurasia
C. trilocularis is similar in appearance to *Sida* sp. (see page 279) but has long pods that split open when ripe to reveal the black seeds within

Corchorus trilocularis

MALVACEAE

Hibiscus cannabinus
wild stockrose * wildestokroos

Hibiscus trionum
bladder hibiscus * Terblansbossie

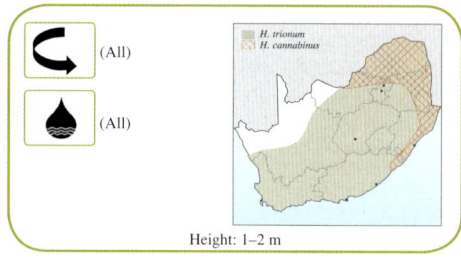

Height: 1–2 m

Hibiscus cannabinus

Hibiscus trionum

Origin and description: These two species are common throughout South Africa and many other warm parts of the world. Although their exact origin is uncertain, they are possibly indigenous, even though *H. trionum* also comes from the Mediterranean region. Improved varieties called 'kenaf' were nevertheless imported from Asia as a fibre crop, the fibre being a substitute for hemp and for making a fine 'tree-free' paper. The hairs on the stem of *H. cannabinus* cause nasty irritations if one attempts

Other common names
H. cannabinus: Deccan hemp * kenaf * wild hollyhock * sosoori (Sh) * umgangampunza (N) * udekane (Z) * umhlakanye (Si)
H. trionum: bladder weed * flower-of-an-hour * iyeza-lentshulube (X) * solwane (S) * uvemvane olukhulu (Z)

Hibiscus cannabinus

to hold the stem with a bare hand or use a combine harvester to cut an infested crop. *H. trionum* does not possess these unpleasant hairs and has a flatter, spreading growth habit. The fruits of these weeds are bladdery and contain the seeds. The leaves of *H. cannabinus* are more finely divided than those of *H. trionum* and are similar to the leaves of *Cannabis sativa* or 'dagga' (see page 257) with which it must not be confused.

Impact: These weeds can be competitive in annual crops and interfere with the harvesting process.

Control: Of the two species, *H. cannabinus* is taller, more aggressive and competitive. It is also a deep germinator and therefore less susceptible to pre-emergence herbicides. More reliable control can be obtained with post-emergence chemicals. *H. trionum* is relatively easy to control.

MALVACEAE

Lavatera arborea
tree mallow * boommalva

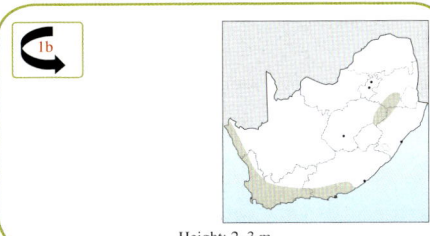

Height: 2–3 m

Origin and description: Introduced from Europe, this plant is now a widespread weed in South Africa, especially in the southwestern parts of the Cape. It looks very similar in terms of leaf shape and flower colour to the small mallow, *Malva parviflora*, but it is much taller and has very large, attractive flowers. The tree mallow can easily reach a height of 3 m and can even develop a woody stem. For these reasons it was probably originally planted as an ornamental shrub.
Impact: It is a biennial and is usually found in waste areas, sand dunes and on roadsides, obstructing vision and appearing generally unsightly.
Control: There are no herbicide registrations for this weed. It should be removed when it is small.

Other common names
rose mallow * mak-kiesieblaa*r*
Similar species recorded as naturalised in South Africa
L. assurgentiflora (mission mallow), from the Channel Islands
L. cretica (smaller tree mallow), from Europe and the Mediterranean region
L. trimestris (annual tree mallow), from the Mediterranean region

Lavatera arborea

MALVACEAE

Malvastrum coromandelianum
prickly malvastrum * stekelrige malvastrum

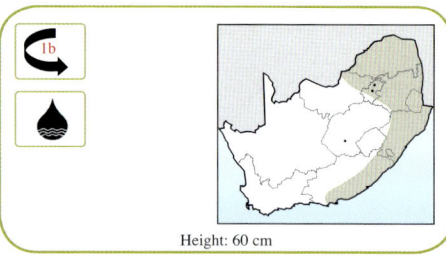

Height: 60 cm

Origin and description: This is a variable perennial or annual weed, native to North America. Despite the common name, it does not actually have any thorns or prickles but is rather tough and leathery. It should not be confused with *Sida cordifolia*, which it resembles strongly.

Impact: *M. coromandelianum* is a common and sometimes serious weed of roadsides, orchards, waste places and perennial crops in the summer-rainfall region, with the exception of the Free State. It is very drought resistant, so it can be found growing on dry road shoulders where other weeds may perish.

Control: There are very few herbicides registered for the control of this weed. It is probably susceptible to conventional herbicides, but only if it is sprayed when young. The seedlings can be removed by shallow cultivation, but the mature plant is very difficult to pull up.

Malvastrum coromandelianum

MALVACEAE

Sida cordifolia **subsp.** *cordifolia*
heart-leaf sida * verdompsterk

Sida rhombifolia **subsp.** *rhombifolia*
arrow-leaf sida * taaiman

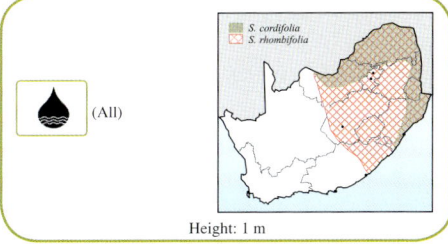

Height: 1 m

Origin and description: Even though these plants are found in other tropical countries, these and other species of *Sida* are probably indigenous and found virtually throughout South Africa. They are perennial shrubs that grow up to about 1 m tall. Some people use the very strong bark as rough cordage.

Impact: Being perennial weeds, these plants do not often infest annual crops, being more weeds of orchards, gardens, roadsides and old lands, etc. They are well known in cotton in Mpumalanga and Limpopo Province. These weeds are characterised by their very strong taproot and stem, making them almost impossible to pull up and tolerant to even the heaviest discing.

Control: These weeds are difficult to control due to their physically resilient nature. They can be controlled by cultivation in the seedling stage, but when mature they require systemic herbicides, which can translocate to the roots. Pre-emergence herbicides give erratic control.

Other common names
S. cordifolia: flannel weed * white burr * hartblaartaaiman * koekbossie * nama (N)
S. rhombifolia: Pretoria sida * smalblaartaaiman * ivivane (Si)

Similar species
These two Sida species should not be confused with Malvastrum coromandelianum
Sida spinosa var. *spinosa* (spiny sida)

Sida cordifolia

Sida rhombifolia

NEPHROLEPIDACEAE

Nephrolepis exaltata (=*Polypodium exaltatum*)
sword fern * swaardvaring
Nephrolepis cordifolia
tuberous sword fern

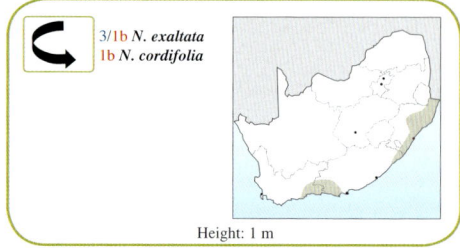

3/1b *N. exaltata*
1b *N. cordifolia*

Height: 1 m

Origin and description: The sword fern is native to tropical regions around the world, being common in humid forests and swamps, especially in Central America and the West Indies. It is, however, considered a serious invader in Florida. It is a popular house and garden plant in South Africa, but it has escaped to flourish in the more humid regions, especially in KwaZulu-Natal. *N. cordifolia* is very similar, and is also very invasive. Both species are spread by spores and stolons, and in Africa both species produce tubers. There are sterile cultivars available but their sterility and non-invasiveness has to be investigated.

Other common names
Boston fern * maidenhair * maidenhair fern
Similar species
Polypodium aureum (=*Phlebodium aureum*): rabbit's-foot fern * haaspootvaring X3

Impact: This plant can be found invading swamps, forests, coastal bush and roadsides. It forms monospecific stands, thereby eliminating indigenous species and threatening biodiversity.

Control: Hand-pulling is possible but it is important to ensure that all parts of the plant have been removed. These ferns are susceptible to glyphosate, but follow-up applications are necessary to control plants regrowing from rhizomes and tubers. Use the sterile cultivars by choice or preferably the indigenous *N. biserrata*.

Nephrolepis exaltata

NYCTAGINACEAE

Mirabilis jalapa
four-o'clocks * vieruurtjies

Origin and description: Introduced from tropical America as a garden flower, *M. jalapa* has escaped into the wild. Known as 'four-o'clocks' because their flowers open at that time or at least in the late afternoon. They remain open all night and collapse in the morning. The plant has yellow, reddish purple and white flowers and all colours can occur on the same plant.
Impact: It is usually found on stream banks, railway embankments and waste places where it competes with indigenous species.
Control: The plant should be removed physically or sprayed with a systemic herbicide.

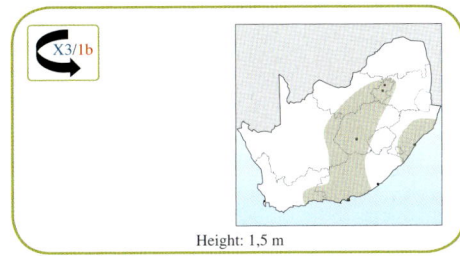

Height: 1,5 m

Other common names
beauty-of-the-night * false jalap * marvel-of-Peru * vieruurblom

Mirabilis jalapa

Mirabilis jalapa

ONAGRACEAE

Oenothera biennis
common evening primrose * nagblom

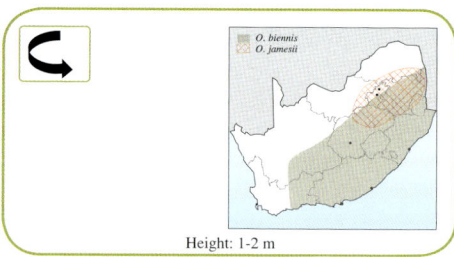

Height: 1-2 m

Origin and description: One of over 14 species of *Oenothera* naturalised in South Africa. This one is from eastern and central North America and arrived as early as 1772. As with other 'evening primroses', the flowers open in the evening and close by noon the next day. Unlike the others though, the stem of *O. biennis* is hairy with red-based, glandular hairs. Also, as the name suggests, it is biennial.

Impact: This weed is commonly found on roadsides, in waste areas and on riverbanks. It is

Similar species
O. jamesii (trumpet evening primrose), also from North America
O. glazioviana, of uncertain origin, but possibly a European bred hybrid of two North American species

becoming much more common and competitive in the moister regions of the eastern part of the country.

Control: The plants are easy to remove by hand when still small and are generally susceptible to contact herbicides when they are seedlings. Once they have become established, however, they become tolerant to most chemicals and therefore control must be initiated early.

Oenothera biennis

ONAGRACEAE

Oenothera indecora
evening primrose * nagblom

Oenothera stricta **subsp.** *stricta*
evening primrose * nagblom

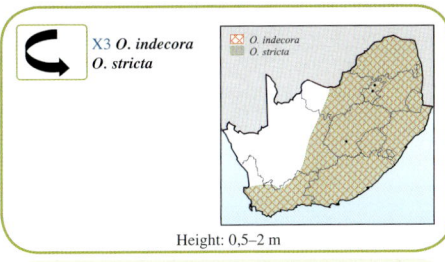

Height: 0,5–2 m

Origin and description: These plants are just two of over 14 species of *Oenothera* that have been introduced to South Africa from the Americas, probably as ornamentals. Most of them come from Central America and have been known as weeds since the end of the 19th century. They are called 'evening primroses' because of their habit of opening just before sunset and closing again before noon the following day. The flowers vary in colour and in some of the yellow-flowered species they turn pink with age. The plants have a thick, stubby taproot.

Similar species
O. laciniata (cut leaf evening primrose). Subject of a herbicide registration, also from the Americas

Oenothera indecora

Impact: They are commonly found on roadsides, in waste areas, old lands and occasionally in crops, especially when under conservation tillage. *O. indecora* and *O. stricta* are serious weeds in

orchards and vineyards in the Western Cape.
Control: The plants are controlled by cultivation when still small and are generally susceptible to contact herbicides when they are seedlings.

Once they have become established, however, they become tolerant to most chemicals and therefore control must be initiated early.

Oenothera stricta

Oenothera indecora

ONAGRACEAE

Oenothera rosea
rose evening primrose * pienkaandblom

Oenothera tetraptera
white evening primrose * witnagblom

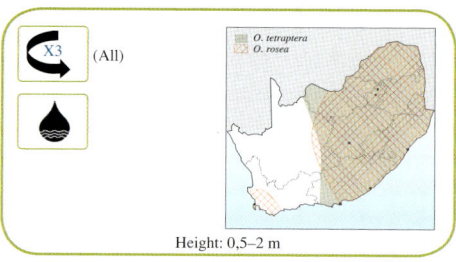
Height: 0,5–2 m

Origin and description: These plants are just two of over 14 species of *Oenothera* that are naturalised in South Africa. They were introduced from the Americas, probably as ornamentals. Most of them come from Central America and have been known as weeds here since the end of the 19th century. They are called 'evening primroses' because of their flowers that open just before sunset and closing again before noon the following day. The flowers vary in colour and in some of the yellow-flowered species they turn pink with age. The plants have a thick, stubby taproot.

Other common names
O. tetraptera: witaandblom * moopeli-o-mosoeu (Ss)

Impact: They are commonly found on roadsides, waste areas and in perennial crops, occasionally in vegetables.
Control: The plants are controlled by cultivation when still small and are generally susceptible to contact herbicides when they are seedlings. Once they have become es-

tablished, however, they become tolerant to most chemicals and therefore control must be initiated early.

Oenothera tetraptera

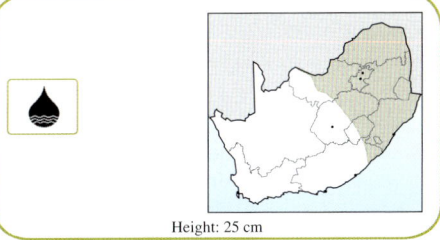

Oenothera rosea

OROBANCHACEAE

Striga asiatica
witchweed * rooiblom

Origin and description: *S. asiatica* is an indigenous, semiparasitic plant. There are several other indigenous species of *Striga* which are not as common. For example *S. gesnerioides*, with more purple flowers, and *S. forbesii*, which is much larger, are two species that occasionally attack commercial crops. *S. hermonthica* is the subject of much attention in many parts of Africa.
Impact: *Striga* germinates only in the presence of the roots of host plants that release 'strigol'. It is encouraged to germinate almost exclusively

Height: 25 cm

Other common names
isona weed * Matabele flower * scarlet lobelia * mieliegif * vuurbossie * bisi (Sh) * isona (Z) * seona (S) * sono (Si) * umnaka (X)

285

by grasses and crops such as maize, sorghum and sugarcane but also, peculiarly, tobacco. The seeds are minute (2 million per gram), with one plant producing up to 500 000 seeds, which can lie dormant in the ground for up to 20 years. Under unfavourable conditions, *Striga* can develop and grow normally – but underground. It takes several weeks to emerge, by which time it might have caused severe damage to a crop. Well-fertilised plants are able to withstand the effects of this parasite, which is why it is a far more significant threat in small-scale farming.

Control: Chemical control is not very successful, although some progress has been made with chemicals applied to the host plants as well as the soil. Various cultural methods of control, such as early planting, are used, so that the roots are strong and can withstand attack. Another cultural method is the planting of susceptible plants that are destroyed before the *Striga* produces seed. Crops such as beans, which release strigol but do not act as hosts to *Striga*, can also be grown. This causes the *Striga* seeds to germinate and die. After several seasons of using this 'suicide germination' as a method of cultural control, the weed-seed reserve should be reduced considerably. Research into using improved strains of promiscuous soybeans specifically for this purpose is significantly improving productivity in many African countries.

Striga asiatica

PAPAVERACEAE

Argemone mexicana* forma *mexicana (=*A. mexicana*)
yellow-flowered Mexican poppy * geel-blombloudissel

Argemone ochroleuca* subsp. *ochroleuca (=*A. subfusiformis*)
white-flowered Mexican poppy * Mexikaanse papawer

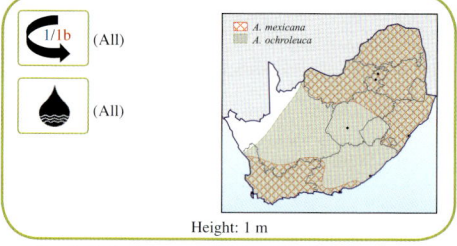

Other common names
devil's fig * Texas poppy * ugud-luthukela (Z)

Origin and description: The Mexican poppies are of Central American origin and are now widespread in South Africa. Because of their pioneering nature, these plants were not originally declared noxious weeds. Pioneer plants often help to prevent erosion, as they are usually the first to become established on bare ground. The flowers of these weeds look like typical poppies, but because of their spiny leaves they are often mistaken for thistles. *A. ochroleuca* exudes a bright yellow sap and they both have a distinctive odour when crushed.

Impact: These plants are known to have caused poisoning, even fatalities, of humans and stock but fortunately the seeds are not readily eaten. They are alternative hosts for verticillium and the tomato spotted wilt virus and the seeds contaminate sheep's wool. They are often found in places such as recently cleared sites and new dam walls as well as in many crops. In maize lands they may appear late in the season, particularly in the maize areas in the western parts of the country. They are sometimes a problem in wheat and care must be taken to avoid contamination of the wheat by the toxic seeds.

Control: These weeds are usually controlled effectively by shallow cultivation. When small, most post-emergence herbicides can be used to control them.

Argemone ochroleuca

Argemone mexicana

PAPAVERACEAE

Papaver aculeatum
wild poppy * doringpapawer

Papaver rhoeas
field poppy * koringpapawer

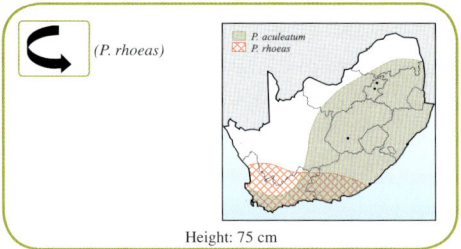

(*P. rhoeas*)

Height: 75 cm

Origin and description: *P. aculeatum* is indigenous, whereas *P. rhoeas* was introduced to South Africa from Europe. *P. rhoeas* is also common in the crops and on old battlefields of northern France and was first planted in South Africa as an ornamental. Both species are annuals.

Impact: *P. aculeatum* has an orange or salmon-coloured flower and spiny stems and is widespread in South Africa. It is mainly a weed of old lands, waste places and roadsides, but also of grain cultivation in winter. *P. rhoeas*, which has a red flower, is a minor weed of wheat in parts of the Cape provinces.

Control: There are no specific herbicides registered for these weeds, but being small-seeded, shallow germinators, they should be susceptible to chemicals and cultivation during the seedling stage.

Other common names
P. aculeatum: thorny poppy * koringpapawer * koringroos * sehlolo (S)
P. rhoeas: Flanders poppy
Similar species recorded as naturalised in South Africa
P. argemone: long pricklyhead poppy
P. hybridum: round pricklyhead poppy

Papaver rhoeas

Papaver aculeatum

PEDALIACEAE

Sesamum triphyllum
wild sesame * wildesesam

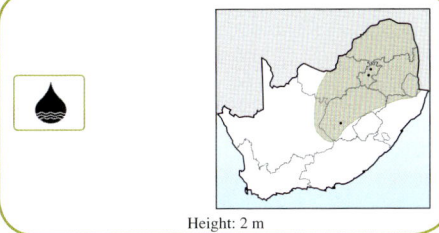
Height: 2 m

Sesamum triphyllum

Other common names
thunderbolt flower * seeroogblaar
Similar species
There are several other similar indigenous species of *Sesamum*. *Sesamum indicum* (sesame * oliebossie) is from Eurasia and has escaped cultivation
Ceratotheca triloba (wild foxglove * vingerhoedblom), common indigenous roadside weed

Origin and description: *S. triphyllum* is an indigenous and widespread annual weed common throughout South Africa's summer-rainfall region.
Impact: This weed is troublesome in various crops in Limpopo Province and Mpumalanga. Also common along roadsides in KwaZulu-Natal and an occasional weed of sugarcane.
Control: It should be controlled when small. Isoxaflutole is registered for its control in sugarcane.

PHYTOLACCACEAE

Phytolacca octandra
inkberry * inkbessie

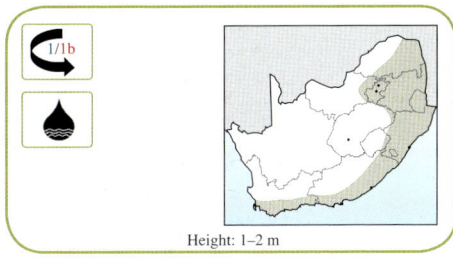
Height: 1–2 m

Origin and description: There are several species of *Phytolacca* occurring as weeds in South Africa. Most of them are natives of the Americas. *P. octandra* is probably exotic and occurs mostly in the eastern parts of the country. It was reported to have come into KwaZulu-Natal when the railway cuttings between Durban and Pietermaritzburg were made in about 1865. Before this it was unknown in the region.

The Afrikaans common name refers to the grapelike fruit, which is favoured by baboons in some areas, even though they are said to be poisonous and capable of causing skin irritations. The seeds are efficiently dispersed by birds.

Impact: *P. octandra* is a major weed of forestry in South Africa as it rapidly invades clear-felled areas and can grow into quite a large bush. It also occurs on roadsides and in waste places and gardens.

Control: The weed is easily controlled by chemical or mechanical means, especially when small.

Other common names
forest inkberry * poke weed * bobbejaandruif * amahashe (X) * umnyandla (Z) * umnanja (X, Z)

Similar species
P. americana ◖ 1b (poke weed * karmosynbos), also from tropical America but is taller (3 m), with pink to red flowers and longer stalks
P. octandra must not be confused with *Cestrum laevigatum* (see page 158), which shares the common name of 'inkberry'

Phytolacca octandra

PHYTOLACCACEAE

Rivina humilis
bloodberry * bloedbessie

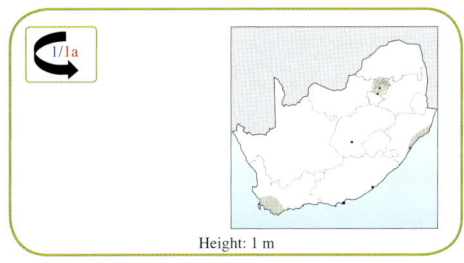

Origin and description: A perennial herb from North America, *R. humilis* was introduced as an ornamental. It has escaped and can be found growing in the wild in the coastal regions of KwaZulu-Natal as well as one or two other places. It is found along the edges of forests, in hedgerows and other sheltered or shady areas. Birds are the main dispersal agent as they eat the fruit.
Impact: Although not yet widespread, this plant could become seriously invasive. It is competitive and replaces indigenous vegetation. It also interferes with the re-establishment of indigenous plants following disturbance of the soil.
Control: Individual plants can be hand-pulled and destroyed, preferably before the formation of fruit. Herbicides may be effective, but there are no registrations.

Other common names
baby pepper * coral berry * rivina

Rivina humilis

PLANTAGINACEAE

Plantago lanceolata
buckhorn plantain * smalweëblaar

Plantago major
broadleaf ribwort * grootweëblaar

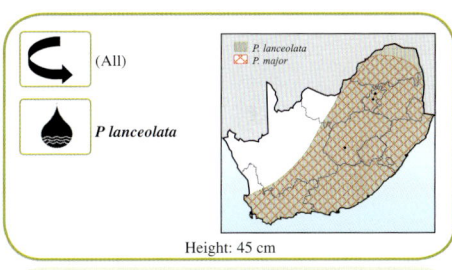

Origin and description: Collectively called 'plantains', these closely related species are natives of Europe and are now cosmopolitan. They have followed humans' pioneering activities into the New World and are called 'White man's footprints' by the North American Indians. The leaves are edible and are used in herbal medicine. These species of *Plantago* are easily recognised by the compact, tough inflorescence at the end of a long stalk commonly producing up to 1 000 seeds. The pollen is a major factor in causing hay fever and the Afrikaans name 'oorpynhoutjie' refers to the old custom of placing small pieces of the inflorescence in the ear as a remedy for earache. *P. major* is shorter, with

Other common names
P. lanceolata: English plantain * German psyllium * ribwort * oorpynhoutjie * smalblaarplantago * bolilanyana (S) * indlebe-kathekwane encane (Z)
P. major: cart-track plant * greater plantain * ripplegrass * rippleseed plantain * wild sago * platvoet * indlebe-ka-tekwane (Z) * bolila (S)

broader, ovate leaves as opposed to the long, narrow leaves of *P. lanceolata*. It also has a relatively larger group of flowers and is considerably less common than *P. lanceolata*.

Impact: They are widespread in South Africa and common along roadsides, on wasteland and in gardens and vineyards. They can become a problem in lucerne, clover fields and orchards, for example, with the seeds being a common impurity of grass and clover seeds. *P. lanceolata* is particularly significant in the winter-rainfall region and in 2003 a specimen from the Breede River Valley was discovered to be resistant to glyphosate. Since then, glyphosate resistance in Cape weeds is a growing and significant problem.

Control: If they are sprayed at an early stage they are normally susceptible to post-emergence herbicides. Badly infested pastures should be cultivated, fertilised and planted with an alternative crop for two years before reseeding.

Plantago lanceolata

Plantago major

PLUMBAGINACEAE

Limonium sinuatum
statice * papierblommetjies

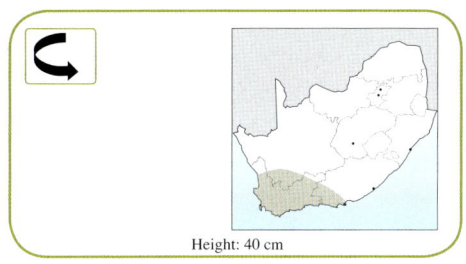

Height: 40 cm

Origin and description: This is a Mediterranean plant species that is cultivated as an ornamental and for use in the cut-flower trade. It is now widely naturalised in the Cape provinces. The usual flower colours are blue and white, although other colours occur. There are about 10 indigenous species of *Limonium*.

Impact: First recorded as naturalised in 1950,

Other common names
sea lavender * marsh-rosemary

statice has since spread into large areas of the Cape. It is most common on roadsides, but can be seen invading grasslands and fynbos, where it pressurises an already fragile and threatened ecosystem.

Control: Unwanted plants should be removed wherever possible.

Limonium sinuatum

POLYGONACEAE

Persicaria lapathifolia *(=Polygonum lapathifolium)*
spotted knotweed * hanekam

Persicaria capitata *(=Polygonum capitatum)*
pink knotweed

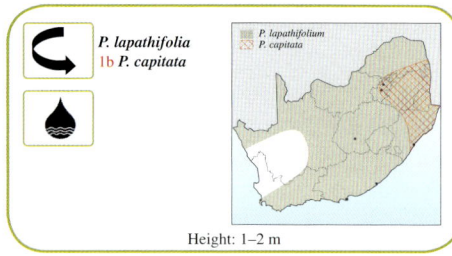

Height: 1–2 m

Origin and description: *P. lapathifolia* is originally from Europe and is now a widespread naturalised weed in South Africa. It has been called a 'follower of man' as it has spread around the world in the footsteps of mankind, with no natural habitat.

Impact: It can tolerate very moist conditions and can be found as a weed in irrigated crops, especially rice. However, it is very versatile and is also found infesting dry land crops, where it will compete for moisture. It is common on river banks, dam walls, in ditches and can even live in water. *P. capitata* is a closely related plant from the Himalayas and Asia that is widely planted as an ornamental groundcover in gardens. However, it is emerging as a potential invader as it is clear that it is becoming established away from gardens where it can quickly replace indigenous species. It can be found in Limpopo, Mpumalanga and Swaziland invading roadsides, forest margins and disturbed grassland.

Control: There are no herbicides registered for these weeds but they are easy to remove by cultivation. In ditches and water courses, *P. lapathifolia* should be removed manually. Care should be taken to ensure that *P. capitata* does

Other common names
P. lapathifolia: pale persicaria * pale smartweed * red shank * viltige duisendknoop * tolo-la-khongoana (S) * umancibikela (Z)

Similar naturalised species
P. limbata (= *Polygonum limbatum*), from central Africa and possibly indigenous. Subject of a herbicide registration
P. hydropiper (water pepper), from temperate Eurasia

not escape from cultivation and any existing infestations should be eliminated before this weed becomes too serious.

Persicaria lapathifolia

Persicaria capitata

POLYGONACEAE

Rumex acetosella* subsp. *angiocarpus *(=R. angiocarpus)*
sheep sorrel * steenboksuring

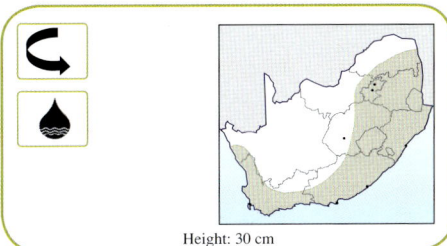

Height: 30 cm

Origin and description: Accidentally introduced from Europe by early Dutch colonists, *R. acetosella* subsp. *angiocarpus* is now a widespread weed in South Africa. It probably arrived with the fist importation of European wheat late in the 17th century and was recorded as a weed of cultivated land in the Cape in 1772.

Impact: It is a common perennial weed of cultivated lands and sometimes, because of its underground rhizomes, can become a serious problem. It is more of a problem on acid soils than in other soils, and is particularly troublesome in the Free State and Griqualand East.

Control: This weed is difficult to eradicate mechanically as small pieces of root left behind are able to regrow. It can be controlled by correct management and with pre- and post-emergence herbicides.

Other common names
dock * field sorrel * rooisuring

Rumex acetosella

POLYGONACEAE

Rumex crispus
curly dock * krultongblaar

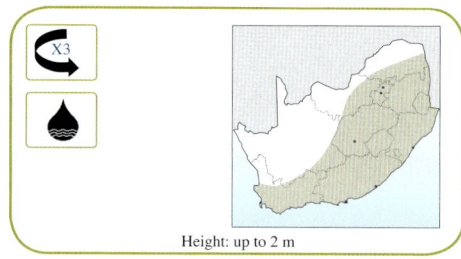

Height: up to 2 m

Origin and description: *R. crispus* is from Europe and is widely distributed in South Africa. The leaves of *R. crispus* are said to have medicinal properties. They are believed to relieve the pain of insect stings and bites, as well as the skin irritation caused by certain plants. The leaves die in winter and the plant hibernates in the rosette.

Impact: *R. crispus* is usually found in ditches and moist, waste places. It can become troublesome in orchards and vineyards, especially in the Western Cape.

Control: Because of its perennial nature, systemic herbicides should be used for the eradication

Other common names
curled sorrel * weëblaar * wildespinasie * idololenkonyane (Z) * ubuklunga (X) (Z)

of this weed. In orchards and vineyards it responds well to systemic, non-selective herbicides. Amitro-le mixtures are registered in the Western Cape for use in vineyards.

Rumex crispus

RESEDACEAE

Reseda lutea
dyer's rocket * katstert

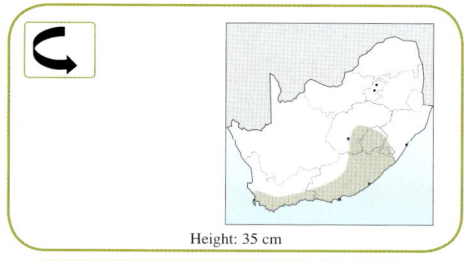

Height: 35 cm

Origin and description: Introduced from Europe and Asia, *R. lutea* is now a weed, especially in the Cape provinces. It is a perennial with a fleshy, tuberous root that survives the annual aerial parts. There are various subspecies and varieties.

Impact: It can cause problems during the harvesting of cereals as it remains green and clogs the combines. It causes particular problems around Riversdale in the southern Cape where it appears late and avoids conventional herbicide programmes.

Other common names
dyer's weed * weld * wild mignonette * wild rocket

Control: *R. lutea* is difficult to control and there are no specific herbicides registered for this weed. It appears to be moderately susceptible to the sulphonyl ureas that are used in cereals.

Reseda lutea

ROSACEAE

Agrimonia procera *(=A. odorata)*
agrimony * geelklits

Origin and description: *A. procera* is a native of Europe that has become a widespread weed of Asia, North America and now South Africa. There are about 15 species worldwide in the genus and they are called 'agrimony'. This species is a perennial plant although the aerial parts will die off in winter and regrow in the spring.

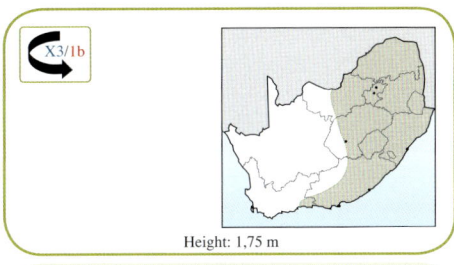

Height: 1,75 m

Other common names
akkermonie * bohome (S) * umakhuthula (Z)

Impact: *A. procera* is found throughout the more temperate regions of the country, on roadsides in particular, but can often be seen invading grassland and perennial pastures. The fruits are covered with hooked bristles that enable them to stick to clothing, hair and sheep's wool. Contaminated wool is downgraded.

Control: As it is a perennial plant, *A. procera* will usually require systemic herbicides unless it can be removed when still a seedling. Established plants should be dug out.

Agrimonia procera

RUBIACEAE

Galium spurium **subsp.** *africanum*
catchweed * kleefkruid

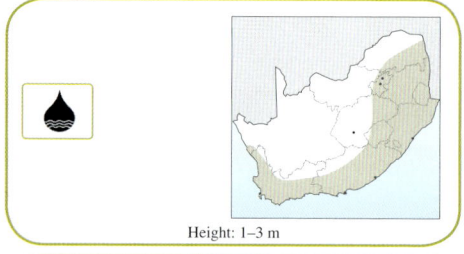

Height: 1–3 m

Origin and description: There are 400 species of *Galium* in the world and 13 of them are native to South Africa. *G. spurium* was originally thought to be indigenous, although it could have formed complexes and hybrids with the North American species *G. aparine*, which is a serious weed in Europe, Japan, Australia and South America.

Impact: In South Africa, this plant is an occasional weed, creeping over hedges, fences and crops, sometimes swamping desired vegetation. It is a weed of grain crops in the Cape. The leaves and stems are covered with little hooks that make them feel distinctly sticky. This makes them stick to clothes and wool, and contaminated wool is consequently downgraded.

Other common names
bedstraw * goosegrass * morarane (S)

Control: It can be controlled post-emergence in wheat and barley in the Cape.

Galium spurium

SOLANACEAE

Datura ferox
large thorn-apple * grootstinkblaar

Datura stramonium
common thorn-apple * olieboom

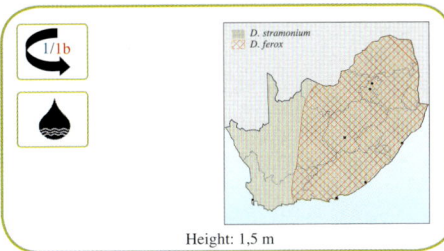

Height: 1,5 m

Origin and description: These two introduced species, *D. ferox* from Eurasia and *D. stramonium* from North America, are now declared weeds in South Africa. *D. stramonium* was reported in the Cape as early as 1714. These plants are widespread and serious annual weeds of many crops.

The main difference between the species is the length of spines on the fruit (*D. ferox* has the large thorns). Often the stems of *D. stramonium* tend to be purple. Forms with purple flowers were once placed in a separate species (*D. tatula*), but this is now considered a natural variation. The seed and seedlings are poisonous and human fatalities have been recorded in cases where people have eaten this plant either accidentally or deliberately. The plant is cultivated in Central Europe and South America for the production of atropine.

Impact: These plants are declared weeds not only because of their poisonous properties, but also because of their tall and aggressive growth habit. In maize, for example, both of these weeds are difficult to control and they contaminate the grain. One seed per 10 kg of maize will cause rejection, this being equivalent to approximately one plant per hectare.

Control: Being deep germinators, these weeds are not adequately controlled by many pre-emergence herbicides. The most reliable control is achieved with post-emergence herbicides. In annual crops, it is advisable to delay their application as long as it is practically possible in order to catch the late germinating individuals.

Other common names
D. ferox: long-spined thorn-apple * white stinkweed
D. stramonium: ditch weed * jimson weed * stinkwort * malpitte * iloqi (Z) * lechoe (S) * umhlavuthwa (X)

Similar species
D. innoxia 1/1b (hairy thorn-apple * moonflower) from Central America, also Category 1b but not very common
D. metel (devil's trumpet * white thorn-apple) of uncertain origin

Datura stramonium

Datura stramonium

Datura ferox

SOLANACEAE

Nicandra physalodes
apple-of-Peru * basterappelliefie

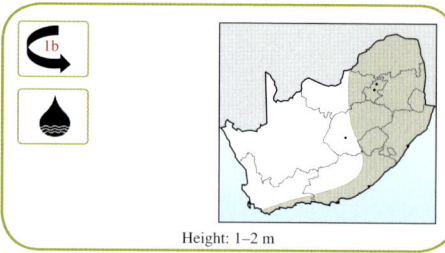
Height: 1–2 m

Origin and description: A native of Peru, *N. physalodes* was first recorded near Durban in 1862. It has since spread into most areas in South Africa, especially in the summer-rainfall regions, with the exception of the Free State. It is an annual plant and the fruit, which is eaten by birds, is a round yellow berry, enclosed in a bladder-like calyx.
Impact: *N. physalodes* is on record as a host to the root knot nematode, *Meloidogyne javanica*. It is a strong competitor and a weed of crops, gardens and waste areas
Control: *N. physalodes* is susceptible to conventional herbicides and shallow cultivation when in the seedling stage.

Other common names
rivabe * shoo-fly plant * bloubitter * wildebitter * linyooko (S) * umgabaganga omncane (Z) * umpungempu (X)

Nicandra physalodes

SOLANACEAE

Physalis angulata
wild gooseberry * wilde-appelliefie

Height: 0,5–1 m

Origin and description: Introduced from the Americas, *P. angulata* is now a widespread weed in South Africa. It is especially prevalent in the summer-rainfall regions north of the Karoo. It can grow to a height of well over 1 m. The fruit, which is like a green berry, is called a bacca and is found in the inflated, bladdery calyx.
Impact: It is a common but rarely a serious an-

Other common names
cutleaf groundcherry * wild physalis * winter cherry * kalkoengif * klapbessie

nual weed in the summer-rainfall areas. It is found in gardens, waste areas and many crops.
Control: *P. angulata* can sometimes be a strong competitor, but is relatively easy to control, being susceptible to conventional herbicides and shallow cultivation when still a seedling.

Similar species
P. peruviana ⟲ (Cape gooseberry) cultivated in the Cape for its fruit since as far back as the 18th century but with the potential to be weedy

Physalis angulata

SOLANACEAE

Physalis viscosa
sticky gooseberry * klewerige appelliefie

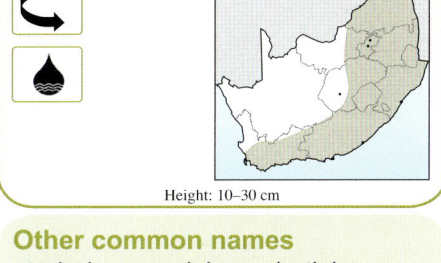

Height: 10–30 cm

Other common names
starhair groundcherry * sticky physalis * wild gooseberry * kusebere (S)

Origin and description: Introduced from the Americas, *P. viscosa* is now a widespread weed in South Africa. It is a perennial plant that spreads into large patches on account of its ramifying underground rhizomes. It never grows very tall (about 30 cm). *P. viscosa* is squatter, tougher and more difficult to remove than *P. angulata*. The sticky fruit is also enclosed in an inflated bladdery calyx. It has been suspected of tainting dairy products.

Impact: It is common in waste areas, on roadsides and in perennial crops in most areas of South Africa. It is a serious weed of sugarcane.

Control: As with most plants with perennial underground systems, *P. viscosa* is very difficult to control, especially in perennial crops. It causes serious problems for South Africa's sugar farmers as it resists the conventional herbicide programmes. Individuals that have escaped or patches of *P. viscosa* should be spot-sprayed with systemic, non-selective herbicides or with triclopyr.

Physalis viscosa

SOLANACEAE

Solanum elaeagnifolium
silverleaf bitter apple * satansbos

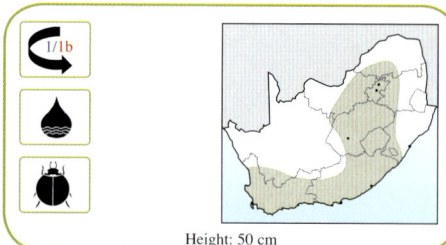

Height: 50 cm

Origin and description: *S. elaeagnifolium* was first recorded in this country in 1952, although some authorities believe it was identified at Wolmaransstad as early as 1919. It was probably introduced from the Americas with hay and has now spread to large parts of the Free State, Mpumalanga and the Eastern and southwestern Cape.

Impact: *S. elaeagnifolium* is an important perennial weed that occurs mainly on disturbed soil, neglected lands, in grazing camps, along roads and in water furrows. Firebreaks that have been ploughed or disked along fence lines provide an ideal environment for the seeds dropped by birds that sit on the fences. When it occurs in cultivated land, it can completely swamp the planted crop. The young fruits and leaves are poisonous and it has been suspected as being a source of potato viruses.

Other common names
bloubos * silwerblaarbitterappel
Similar species
S. panduriforme (bitter apple) is indigenous and found in the summer-rainfall region. It does not have the reddish prickles on the stems and stalks

Control: In recent years the government has spent large sums of money on the control of *S. elaeagnifolium*, without significant success. Its very extensive root system, which penetrates to depths up to 3 m or more, and its ability to propagate from its roots, make this an extremely difficult weed to control. Fluroxypyr is registered as a foliar application. Biological control is showing promise and several defoliating beetles are being studied by the Department of Agriculture. The plants, with as much of the root as possible, should be removed before seeds are formed. Continuous removal will debilitate the plant and prevent the roots from forming shoots.

Solanum elaeagnifolium

SOLANACEAE

Solanum nigrum (=*Solanum retroflexum*)
nightshade * nastergal

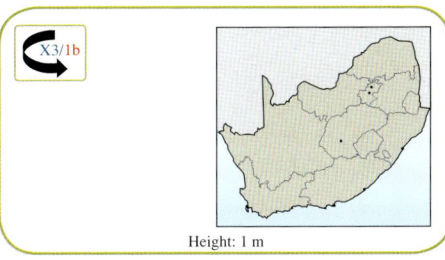

Height: 1 m

Origin and description: *S. nigrum* is probably from Europe and is now a cosmopolitan weed common throughout South Africa. The ripe fruit can be eaten when cooked, but the unripe fruits are poisonous. It is when the berries are green that the plant can be a danger to livestock.
Impact: *S. nigrum* is not usually very troublesome, although it is frequently found in fields under cultivation and like many solanaceous plants, it is a host to nematodes.
Control: This weed is easy to control by cultivation and with conventional pre- or post-emergence herbicides. It is also easy to remove by hand.

Other common names
black nightshade * inkberry * sobosobo berry * galbessie * ixabaxaba (N) * musaka (Sh) * seshoabohloko (S) * umsobosobo (X) (Z) * umsobo (Si)

Solanum nigrum

SOLANACEAE

Solanum pseudocapsicum
Jerusalem cherry * wilderissie

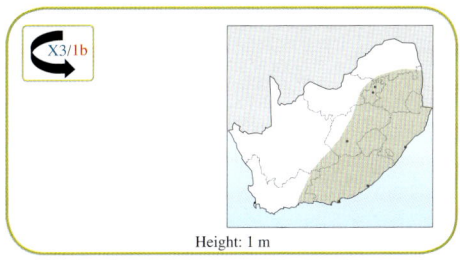

Height: 1 m

Origin and description: *S. pseudocapsicum*, one of maybe 2 000 species of *Solanum*, is an invasive, perennial, evergreen and alien plant,

303

Other common names
false capsicum * Madeira winter cherry * Natal cherry * winter cherry * bosgifappel * gifbessie * Jerusalemkersie

having originated in Mediterranean Europe, probably Madeira, and was introduced as an ornamental. The bright red berries, which persist for many months, are poisonous. The plant usually occurs in the shade of large trees.

Impact: *S. pseudocapsicum* is poisonous to animals and can replace indigenous vegetation. It should not be confused with the cherry tomato, since the attractive red berries can cause unpleasant, but not fatal symptoms in humans as well.

Control: These plants are relatively easy to control by cultivation and with conventional pre- or post-emergence herbicides. They are also easy to remove by hand.

Solanum pseudocapsicum

URTICACEAE

Urtica dioica
common stinging nettle * gewone brandnetel
Urtica urens
small stinging nettle * kleinbrandnetel

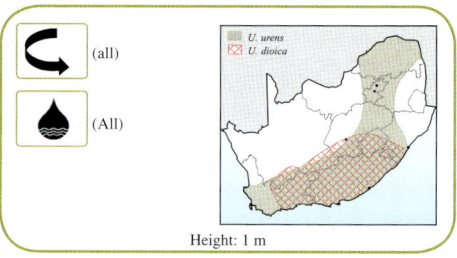

Origin and description: These two species of *Urtica* originated from Europe. Despite the names, *U. urens* is far more common than *U. dioica*.

Impact: *U. urens* is smaller and an annual. It is found throughout South Africa, but is more common in the Western Cape, occurring in waste places, damp areas and in orchards and vineyards. The perennial *U. dioica* is restricted to the Cape provinces. It has a creeping rootstock from which stems are produced. Although both species cause painful stings, they are in fact edible, especially when young.

Control: *U. urens* is very susceptible to contact herbicides such as paraquat and also to pre-

Other common names
U. dioica: Swedish hemp * tall nettle * brandneuker * bobatsi (S)
U. urens: bush stinging nettle * bubati (P) * imbathi (Z) * isibathi (Si) * umbabazane (Si) (X)

Urtica urens

emergence herbicides. *U. dioica* requires systemic herbicides for effective control due to the underground rhizomes of the plant.

Urtica urens

VERBENACEAE

Stachytarpheta urticifolia
nettleleaf velvetberry

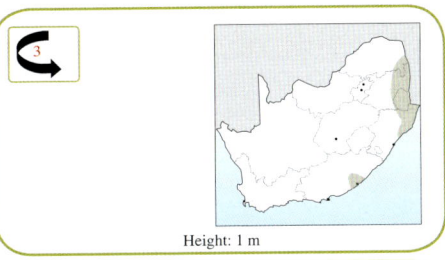

Height: 1 m

Origin and description: Of South American origin, *S. urticifolia* is one of several species of this genus that are popular garden plants throughout the world and two of them have now escaped and have become naturalised in South Africa.
Impact: The flowers of this plant are very attractive to bees, butterflies and birds, creating the potential of being a distraction from indigenous species. It can be found invading forest margins,

Other common names
blue porter weed * blue rat's tail * snakeweed
Similar species
S. mutabilis (changeable velvetberry * red porter weed), also from the Americas and also naturalised in South Africa

roadsides and waste areas in the subtropical eastern areas. Seeding profusely, it has escaped from tourist camps and invaded nearby vegetation in the Kruger Park.
Control: There are no herbicides recommended for this weed. Unwanted plants should be removed by hand.

Stachytarpheta urticifolia

VERBENACEAE

Verbena bonariensis
purple top * blouwaterbossie
V. officinalis
common vervain * verbain

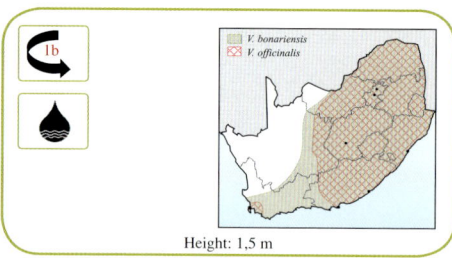

Height: 1,5 m

Origin and description: *V. bonariensis* is a native of South America and was naturalised in the Cape more than a century ago. *V. officinalis* is very similar in habit and distribution, but with much smaller flowers. It is now widespread, occurring in all provinces.

Impact: It is a weed of gardens, roadsides, waste places and fallow lands. It can become a particular problem in lands under conservation tillage. Its presence on a site indicates a suitable habitat for the very dangerous pompom weed (*Campuloclinium macrocephalum*).

Control: When young, *V. bonariensis* can easily be controlled by cultivation and with the usual broadleaf weed herbicides. The mature plant, however, is tough, wiry and more tolerant to herbicides.

Other common names
V. bonariensis: tall verbena * wild verbena
V. officinalis: European verbena * seona-se-seholo (Ss)
Similar species
V. brasiliensis 1b (from Brazil), very similar but less common

Verbena bonariensis

Verbena bonariensis

Verbena officionalis

ZINGIBERACEAE

Hedychium coronarium
white ginger lily * witgemmerlelie

Hedychium gardnerianum
kahili ginger lily * kahiligemmerlelie

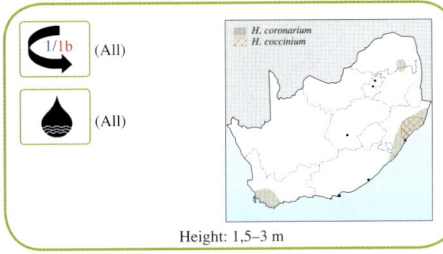

Height: 1,5–3 m

Origin and description: There are several species and cultivars of *Hedychium* from Asia that were introduced as ornamentals. These perennial plants have now become established in parts of KwaZulu-Natal and recently in the Kruger National Park and parts of Limpopo. The closely related *Alpinia zerumbet* from Eastern Asia is also potentially invasive and is found wild in parts of the East Coast and Kruger Park, but is common in gardens.

Impact: The ginger lilies are found on shady and damp roadsides and in waste places, with some dense infestations in the forests near Pietermaritzburg. The plants grow from tough rhizomes, which make them extremely difficult to eradicate once they have become established. *H. gardnerianum* is No. 40 on the list of the *World's Worst Invasive Species*.

Control: There are no herbicides registered for these plants and none appear effective. The plants should be dug up and the rhizomes removed in their entirety.

Similar species
H. coccineum 1/1b (red ginger lily * rooigemmerlelie)
H. flavescens 1/1b (yellow ginger lily * geelgemmerlelie)
Alpinia zerumbet 3 (pink porcelain lily * shell ginger * pienkporseleinlelie * skulpgemmer)

Hedychium coronarium

Alpinia zerumbet

Hedychium gardnerianum

SPREADING OR FLAT-GROWING HERBS

Within the section, the plants are arranged alphabetically according to family and genus.

SPECIES	Exudes white sap	Fruits that are burs or that stick	Fluffy seeds that blow away easily	Seeds in multiseeded capsules	Red or pink flowers	Yellow flowers	Purple flowers	White flowers	Blue flowers	Green or inconspicuous flowers
Taraxacum officinale, p. 320	■		■			■				
Euphorbia spp., p. 331	■			■						■
Medicago spp., p. 332		■				■				
Tribulus terrestris, p. 349		■				■				
Acanthospermum spp., p. 316		■						■		■
Alternanthera pungens, p. 311		■								■
Emex australis, p. 340		■								■
Tridax procumbens, p. 319			■			■		■		
Anagallis arvensis, p. 343				■	■				■	
Oxalis corniculata, p. 338				■		■				
Oxalis pes-caprae, p. 338				■		■				
Spergula arvensis, p. 322								■		■
Stellaria media, p. 323				■				■		
Veronica persica, p. 346				■					■	
Modiola caroliniana, p. 336					■					
Tropaeolum majus, p. 346					■					
Tradescantia spp., p. 327					■			■		
Vinca spp., p. 315, 316					■					
Oxalis latifolia, p. 339					■					
Cotula australis, p. 317						■				
Duchesnea indica, p. 344						■				
Gisekia pharnacioides, p. 334						■				
Lotus subbiflorus, p. 332						■				
Portulaca oleracea, p. 341						■				
Citrullus lanatus, p. 329						■				
Cucumis myriocarpus, p. 330						■				

Spreading or flat-growing herbs

	Exudes white sap	Fruits that are burs or that stick	Fluffy seeds that blow away easily	Seeds in multiseeded capsules	Red or pink flowers	Yellow flowers	Purple flowers	White flowers	Blue flowers	Green or inconspicuous flowers
Spilanthes decumbens, p. 318						■				
Boerhavia spp., p. 336							■	■		■
Verbena aristigera, p. 348							■	■		
Trifolium spp., p. 333							■	■		
Polygonum aviculare, p. 340							■			■
Malva parviflora, p. 334							■			
Guilleminea densa, p. 313								■		■
Richardia brasiliensis, p. 344								■		
Eclipta prostrata, p. 318								■		
Gomphrena celosioides, p. 312								■		
Commelina benghalensis, p. 326									■	
Callisia repens, p. 324										■
Chenopodium carinatum, p. 324										■
Coronopus didymus, p. 321										■
Atriplex semibaccata, p. 324										■
Dichondra micrantha, p. 328										■
Hydrocotyle americana, p. 313										■
Hypoestes phyllostachya, p. 310										■

ACANTHACEAE

Hypoestes phyllostachya
polka-dot plant

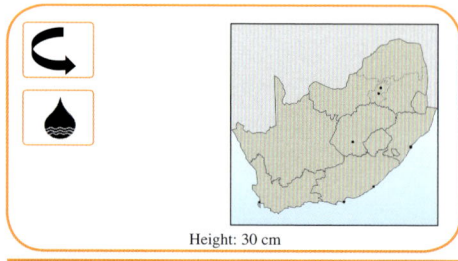
Height: 30 cm

Origin and description: The polka-dot plant is a tropical herb, native to Madagascar. It was introduced as an ornamental groundcover.
Impact: This plant has become naturalised and is now found invading moist and shady areas in ditches, along roadsides and in waste areas, usually near gardens. It is considered an emerging weed by the Department of Agriculture and it should be cultivated with caution. It should not be planted at all in conservation areas as it can spread into nearby forest. It can smother and replace indigenous vegetation.
Control: This plant should be cultivated with caution and unintended plants removed.

Hypoestes phyllostachya

AIZOACEAE

Zaleya pentandra
African purslane * muisvygie

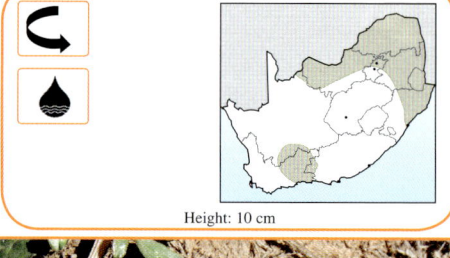
Height: 10 cm

Origin and description: Of uncertain origin but probably an exotic, *Z. pentandra* arrived in southern Africa centuries ago and is found throughout tropical Africa and into the Middle East. It is a perennial, evergreen plant and is a widespread weed in the subtropical parts of South Africa. It has been recorded in all provinces except the Free State.
Impact: *Z. pentandra* is commonly found in waste areas, old lands and disturbed soil, especially sandy soil. It can become dense and competitive on occasions as it has a strong pioneering nature.
Control: There was one herbicide registered for its control, that is used in cotton fields, but it appears to have lapsed. Cultivation will destroy this weed quite easily.

Zaleya pentandra

AMARANTHACEAE

Alternanthera pungens
(=A. repens)
khakiweed * kakiedubbeltjie

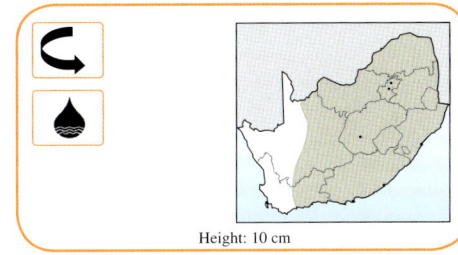

Height: 10 cm

Other common names
khaki burweed * paperthorn * ikhungele (Z)

Origin and description: *A. pungens* is a native of South America and has spread to most areas of southern Africa. It is said that *A. pungens* was introduced as an impurity in fodder brought in for the horses of British troops ('khakis': hence the common name) during the Anglo-Boer War. The species followed the railway system and was usually first seen in areas near the stations, being easily spread on grain sacks as they were loaded and unloaded. The common name of 'khaki weed' is also commonly used for *Tagetes minuta* and *Inula graveolens*.

Impact: *A. pungens* is a very unpleasant weed, as the seeds can penetrate bare feet and even stick to rubber-soled shoes. It occurs in such places as lawns, gardens, pathways and playing fields and may not be noticed until trodden or sat upon, hence the names referring to 'thorn' or '*dubbeltjie*'. (The old name '*repens*' refers to its low, creeping growth habit.) The burrs also cause downgrading of wool. It has a large taproot and roots at the nodes. It forms large mats, which are difficult to remove. It is of agronomic importance in the Mpumalanga lowveld and the Limpopo Province where it is a weed in many different crops.

Control: This weed is controlled effectively by many pre-emergence herbicides but becomes tolerant to post-emergence herbicides as it matures. It should be removed when small.

Alternanthera pungens

Spreading or flat-growing herbs

AMARANTHACEAE

Gomphrena celosioides
prostrate globe amaranth * mierbossie

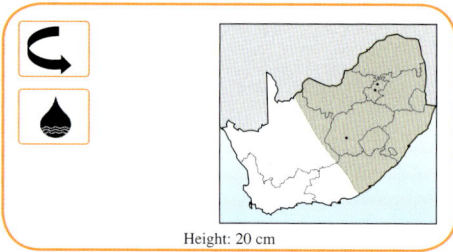

Height: 20 cm

Origin and description: *G. celosioides* is a native of tropical South America. It is now a cosmopolitan weed and widely distributed in the summer-rainfall region of South Africa.

Impact: It is commonly found in waste places and on roadsides but also in crops and gardens, particularly in neglected lawns and grassy areas. This plant is quite different from the cultivated bachelor's button (*G. globosa*), which has purple flowers.

Control: It is a weak competitor that rarely requires chemical control, but being a perennial, may need a systemic herbicide once it is well established. Seedlings are easy to remove by cultivation.

Other common names
bachelor's button * kruip-knopamarant * intandangulube (Z) * unyawolwengulube (Z)

Gomphrena celosioides

AMARANTHACEAE

Guilleminea densa (=*Brayulinea densa*)
carrot weed

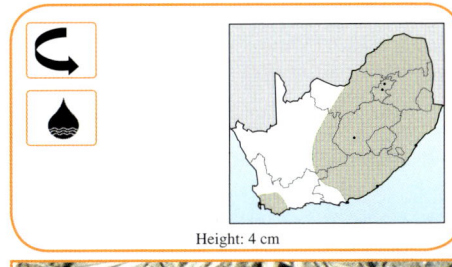
Height: 4 cm

Origin and description: *G. densa* was introduced from Central America. It is now a very troublesome weed in various parts of South Africa, but especially in Mpumalanga. It was first recorded in 1909 in the former Transvaal.

Impact: *G. densa* is an annual plant, reproducing only from seed. The large, fleshy underground parts, however, can survive from year to year and for this reason it is a difficult weed to control. It causes serious problems on sports turf and golf courses, but is also found on roadsides, disturbed veld and on other sites.

Control: Chemical control in grass swards will require repeated applications of selective herbicides. Removal by hand is difficult, because the plant can regrow from roots left behind. When using herbicides, it is always better to fertilise and avoid mowing for as long as possible prior to application. This allows the weed to become large, lush, vigorous and more capable of absorbing and translocating the herbicide.

Guilleminea densa

APIACEAE

Hydrocotyle americana
navelwort * perdekloutjies

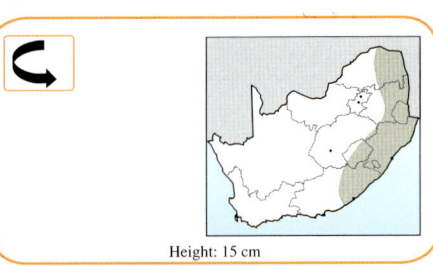
Height: 15 cm

Origin and description: *H. americana* is a cosmopolitan weed of uncertain origin. It is probably exotic. It is a perennial, evergreen plant with shiny leaves that are often yellow-green. It is widely distributed in the summer-rainfall areas and commonly found in damp places in gardens, lawns, golf greens and in any moist and sheltered spot. It can grow up to about 15 cm, but will flourish even under heavy mowing, such as on golf greens, although the leaves often become much smaller. It flowers when allowed to grow lank, and will produce seeds. Once established, spreading is accomplished by running stems that root at the nodes and by rhizomes.

Impact: This plant can become a serious problem in shaded and damp areas of lawns and golf

Similar species

There are several species of plant that look very similar and are also problems in gardens and turf These include *Centella asiatica* (marsh pennywort – pantropical) and *Dichondra micrantha* (kidney weed or wonderlawn – possibly indigenous). The latter is sometimes used as a lawn itself (see page 328).

greens (for example). It can completely replace the grass species.
Control: Because of its perennial nature, control is difficult. Repeated applications of selective herbicides may be necessary in grass swards such as golf greens.

Hydrocotyle americana

Centella asiatica

APOCYNACEAE

Vinca major
greater periwinkle * maagdepalm

Catharanthus roseus (=Vinca rosea)
Madagascar periwinkle * begraafplaasblom

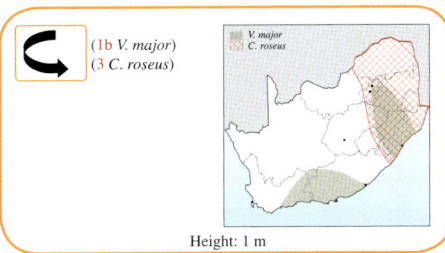

(1b *V. major*)
(3 *C. roseus*)

Height: 1 m

Origin and description: These periwinkles are well-known perennial garden plants used as an ornamental groundcover that have now escaped into the wild. *Vinca* species are originally from Europe and Asia and are now found throughout South Africa. *V. major* has slightly larger flowers, as opposed to *V. minor*, with minutely hairy leaf margins. Otherwise they are very similar in origin and appearance. *C. roseus* is in fact an endangered species in its home range in Madagascar on account of habitat destruction caused by slash-and-burn agriculture. It has been and still is grown commercially in many parts of the world, including KwaZulu-Natal, for the extraction of medicinal compounds. The flowers are pink or white. From the Afrikaans name, it is clearly a popular plant for putting on graves, from where it can drop seed and establish itself.

Impact: These plants invade moist, shaded places and, once established, they form extensive mats that smother all other vegetation and eventually replace it. *V. major* is particularly common in the resorts of the Drakensberg, where it invades nearby forests and shady waste areas.

Control: These plants do not have a very strong root system and can be easily removed by hand or other physical methods. The sterility or non-invasiveness of cultivars has to be investigated.

Other common names
V. major: band plant * blue buttons * gewoneopklim
C. roseus: oldmaid * pink periwinkle * vinca * maandrosie * soldatenblom * isishushlungu (Z)

Similar species
V. minor 1b (myrtle * running myrtle), also from Eurasia

Spreading or flat-growing herbs

Vinca major

Catharanthus roseus

Vinca major

ASTERACEAE

Acanthospermum australe
prostrate starbur * kruipsterklits

Acanthospermum hispidum
upright starbur * regopsterklits

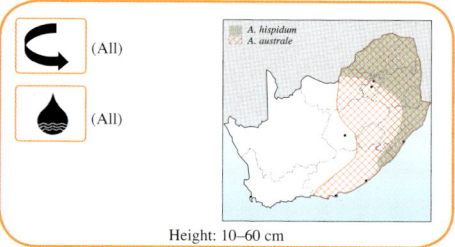
Height: 10–60 cm

Origin and description: These 'starburs' are of South American origin and were probably introduced in imported fodder during the Anglo-Boer War. They are now common weeds of crops in the warmer parts of South Africa and are of particular importance in the northern and eastern regions. *A. australe* now includes what was *A. brasilium*, which was always very similar but had eight or nine seeds per inflorescence instead of five. *A. hispidum* is a closely related annual species with a more upright growth habit. Because of this, it is more likely to contaminate a crop.

Impact: The weed's burs contaminate sheep's wool and cause subsequent downgrading. *A. hispidum*, in particular, can contaminate cotton lint and cause the downgrading thereof.

Control: Shallow cultivation and most broadleaf herbicides are successful in controlling these weeds.

Other common names
A. australe: eight-seeded starbur * Paraguay bur * donkieklits * Jode-luis * setla-bocha (S)
A. hispidum: goat's head * kleinkankerroos * Tsumeb onkruid * bima (N) ubima (Sh)

Acanthospermum hispidum

Acanthospermum australe

Acanthospermum brasilium

ASTERACEAE

Cotula australis
cotula
Cotula turbinata (=*Cenia turbinata*)
goose daisy * ganskos

Origin and description: Originating in Australia, *C. australis* now occurs throughout most of the southern hemisphere. In South Africa, it was originally observed at Grahamstown in 1868; since then it has gradually spread northwards, possibly as far as Swaziland. It flowers throughout the year and for this reason it is noticed more in winter than at other times. *C. turbinata* is an indigenous weedy plant found mainly in the winter-rainfall areas in the Western Cape. The plant has a distinctive smell and the flowers can be yellow or white.

Impact: *C. australis* is usually a weed of damp and shady places but also appears able to withstand direct sunlight and heavy mowing. It is therefore frequently found in lawns and golf greens where it can be a problem. *C. turbinata* occurs as a weed in gardens, orchards, waste places, along roadsides and in agricultural crops, occasionally becoming a problem in cereals.

Control: No herbicides have been registered to control *C. australis*, but in most situations it

(All)
(*C turbinata*)

Height: 20 cm

Similar species
Of the 45 species of *Cotula* found in South Africa and South America, 43 are native to South Africa
C. tenella is also covered by a herbicide registration
Do not confuse *C. australis* with *Coronopus didymus* (see page 321)

Cotula turbinata

Cotula-spesies

can easily be removed by cultivation. In lawns and on golf greens, however, it appears fairly hardy and repeated applications of selective broadleaf herbicides may be necessary. *C. turbinata* is usually only of nuisance value and it can easily be removed by cultivation, although it is the subject of a herbicide registration.

Cotula australis

ASTERACEAE

Eclipta prostrata (=*E. alba*)
eclipta

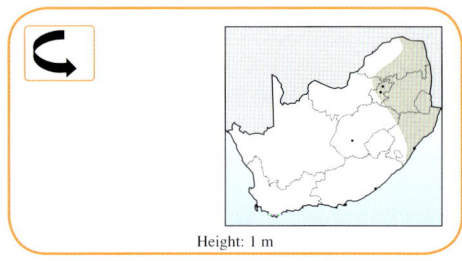

Height: 1 m

Origin and description: This plant originates from Europe and Asia and is now cosmopolitan in the warmer parts of the world. It has been known in South Africa for over 80 years. It is distributed mainly in the summer-rainfall region.

Impact: It is an annual weed of many situations, but preferring damp places and can become dense and competitive. Commonly used as a traditional love-charm and in sorcery. It is also reputed to have strong medicinal properties.

Control: There are no registered control measures but this weed is unlikely to become eco-nomically significant. Cultivation and conventional herbicides should give adequate control.

Other common names
false daisy * iphamphuce (Z) * ikhambi lakwangcolosi (Z) * ungcolozi (Z)

Eclipta prostrata

ASTERACEAE

Spilanthes decumbens
spilanthes * madeliefie

Origin and description: A native of Uruguay, *S. decumbens* is a relatively recent introduction to South Africa but it is now established and spreading. It was first recorded at East London in 1926 but has spread to most parts of the eastern half of the country. It often forms dense colonies, especially on roadsides. The other species of invasive daisies (see below) have a similar history and impact.

Impact: These are good examples of exotic species establishing themselves and having the potential to spread into a variety of niches and displace indigenous vegetation.

Control: *S. decumbens* is controlled effectively by triazines, but only pre-emergent. The best method of control of all these daisies is probably to first clear established plants by hand and then to use systemic chemicals on the regrowth. Unwanted plants should be physically removed and further plantings should be avoided.

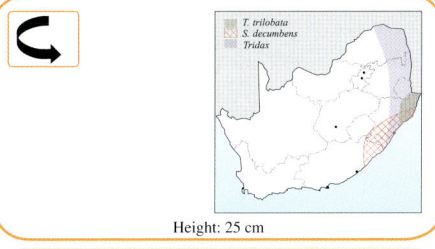

Height: 25 cm

Other common names
isisinini (Z)

Similar species of invasive daisies
Spilanthes mauritiana ↻ from the tropics
Sphagneticola trilobata ↻ 1/1b (=*Wedelia trilobata, Thelechitonia trilobata*) (Singapore daisy * Singapoer-madeliefie) from tropical America is an invasive weed in KwaZulu-Natal
Tridax procumbens ↻ (tridax daisy * aster) from Central America is invading orchards and waste places in the eastern parts

Spreading or flat-growing herbs

Spilanthes decumbens

Tridax procumbens

Sphagneticola trilobata

ASTERACEAE

Taraxacum officinale
common dandelion * perdeblom

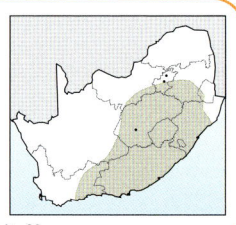
Height: 30 cm

Origin and description: Of European origin, *T. officinale* arrived in the Cape as early as 1700 and is now a common perennial weed throughout South Africa. The leaves are used as a salad and the roots are used as a medicine. The roots can also be roasted and used as a coffee substitute when coffee is unavailable, as was the case during World War I and II. There are 12 indigenous species of *Taraxacum* that occur in South Africa that look very similar, and one other alien species, *T. serotinum*, but none of them are as widespread or as common. They all have the common name 'dandelion' and all exude a white latex when broken.

Impact: *T. officinale* is found on roadsides, in waste places, in gardens and occasionally in crops and orchards. It is a weak competitor but can be particularly troublesome in lawns and occasionally in maize under conservation tillage. It is easily spread by the seeds that are blown around by even a slight breeze.

Control: Young plants are easy to control with post-emergence, systemic broad-leaf herbicides

Other common names
lion's tooth * platdissel * irwabe lenyoka (X) * umkhothane (X) * umashwababa (Z)

Similar species
Do not confuse it with *Hypochaeris* spp. (cat's ears * wild lettuce), sometimes called 'false dandelions'

but, once established, they become more difficult to control on account of their strong tap roots. Mature plants can be hand-pulled but will regenerate if even just a broken piece of the root is left behind. The soil should be watered before this is attempted to minimise this risk. It is said that vinegar will also kill dandelions.

Taraxacum officinale

BRASSICACEAE

Coronopus didymus
carrot weed * peperkruid

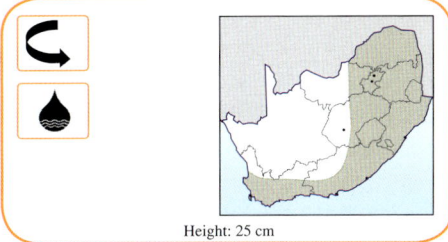
Height: 25 cm

Other common names
swinecress * wild carrot * peperbossie

Origin and description: Introduced from South America, this is now a common annual weed throughout South Africa. Although generally growing fairly flat, it can become about 30 cm tall under favourable conditions and can create a tangled mass. The flowers are indistinct and produce what look like small bunches of grapes. These are in fact lines of paired seeds, hence the name '*didymus*', which is Latin for 'twin'. When crushed, the leaves smell like a pigsty, hence the alternate common name 'swinecress'.

Impact: It can become quite a problem in winter-grown, irrigated vegetables, especially along the KwaZulu-Natal escarpment. It is also well known in gardens, orchards and winter wheat. In maize lands, winter weeds such as this provide cutworm moths with a place to lay their eggs, thereby increasing the potential cutworm problem the following season.

Control: *C. didymus* is relatively tolerant to selective herbicides, especially those used in wheat. Particular care should therefore be taken with the choice and application of cereal herbicides; otherwise young plants are susceptible to most of the contact herbicides.

Coronopus didymus

Spreading or flat-growing herbs

Coronopus didymus

CARYOPHYLLACEAE

Spergula arvensis
corn spurry * sporrie

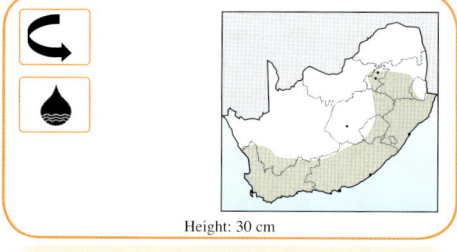
Height: 30 cm

Origin and description: *S. arvensis* is a major annual weed of wheat in Europe and was introduced to South Africa by Jan van Riebeeck as a fodder plant in about 1653. It is now widespread in the country. Seeds excavated from Iron Age sites in Denmark, dating back 2 000 years, were found to be still viable. Livestock and poultry are said to eat it readily.

Impact: It is an occasional problem plant in pastures, cereals and gardens. It is a weak competitor and has not achieved the same pest status as in Europe.

Control: *S. arvensis* is susceptible to pre-emergence herbicides, but effective wetting with post-emergence herbicides is difficult on account of the plant's fine, waxy leaves. This weed is well controlled by the sulphonyl urea group of herbicides and by heavy liming.

Other common names
Hollandsche gras * perdegras * bolepo-ba-seokha-sa-merung (S)

Spergula arvensis

322

Spergula arvensis

CARYOPHYLLACEAE

Stellaria media
chickweed * sterremuur

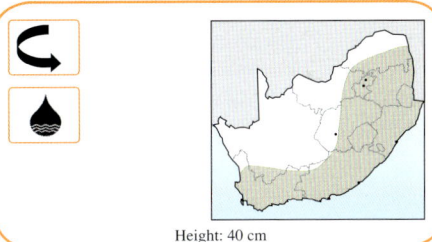
Height: 40 cm

Origin and description: A native of Europe, *S. media* probably arrived with the early settlers. It is a common annual weed in gardens, particularly in damp and shaded areas. It is also common in vineyards in the Western Cape.

Impact: It can form large mats with its sprawling growth habit. It is especially common in the Western and southern Cape and can cause serious problems in wheat as it interferes with harvesting by clogging the combine harvesters. Because of the more favourable climate than in Europe, it is generally a more serious problem here than in that region.

Control: It is easily controlled by cultivation and is susceptible to many herbicides.

Other common names
satin flower * starwort * stitchwort * tonguegrass * white bird's eye * winterweed * mier * muggiesgras * qoqobala (S)

Similar species
Anagallis arvensis (pimpernel) see page 343
Euphorbia peplus (stinging milkweed) see page 266

Stellaria media

CHENOPODIACEAE

Atriplex semibaccata
Australian salt bush * Australiese brak(bossie)

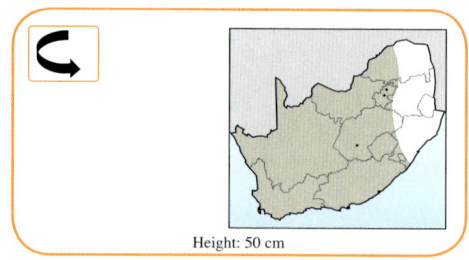

Height: 50 cm

Origin and description: Introduced from Australia at the beginning of the 20th century as a fodder crop, this plant is now common as a weed on disturbed ground throughout the coastal and dry areas of the Cape and up into southern KwaZulu-Natal. *A. semibaccata* is readily grazed by sheep.
Impact: It replaces indigenous vegetation.
Control: It is a perennial woody and semideciduous bush when mature and is not effectively controlled by contact herbicides. For this reason, it can be a problem in places such as orchards where these chemicals are widely used. Control should be initiated when the plant is young and tender.

Other common names
berry saltbush * wild lucerne * brak * kruipsoutbos * Robertsonbrak

Atriplex semibaccata

CHENOPODIACEAE

Chenopodium carinatum (=C. bontei)
green goosefoot * groenhondebossie

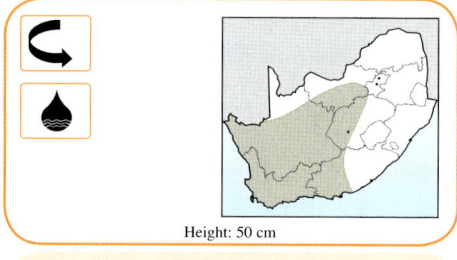

Height: 50 cm

Origin and description: *C. carinatum* came from Australia at the beginning of the 20th century and is widespread throughout South Africa. Young plants and leaves are edible.
Impact: It is a serious weed of many crops, especially vegetables, in Mpumalanga and the Cape Provinces.

Other common names
creeping goosefoot * keeled goosefoot

Chenopodium carinatum

Control: *C. carinatum* is easily controlled by cultivation while still in the seedling stage. It is susceptible to many pre- and post-emergence herbicides.

Chenopodium carinatum

COMMELINACEAE

Callisia repens
creeping inch plant

Origin and description: A native of Central and South America and the West Indies, this plant was introduced into South Africa as an ornamental ground cover and for growing in hanging baskets.

Impact: Having escaped into the wild, the creeping inch plant is in fact a very aggressive ground cover and can quickly multiply and spread, totally eliminating all other vegetation over which it spreads. It is now commonly found in disturbed sites around human habitation, as well as on sandy river banks and in riverbeds. It can also invade and block ditches and drains.

Control: Control is best achieved manually, with chemical follow-ups, if necessary, on seedlings that reappear in cleared areas. The carpet it produces is not too difficult to remove manually. Proposed as a Category 1b plant in order to prevent further invasions.

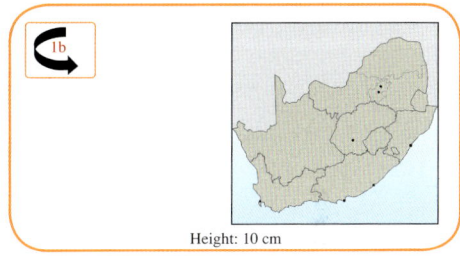

Height: 10 cm

Other common names
baby's tears * itsy bitsy inch vine * turtle vine

Callisia repens

COMMELINACEAE

Commelina benghalensis
wandering Jew * wandelende Jood

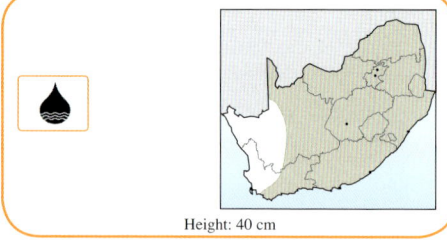

Height: 40 cm

Origin and description: *C. benghalensis* is one of the most serious and widespread weeds of South Africa. It is probably indigenous and, although common throughout South Africa, it is most common in the summer-rainfall regions. It has small blue flowers that only last one or two hours. Because of its vigorous growth and climbing habit, *C. benghalensis* is often grown as an indoor pot plant. In times of famine, the leaves are eaten by some people.

Impact: *C. benghalensis* is a significant weed of many crops. It competes strongly for moisture and in maize it is a major late-season weed. In other crops, such as tea, it can climb into the bush, interfering with growth and harvesting. Because of its creeping growth habit, one single plant can become very large and competitive.

Control: Despite the fact that most of the vegetative growth and aerial seeds are produced above ground, the plant also has burrowing runners, that are capable of producing underground flowers and seeds which are considerably larger than those produced above ground. This is the main reason *C. benghalensis* is so difficult to control, as plants that grow from underground seeds cannot be killed by surface-applied herbicides until they emerge. This weed can also produce vegetatively, regenerating from small pieces that are broken from the parent plant. Control is best achieved with post-emergence herbicides and deep cultivation. If control is attempted by means of cultivation, the cut stems must be buried deeper than 5 cm to kill them. Shallow cultivation only tends to spread the weed. It can be controlled post-emergence with certain herbicides and the addition of 2,4-D or MCPA to glyphosate will help.

Other common names
Benghal wandering Jew * commelina * blouselblommetjie * damba (V) * idambizo (Z) idlebendlele (Z) * khotswana (S) * uhlotshane (X)

Similar species
Several other indigenous species of *Commelina* with both yellow and blue flowers but they are not considered weedy

Commelina benghalensis

COMMELINACEAE

Tradescantia fluminensis
wandering Jew * wandelende Jood

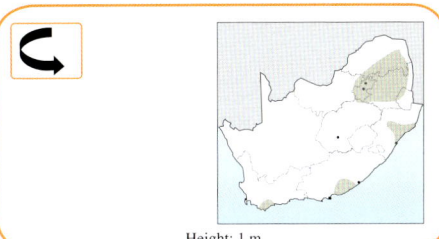
Height: 1 m

Origin and description: *T. fluminensis* is endemic to the tropical rain forests of Brazil. It was introduced into South Africa as an ornamental and has now become widely naturalised. It can produce seed, but is more commonly spread by being dragged around by stock, other animals and humans. It propagates easily from fragments.

Impact: *T. fluminensis* thrives in areas of low light intensity and can form dense mats under forest canopies. This mat smothers low-growing plants and prevents the natural regeneration of the taller forest species. If left unchecked, it has the potential to destroy native woodland and can take over gardens if not attended to carefully. The succulent stems break easily at the nodes and establish themselves wherever they land on moist soil. It does not seem to be a significant weed of crops. It is recognised as an emerging weed along the Garden Route.

Control: The foliage is easy to remove but this should be done at frequent intervals. This can lead to eventual destruction. Herbicides would also have to be applied repeatedly, but this is not desirable in a wild environment.

Similar species
T. pallida (purple heart)
T. zebrina 1b (wandering Jew * wandelende Jood)

Tradescantia zebrina

Tradescantia pallida

Tradescantia fluminensis

Spreading or flat-growing herbs

CONVOLVULACEAE

Dichondra micrantha (=*D. repens*)
wonderlawn * wondergrasperk

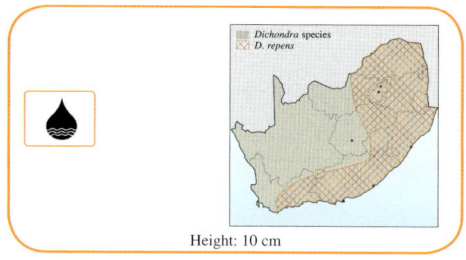
Height: 10 cm

Origin and description: One of only a few species of *Dichondra* that occur throughout the world and the only one to occur in South Africa. They are all native to tropical and warm temperate regions and although the origin of *D. micrantha* is not certain, it is thought to be indigenous to southern Africa. It is also considered as indigenous in Australia and some authorities maintain it is from North America and Asia.

Impact: These plants are creeping members of the convolvulus family and are commonly used as grass substitutes for lawns in warm climates. Unfortunately they are very aggressive competitors and if not wanted as a lawn, they out-compete the planted grass and are difficult to eradicate. Even if they have been planted as a lawn, they spread to flower beds and can become a serious nuisance. *D. micrantha* is also a significant weed in lucerne in the southern Cape. This

Other common names
dewdrop lawn * kidney grass * kidney weed

is especially true during the lucerne phase of a crop rotation cycle, where the lucerne is cultivated under 'minimum tillage', which allows the underground rhizomes to survive and resist conventional herbicides.

Control: A selective herbicide is registered for use on lawns, but in flower beds this weed will need continuous and thorough cultivation to remove it as it can regrow from broken pieces of the underground rhizomes.

Dichondra micrantha

CUCURBITACEAE

Citrullus lanatus
wild watermelon * wilde waatlemoen

Origin and description: *C. lanatus* is an indigenous annual plant that is widespread in South Africa and a common weed. Although resembling the cultivated watermelon, the wild watermelon has a very bitter taste. Despite this, the skin and flesh is made into a very tasty preserve when processed correctly.

Impact: It is found wherever the soil has been disturbed, especially in cultivated land, orchards and vineyards. It is not usually a serious weed, but because of its climbing growth habit, one plant can become large and competitive. It is a problem where watermelons are grown as the two plants can cross pollinate. It favours sandy soils and is particularly noticeable in the winter months.

Control: Because the seed of this plant is large and resilient, it can withstand many of the pre-emergence herbicides. Control is best achieved with post-emergence herbicides or physical methods.

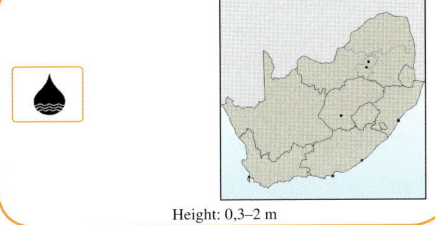

Height: 0,3–2 m

Other common names
bitter melon * colocynth * tshamma * karkoer * ikhabe (Z) * kaate (S) * litjoti (Si)

Similar species
C. lanatus should not be confused with the striped wild cucumber, *Cucumis myriocarpus* (see page 330)

Citrullus lanatus

Citrullus lanatus

Spreading or flat-growing herbs

CUCURBITACEAE

Cucumis myriocarpus
striped wild cucumber * streepwildekommer

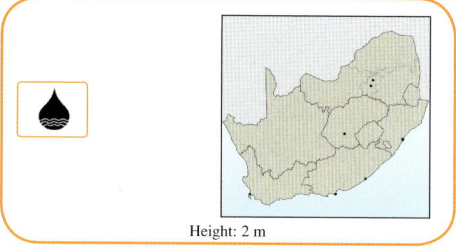
Height: 2 m

Origin and description: *C. myriocarpus* is an indigenous weed that is common in the summer-rainfall regions of southern Africa.
Impact: The fruits are poisonous and are responsible for cattle deaths, but the leaves are eaten by some people. It can become troublesome in a variety of crops because of its tangled, much-branched stems.
Control: Because it is a deep germinator with large seeds, pre-emergence chemical control is not always successful. The weed can be controlled by shallow cultivation in the seedling stage. It is susceptible to post-emergence herbicides.

Other common names
bitter apple * gooseberry cucumber * agurkie * bitterboela * monyaku (S) * sendelenja (Z) * thlare-sampja (T)

Similar species
C. myriocarpus should not be confused with the wild watermelon, *Citrullus lanatus* (described page 329)

Cucumis myriocarpus

EUPHORBIACEAE

Euphorbia hirta (=*Chamaesyce hirta*)
red milkweed * rooimelkkruid

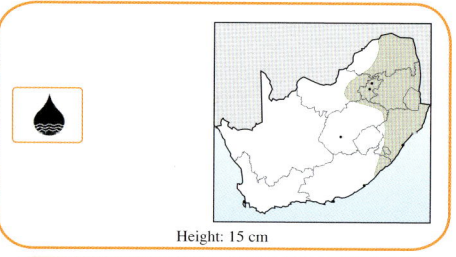
Height: 15 cm

Origin and description: This plant is probably indigenous, although there is some doubt. It is more common in the warmer, subtropical areas than elsewhere in South Africa. It exudes poisonous latex when broken and has been recorded as an alternate host to some important nematode species. The leaves go red as the plant matures.
Impact: It is a weed of gardens and occurs in exposed areas such as pathways, lawns and crops.

Other common names
asthma plant * garden spurge * hairy spurge * pill-pod * sand-mat * snakeweed * bloubekruip * rooimelkbossie

Control: This weed is easy to remove by cultivation and is susceptible to conventional herbicides.

Euphorbia hirta

EUPHORBIACEAE

Euphorbia inaequilatera* var. *inaequilatera
(=*Chamaesyce inaequilatera*)
smooth creeping milkweed * gladde kruipmelkkruid

Euphorbia prostrata (=*Euphorbia chamaesyce*, *Chamaesyce prostrata*)
hairy creeping milkweed * harige kruipmelkkruid

Origin and description: *E. prostrata* is a native of tropical America, whereas *E. inaequilatera* is indigenous. The main difference between these two species is that *E. prostrata* tends to be hairy

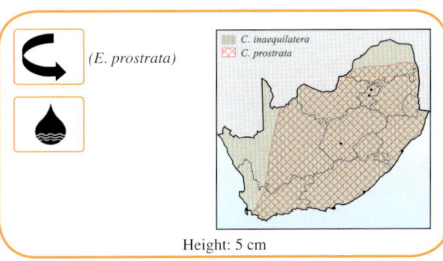

(*E. prostrata*)

Height: 5 cm

Other common names
E. inaequilatera: smooth prostrate euphorbia * gladde-rooi-opslag * hlabo (S)
E. prostrata: blueweed * prostrate spurge * red caustic creeper

331

and have reddish stems, whereas *C. inaequilatera* has smooth, green or yellow stems. Otherwise these two are remarkably similar in appearance, habit and distribution, which is widespread. These plants are annuals and exude a white latex when the stems are broken, as is the case with all Euphorbiaceae.

Impact: They are weeds of gardens and occur in bare, exposed areas such as pathways, lawns and crops. These weeds are sometimes a particular problem in citrus orchards, rapidly spreading onto the bare earth under the trees and competing for water and nutrients.

Control: They are easy to remove by cultivation and are susceptible to conventional herbicides.

Euphorbia inaequilatera

FABACEAE

Lotus subbiflorus
lotus

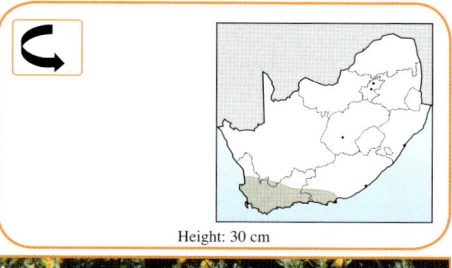

Origin and description: This is probably a fairly recent introduction from Europe. It is widespread in the southern and Eastern Cape where it grows in dense mats in moist places.

Impact: It causes particular problems in lawns as it can swamp many lawn grasses.

Control: There are no specific recommendations for the control of this weed.

Lotus subbiflorus

FABACEAE

Medicago laciniata (=*M. aschersoniana*)
little burweed * klitsklawer

Medicago polymorpha (=*M. hispida*)
bur clover * klitsklawer

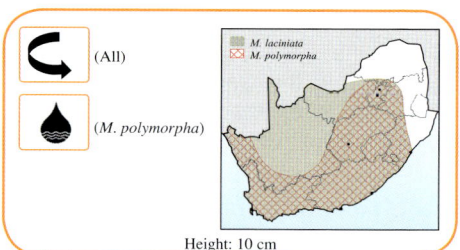

Origin and description: These plants are natives of Europe and are now cosmopolitan annual weeds. *M. polymorpha* has been known in South Africa for over 130 years and is now fairly widespread. Although *M. laciniata* is very similar to *M. polymorpha* in habit and distribution,

Other common names
M. laciniata: burclover * veld shamrock * karooklits * klawergras * bohomenyana (S)

332

it has smaller leaves and burs.

Impact: *M. polymorpha* is a good fodder plant as long as it is kept grazed so that the burs do not develop, otherwise they can become entangled in the wool of sheep. It is, however, most common in the Cape provinces where it is a nuisance in orchards, vineyards and other disturbed areas.

Control: These weeds are easy to control by cultivation or with nonselective post-emergence herbicides.

Other common names
M. polymorpha: rough medic * toothed medic * growwe medicago * stekelklawer

Similar species recorded as naturalised in South Africa
M. falcata ↻ (sickle medic), from Eurasia
M. sativa ↻ (lucerne), from Eurasia
M. lupulina ↻ (black medic * yellow trefoil * shamrock), from Eurasia and North Africa

Medicago polymorpha

FABACEAE

Trifolium angustifolium
narrow-leaved clover

Trifolium repens
white clover * witklawer

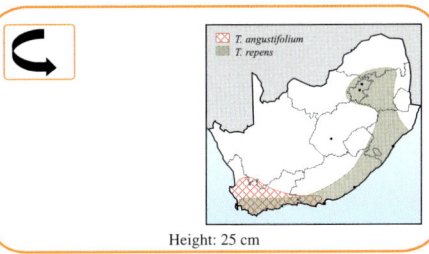

Height: 25 cm

Origin and description: These are two of several clovers introduced from Europe and Asia as fodder crops. *T. angustifolium* was introduced from southern Europe before 1862 but is now found only in the Cape provinces where it is very common on roadsides. It is also a weed in Australia and North America, amongst many others. *T. repens* looks like a typical clover, whereas *T. angustifolium* is often mistaken for a grass on account of its narrow, grass-like leaves.

Impact: Although still valuable as fodder, many 'clovers' are now naturalised in South Africa and are common and widespread weeds, especially in lawns. In turf areas they are unsightly and will replace the grass where they occur.

Control: Most of these plants, being annuals

Other common names
T. repens: white Dutch clover * Brabantse klawer

Some similar species
T. africanum (African wild clover), indigenous
T. arvense ↻ (hare's-foot clover), from Eurasia
T. campestre ↻ (hop trefoil * yellow clover * geelklawer), from Europe
T. dubium ↻ (cowhop clover * Irish shamrock * lesser trefoil), exotic – from Europe

or biennials, will require systemic herbicides for eradication. They are susceptible to 2,4-D but not to 2,4-DB. The latter is used in pastures for broadleaf weed control when these trifoliums are a desirable component. They can easily be removed by hand, especially when the soil is moist.

> **Some similar species**
> *T. pratense* 🍃 (purple clover * red clover * Brabantse klawer), exotic – from Europe
> *T. tomentosum* 🍃 (woolly clover), exotic – from Europe

Trifolium angustifolium

Trifolium repens

GISEKIACEAE

Gisekia pharnacioides
gisekia

Height: 50 cm

Origin and description: *G. pharnacioides* is an indigenous, annual plant that is an occasional weed in crops and gardens. It is found in the summer-rainfall regions, mainly in the warmer areas. There are different varieties of this species.
Impact: Gisekia is often found in sugarcane, although it is more often seen in overgrown areas of gardens and waste areas in farmyards. It rarely becomes dense or competitive.
Control: *G. pharnacioides* is susceptible to the usual herbicides, especially those used in sugarcane and is the subject of several registrations.

Gisekia pharnacioides

MALVACEAE

Malva parviflora
small mallow * kiesieblaar

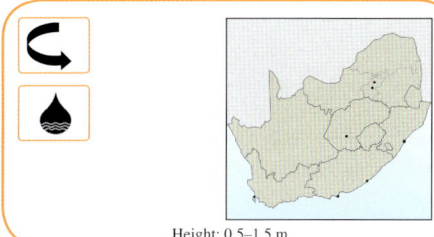

Height: 0,5–1,5 m

Origin and description: This exotic weed originated in Europe and Asia and is now widespread in South Africa. It was possibly first introduced by early settlers for medicinal purposes. If eaten in sufficient quantities, it is known to be poisonous to livestock and in particular horses, especially after exercise. If hens eat it, it is said to cause pink yolks in the eggs. Some indigenous people eat the weed with no apparent ill effects. It has small, inconspicuous white or pink flowers. Recently divided into varieties.

Impact: *M. parviflora* can now be found in all areas of South Africa where it is usually seen on roadsides and in waste places. However, in the Cape provinces it is a serious competitive weed of orchards and vineyards where, under favourable conditions, it can grow into quite a large bush. It has been implicated as an alternative host for the 'kromnek' or tomato spotted wilt virus, which causes a serious viral disease in potatoes, tomatoes, tobacco and peas.

Control: It is a perennial plant with the aerial parts being annual. For this reason it can be easily controlled with contact herbicides if it is sprayed while still young. Large, established plants will require a systemic chemical for effective long-term control. On account of the hairy leaves, it is recommended that a suitable wetting agent is added to all foliar sprays.

Other common names
bread-and-cheese * cheeseweed * brood-en-botter * kasies * mo-ora-tsatsi (S) * thibapitsa (Sh) * unomolwana (X)

Similar species recorded as naturalised in South Africa
M. verticillata 1b (Chinese mallow * kiesieblaar), from eastern Asia
M. aegyptia 1/1b
M. pusilla 1/1b (small mallow)

Malva parviflora

Spreading or flat-growing herbs

Malva parviflora

MALVACEAE

Modiola caroliniana
red-flowered mallow

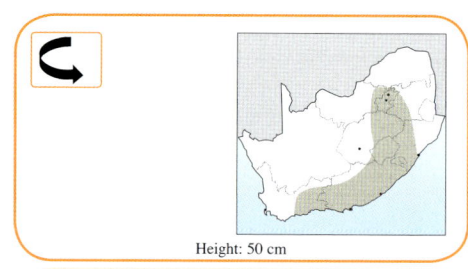

Height: 50 cm

Origin and description: Introduced from tropical America probably as a forage plant or as a contaminant of other forage types, the red-flowered mallow is now widespread in South Africa.

Impact: *M. caroliniana*, being both an annual and a perennial, is a prostrate herb and is commonly found in pastures, gardens, on golf courses and in utility areas. It is not considered a significant weed of crops, even though it can become quite competitive. Although a useful forage plant, many members of the Malvaceae can sometimes be toxic if consumed in large quantities.

Other common names
bristle mallow * creeping mallow

Control: No herbicides are registered for this weed, but it should be easily controlled by most commonly used herbicides. It can also be easily removed by physical means.

Modiola caroliniana

NYCTAGINACEAE

Boerhavia diffusa
spiderling

Origin and description: *B. diffusa* is an exotic, annual weed from South America, that is now common throughout South Africa, especially in the warmer, northern areas. Subspecies exist.
Impact: It is found in most situations, including croplands and gardens, frequently becoming a nuisance. It is especially common on roadsides, along railways and in bare areas.
Control: *B. diffusa* is susceptible to the usual herbicides.

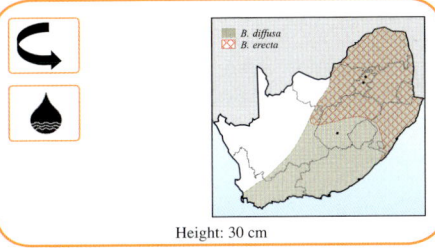
Height: 30 cm

Similar species
Boerhavia erecta (erect spiderling), very similar but does not have sticky seeds

Boerhavia diffusa

337

OXALIDACEAE

Oxalis corniculata (=*O. repens*)
creeping sorrel * tuinranksuring

Oxalis latifolia
red garden sorrel * rooituinsuring

Oxalis pes-caprae
yellow sorrel * geelsuring

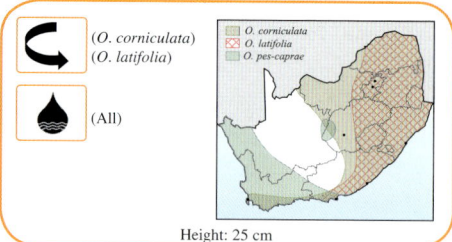

Height: 25 cm

Origin and description: There are approximately 700 species of *Oxalis*, of which about 200 are indigenous to southern Africa, most of which are found only in the Western Cape. *O. corniculata* is from Europe and *O. latifolia* from South America, while *O. pes-caprae* is indigenous. It is significant that of all the species of *Oxalis* that occur in South Africa, the two that are not indigenous are two of three that are considered troublesome weeds. *O. latifolia* and *O. pes-caprae* have bulbs, whereas *O. corniculata* has underground stolons. They all survive as perennials by means of these underground parts.

Impact: These plants sometimes contain enough oxalic acid to cause poisoning in livestock. They can also act as hosts for plant diseases such as various kinds of rust. *O. corniculata* is a problem of lawns and gardens as well as crops and orchards, but the other two, especially *O. latifolia*, are more common in cultivated crops. *O. pes-caprae* is a very colourful roadside plant in the Cape.

Control: These plants are not very competitive and do not usually cause crop losses. They are, however, difficult to eradicate as the bulbs or stolons are very tolerant to herbicides even though surface growth can be controlled easily with 2,4-D and other hormone-type herbicides. If attempts are made to dig them up, all the bulblets or stolons must be removed to prevent regeneration. It is claimed that *O. corniculata* in lawns can be controlled by an application of lime as it is known to be intolerant of alkaline conditions.

Other common names
O. corniculata: jimson weed * yellow sorrel * steenboksuring * bolila (S)* isithathe (Z)
O. latifolia: (pink) garden sorrel * rooisuring
O. pes-caprae: Bermuda buttercup * sour sobs * klawersuring * varksuring

Similar Species
O. luteola (pink sorrel * pienksuring), indigenous – pink flowers
O. polyphylla (finger sorrel * vingersuring), indigenous – pink flowers

Oxalis latifolia

Oxalis latifolia

Oxalis corniculata

Oxalis pes-caprae

Spreading or flat-growing herbs

POLYGONACEAE

Emex australis
spiny emex * Kaapse dubbeltjie

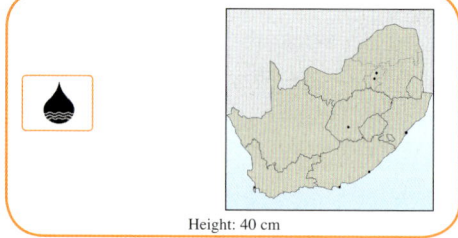
Height: 40 cm

Origin and description: Indigenous to the Cape provinces, *E. australis* is now widespread in South Africa and elsewhere in the world. It is now a serious weed in Australia, having been accidentally introduced there from South Africa aboard early sailing ships crossing the Pacific. The plant was cultivated on these ships as a source of vitamin C.

Impact: It is usually a spreading plant but when dense, it can grow to about 60 cm, becoming very competitive under favourable conditions; this often happens in citrus orchards. The fruit, which is present throughout the year, has three strong spines and can injure the feet of bare-footed people and animals. In cultivated land this weed is more important in the winter-rainfall region but during recent years it has spread to irrigated and dryland wheat in the summer-rainfall areas.

Other common names
cat's head * devil's thorn * prickly jack * duiwelsdis * emexdubbeltjie * volstruisdoring * doublegee (Australia)

Although *E. australis* occurs in wheat fields in the Free State, it is not an important weed in this region. It also occurs on roadsides and in waste areas as an untidy and unpleasant weed.

Control: *E. australis* is susceptible to shallow cultivation and conventional broadleaf weed herbicides.

Emex australis

POLYGONACEAE

Polygonum aviculare
prostrate knotweed * voëlduisendknop

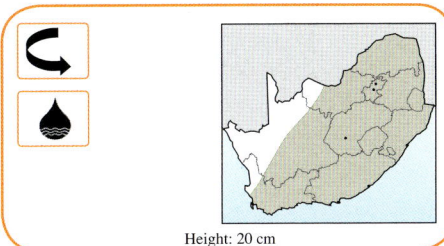
Height: 20 cm

Other common names
bird knotgrass * dooryard knotweed * hogweed * knotweed * mat grass * wire weed * armmanslusern * kanferfoelie * koperdraad * litjiesgras * lira-ha-li-bone (S)

Origin and description: Naturalised from Europe, this annual weed is widespread in South Africa. It is found on roadsides, in waste places and croplands. It is usually recognised by its small, pink flowers and is likely to sprawl along the ground in the absence of a dense crop. It has a tough, resilient stem and the seedling resembles a grass, hence the common names 'wire grass' and 'koperdraad'.

Impact: *P. aviculare* is a major weed of winter wheat and vegetables, particularly in the Free State and Cape provinces. In wheat, for instance, if conditions are favourable, it can clamber over the crop, severely reducing the crop potential and hampering the harvesting operation.

Control: *P. aviculare* is difficult to control with some wheat herbicides as it has an extended period of emergence, adding to its status as a problem plant. Particular care must be taken with the choice and application of herbicides. Nevertheless, many farmers in the Cape provinces value this plant as grazing after their cereal crop. Since 100% control is not considered desirable, these farmers choose chemicals that are less effective and only suppress the weed.

Polygonum aviculare

POLYGONACEAE

Portulaca oleracea
purslane * porslein

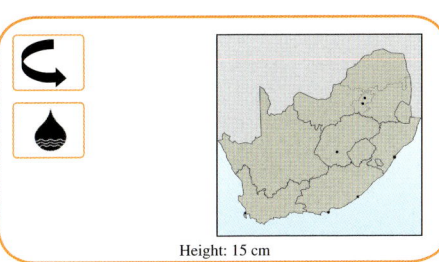
Height: 15 cm

Other common names
pigweed * bobbejaandraad * kanniedood * misbredie * snotterbel * varkkos * igwanitsha (X) * lenjana (Z) * selatsi (S) * silele (Si)

Origin and description: *P. oleracea* is a creeping, succulent annual, introduced into Africa from Europe during the early settler days. In Kenya, the leaves are eaten as spinach. It occurs in all areas of South Africa, often growing vigorously into dense mats in crops and gardens. The common names 'pigweed' and 'misbredie' are confusing, as various members of the *Amaranthus* family are also referred to by these names.

Impact: *P. oleracea* is a common and competitive

Spreading or flat-growing herbs

crop weed and has been implicated as a host plant for the overwintering of the pathogen that causes *Verticillium* wilting disease in cotton.

Control: Because of their succulent nature and having large moisture reserves, the plants can survive for some time after being uprooted or broken. Small pieces can then set root and produce new plants, especially under moist conditions. This makes it difficult to eradicate by cultivation. *P. oleracea* is relatively easy to control with pre-emergence herbicides, as it is small-seeded and a shallow germinator. Good wetting agents should be used with post-emergence herbicides, however, to ensure good coverage and penetration of the waxy leaves.

Similar species
P. quadrifida (wild purslane * porselein * amelanyane (Si) (Z)), uncertain origin

Portulaca quadrifida   *Portulaca oleracea*

Portulaca oleracea

PRIMULACEAE

Anagallis arvensis
pimpernel * rooimuur

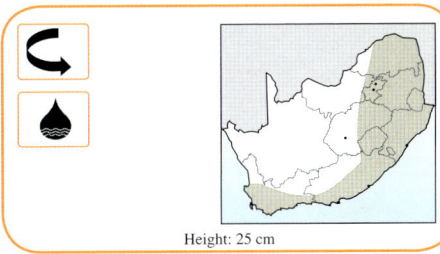
Height: 25 cm

Origin and description: *A. arvensis* is an exotic annual from Eurasia that has been known as a weed in South Africa since the middle of the 19th century. It was introduced as a contaminant of crop seeds by European settlers. In South Africa the most common flower colour is blue, but in Europe it is red, and it is commonly called the 'scarlet pimpernel'. Even though the two flower colours can often be seen growing together, they have recently been ascribed to different subspecies.

Impact: It is usually found in damp situations, being capable of flourishing in lower light densities. It is common in cereals in the southern and Western Cape. *A. arvensis* is poisonous to birds and other animals.

Other common names
bird's eye * poor man's weather glass * red chickweed * scarlet pimpernel * shepherd's calendar * blouselblommetjie

Similar species
Stellaria media (see page 323)

Control: It does not often grow profusely and specific control measures are not normally required. It is easy to control by cultivation and is susceptible to normal herbicides.

Spreading or flat-growing herbs

Anagallis arvensis

Anagallis arvensis

ROSACEAE

Duchesnea indica *(=Potentilla indica)*
false strawberry * wilde-aarbei

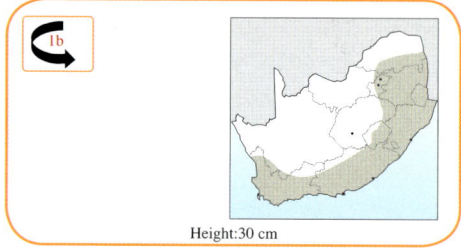

Height: 30 cm

Origin and description: A native of India and Sri Lanka, *D. indica* has spread as a weed in the warm, high-rainfall areas of South Africa. It is a creeping perennial herb spreading by slender stolons. The fruit look like strawberries but taste very insipid.

Impact: This plant favours damp, shady places and is often found in gardens, forest undergrowth, ditches and urban waste areas. As a ground cover, the creeping vines are capable of spreading rampantly, blocking drainage lines and replacing intended species.

Other common names
Indian strawberry

Control: Shallow cultivation and most broadleaf herbicides are successful in controlling this weed. Unintended plants should be removed.

Duchesnea indica

RUBIACEAE

Richardia brasiliensis
Mexican richardia * Meksikaanse richardia

Origin and description: As the botanical name suggests, this weed originated somewhere in South America. It has been known in South Africa for over a century, however, and is common in the warmer parts of the eastern half of the country and up into Zimbabwe.

Impact: It is common in many crops and situations and is particularly well known by sugar farmers and gardeners. It seldom becomes a very serious problem, however. It is occasionally a problem on golf course fairways as it can survive the effects of continuous mowing and cannot be easily controlled when conditions for growth are poor and the uptake of systemic herbicides is slow. Under such conditions it can totally replace the fairway grass over large areas.

Control: It is not otherwise a very competitive weed and is generally susceptible to all types of herbicides. It can also be controlled effectively by cultivation.

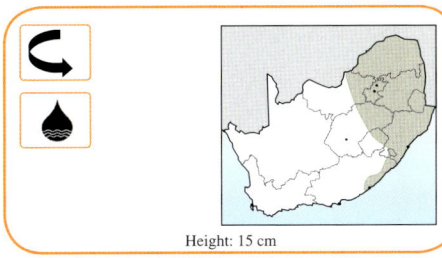
Height: 15 cm

Other common names
Mexican clover * tropical richardia * white eye * Meksikaanse klawer

Richardia brasiliensis

345

SCROPHULARIACEAE

Veronica persica
field speedwell * akker-ereprys

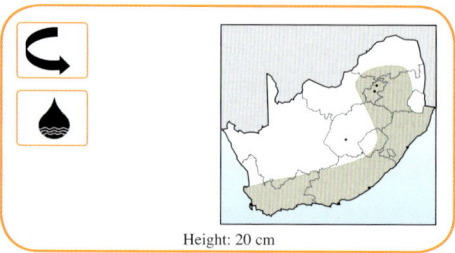
Height: 20 cm

Origin and description: Originally from Europe and Asia, *V. persica* was introduced to South Africa at the end of the 19th century. This annual weed is well known in crops, lands and gardens throughout South Africa, particularly in damp corners.
Impact: It is often found in winter wheat, irrigated vegetables and is one of the main weeds in many citrus orchards of the Eastern Cape. It seldom becomes a serious problem in these places, as it is not particularly vigorous or competitive. It is an untidy weed of lawns and turf areas.
Control: It is controlled effectively by shallow cultivation at the seedling stage and is susceptible to the normal herbicides. In lawns it is usually removed with a low mowing.

Other common names
bird's eye speedwell
Similar species
V. agrestis (field speedwell)
V. anagallis-aquatica (water speedwell * water-ereprys * mo-qhobo-o-monyenyane (S))

Veronica persica

TROPAEOLACEAE

Tropaeolum majus
nasturtium * kappertjie

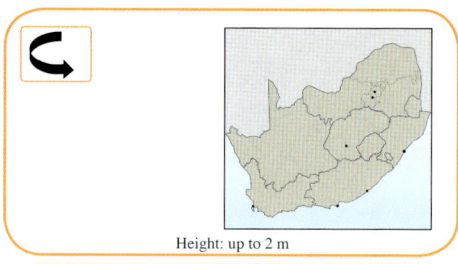
Height: up to 2 m

Origin and description: Originally from the Andes region of South America, the nasturtium has been a popular garden plant in South Africa for many years. It probably exists as a range of hybrids, even in the wild. The plant is usually grown for the attractive flowers but the leaves are also used in salads as they contain mustard oils, which give them a delicate peppery taste.

Other common names
Indian cress * monk's cress

The taste is very similar to that of watercress and it could be a reason that the names are connected. The Latin name for nasturtium is *T. majus*, and that for water cress used to be *Nasturtium officinale*, but is now *Rorippa nasturtium-aquaticum* (see page 385).

Impact: *T. majus* is now naturalised in several regions of South Africa. It is particularly invasive in the Western Cape, especially around Cape Town. It can be very aggressive and can smother and replace indigenous vegetation. It invades roadsides, river banks and forest margins.

Control: Unwanted plants in the garden should be removed and naturalised infestations eradicated with nonselective contact herbicides. Care should be taken with its cultivation, especially in the Western Cape.

Tropaeolum majus

VERBENACEAE

Verbena aristigera *(=V. tenuisecta)*
fine-leaved verbena * fynblaarverbena

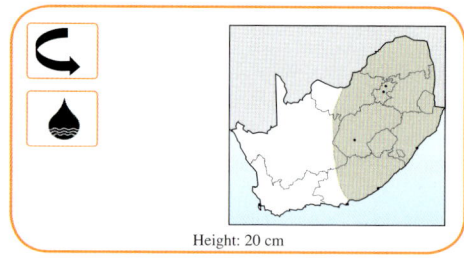

Height: 20 cm

Origin and description: *V. aristigera* is a perennial plant from South America that was introduced to South Africa for cultivation as an ornamental. It has now escaped from gardens and has become naturalised throughout the country. The attractive white and purple flowers are commonly seen on roadsides and in waste places.

Impact: Although in many ways an attractive plant, it often grows in such dense profusion that all indigenous vegetation is replaced.

Other common names
moss verbena * tuber vervain * wild verbena

Control: *V. aristigera* would be susceptible to herbicides normally used on roadsides.

Verbena aristigera

ZYGOPHYLLACEAE

Tribulus terrestris
devil's thorn * dubbeltjie

Origin and description: *T. terrestris* is an indigenous weed that is widespread in South Africa. It does not occur in the cold or wet areas. It is a creeping annual, which can spread horizontally into a plant of over a metre in diameter. It is spread easily by the spiked fruits that stick to the feet of animals and tyres of vehicles.

Impact: The vicious spikes of the fruits can cause damage to the feet of stock animals. The plants can be toxic, causing 'geeldikkop' or tribulosis in sheep, but only appear to be so when they are in a wilted condition, such as during a hot, dry spell following summer rain. In the Karoo, farmers consider it an essential and life-saving fodder plant, but in other regions it is a serious weed. It is a severe weed of maize in parts of the western Free State, particularly on sandier soils where it is a strong competitor for the already limited moisture. It is able to germinate even under very arid conditions.

Control: *T. terrestris* can easily be controlled by cutting the taproot, but the use of herbicides needs special attention. Control with pre-emergence herbicides is erratic because of the size of the seed and the depth of germination. In maize, for example, *T. terrestris* is easily controlled by post-emergence herbicides when very small. When it has passed about the six-leaf stage, however, specific herbicide mixtures are required for successful control.

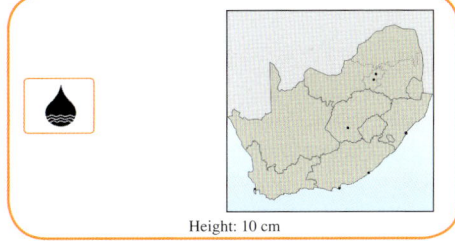

Height: 10 cm

Other common names
burnut * common dubbeltjie * plat-dubbeltjie * volstruisdoring * inkunzane (Z)

Tribulus terrestris

Spreading or flat-growing herbs

CREEPERS AND CLIMBERS

Within the section, the plants are arranged alphabetically according to family and genus.

SPECIES	Exudes white sap	Thorny stems	Parasitic (not green)	Fruit is a bean-like pod	Fruit is like a berry	Tendrils or claws	Succulent	Red flowers	Yellow flowers	Purple flowers	White flowers
Araujia sericifera, p. 353	X										X
Pereskia aculeata, p. 356		X			X		X				X
Cuscuta campestris, p. 360			X								
Macfadyena unguis-cati, p. 355				X		X			X		
Vicia species, p. 365				X		X				X	
Pueraria lobata, p. 364				X						X	
Aristolochia elegans, p. 352				X							
Cardiospermum grandiflorum, p. 369					X	X			X		X
Ipomoea alba, p. 361					X		X				
Ipomoea purpurea, p. 363					X					X	
Solanum seaforthianum, p. 370					X					X	
Hedera helix, p.352						X					X
Passiflora spp., p. 366						X					
Anredera cordifolia, p. 354							X				
Ipomoea hederifolia, p. 362								X			
Syngonium podophyllum, p. 351									X		
Convolvulus spp., p. 358, 359											X
Lonicera japonica, p. 358											X
Fallopia convolvulus, p. 368											X

ARACEAE

Syngonium podophyllum
arrow-head vine * gansvoete

Origin and description: *S. podophyllum* was introduced from the tropical rain forests of Central and South America as a pot ornamental and as a useful and decorative groundcover.

Impact: Under favourable growing conditions of a minimum winter temperature of 16°C, this plant will start to climb. The leaf shape changes and it becomes an aggressive climber capable of swamping indigenous trees. It is also capable of self-regeneration and can be found invading subtropical urban roadsides and waste areas. Not only is it widely naturalised in the warmer regions, but it is planted as a groundcover by people who are totally unaware of its ability to change into an aggressive climber.

Control: Unwanted plants should be removed and new plants only planted if the full capabilities of the species are understood.

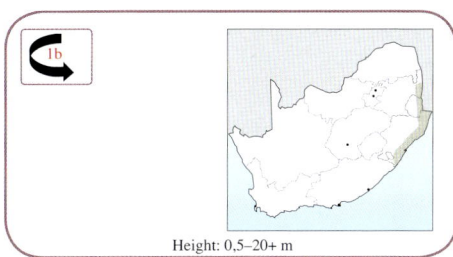

Height: 0,5–20+ m

Other common names
goosefoot plants * arrow-head philodendron

Syngonium podophyllum

ARALIACEAE

Hedera helix
English ivy * Engelse hedera

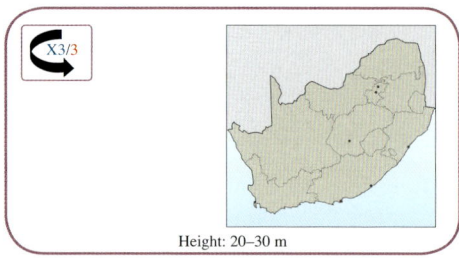

Height: 20–30 m

Origin and description: English ivy is native to most of Europe and was introduced into South Africa as a groundcover and climbing ornamental. There are many cultivars with colourful properties such as variegated leaves as well as various subspecies, which are usually difficult to differentiate – as are the two subspecies named as invasive in South Africa, *H. helix* subsp. *canariensis* and *H. helix* subsp. *helix*. The evergreen foliage of ivy is useful for covering unsightly walls and fences. The plant attaches itself to the wall, rock or tree substrate with strong stems that root from the nodes. The plant can be spread by the seeds in the fruit or by broken pieces of stem.

Impact: In regions of the world where winter is less severe, *H. helix* has become an invasive weed. In South Africa it is widely naturalised and can on occasion be seen climbing trees in urban areas and these trees are likely to eventually collapse from the weight of the ivy. It also invades roadsides and waste areas, being capable of swamping and eliminating indigenous plants. If allowed to grow uncontrolled over buildings, ivy can damage walls and roofs with the roots being able to penetrate the grain of wooden structures, allowing moisture and fungi to enter and cause rotting. The plant is poisonous.

Control: On trees, the stems of mature ivy plants should be cut at ground level and removed from

Hedera helix

the tree for 1 or 2 m. The upper vines should be left to rot, as they will die if they are not rooted in the ground. The remaining stems and shoots can be treated with herbicide. The sterility or non-invasiveness of cultivars has to be investigated.

ARISTOLOCHIACEAE

Aristolochia elegans
calico flower * sisblom

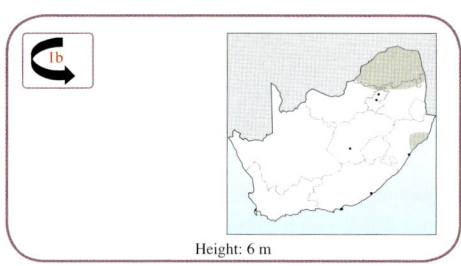

Height: 6 m

Origin and description: Just one of over 500 species of pipevines, which was introduced from Brazil and South America as an ornamental on account of its very unusual flowers. However, it has found a receptive environment in subtropical parts of South Africa and has become naturalised in the wild. It can be found in riverine forests and

Other common names
Dutchman's pipe * pipevine * oupa-se-hoed * pypblom

waste areas where it clambers up trees and over fences.
Impact: *A. elegans* can smother and replace indigenous vegetation.
Control: There are no herbicides registered for this weed, but because of its scrambling nature it should be cut back, uprooted or if necessary, cut down and the remaining stump treated with a suitable herbicide.

Aristolochia elegans

ASCLEPIADACEAE

Araujia sericifera
moth catcher * motvanger

Origin and description: A native of Peru that was introduced into South Africa as an ornamental, *A. sericifera* is now found in many parts of South Africa as a weed of gardens, waste areas and plantations. Having escaped from cultivation, it is closely associated with urban areas, especially Johannesburg, Pretoria and Pietermaritzburg. It is a robust, semiwoody, perennial creeper that can reach the top of tall trees. It exudes a white juice, which will leave sticky patches on skin and clothes, but is not considered dangerous. The fruit is a large spongy capsule, that splits open when mature, releasing the black seeds that are attached to fluffy hairs and blown away by the wind.

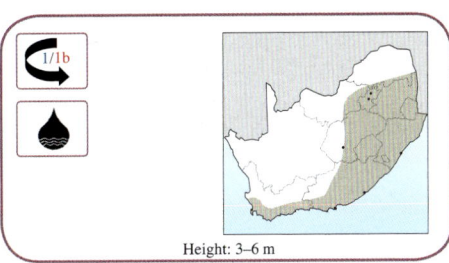

Height: 3–6 m

Other common names
bladder flower * cruel plant * milkweed * moth-vine * stranglehold plant

Araujia sericifera

353

Impact: It smothers desirable plants and causes them to collapse under the extra weight.
Control: Triclopyr is registered as a spot treatment, but this plant is easily uprooted when small and growing in soft soil. The plants should be destroyed before they produce seeds.

Araujia sericifera

BASELLACEAE

Anredera cordifolia (=*Boussingaultia baselloides*)
Madeira vine * Madeira-ranker

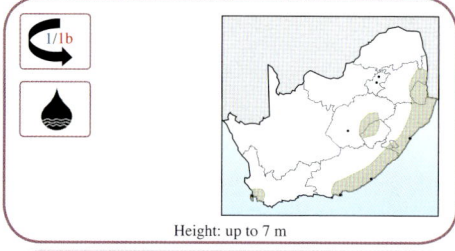

Height: up to 7 m

Origin and description: Originating from sub-tropical South America, *A. cordifolia* was cultivated in gardens and it is probable that from here it escaped into the wild. It is found mainly in the warmer eastern parts of South Africa. It is a perennial, succulent creeper that can become longer than 7 m, crawling along the ground and over crops, hedges and fences. The vine does produce seeds but is spread mainly by means of brittle underground as well as aerial edible tubers, which drop to the ground and take root. It is also spread around by human activities.

Impact: *A. cordifolia* is widespread in South Africa, causing particular problems in the coastal regions of KwaZulu-Natal where it is rapidly becoming a major problem in sugarcane. The combination of fleshy leaves and the thick aerial tubers makes this a very heavy vine and it can smother entire trees and cause them to collapse. It is also troublesome in Mpumalanga and in parts of the Western Cape such as Ceres, Cape Town and Tulbagh. It is also a weed of urban waste areas and gardens.

Control: This weed is difficult to control in most situations. Triclopyr is registered as a spot spray

Other common names
bridal wreath * cascade creeper * lamb's tails * white shroud * intandela (Z)
Similar species
A. baselloides: There is some confusion as to whether this is a very similar but separate species or a misapplication of the name

Anredera cordifolia

in sugarcane and as a full cover spray elsewhere. Any cut plant material must be destroyed and not added to the compost!

Anredera cordifolia

BIGNONIACEAE

Macfadyena unguis-cati
cat's claw creeper * katteklouranker

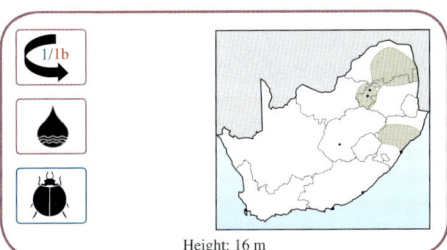

Height: 16 m

Origin and description: Probably a relatively recent introduction from South America, *M. unguis-cati* has rapidly attained the status of a major threat. It is cultivated in most areas of South Africa as a useful and attractive garden plant. It is fast growing, frost and drought tolerant and is ideal as a cover for hedges and walls. In spring it produces masses of large, yellow, trumpet-shaped flowers and in summer, long, slender, pod-like seed capsules that split open to release the winged seeds. The tendrils, which it uses to cling to objects, end in three small, hooked claws, from which the common name is derived. It also develops resilient, tuberous roots as it matures, from which new plants can grow.

Impact: It has escaped cultivation and invaded woodland, forests, orchards, roadsides and open urban spaces in many areas of South Africa. It has become a particular problem in areas of Limpopo Province such as Magoebaskloof, and around Pietermaritzburg in KwaZulu-Natal and is threatening areas in Swaziland and the Mpu-

Macfadyena unguis-cati

355

malanga lowveld. The plant can grow to enormous lengths and can climb in dense masses up the tallest trees. A combination of weight and shade can kill these trees and suppress the regeneration of indigenous vegetation.

Control: Chemical and mechanical control are not very successful, although painting cut stems with a systemic herbicide can work and older plants will die completely if cut at ground level. However, the plants can quickly regenerate from the tubers below ground, even broken pieces. These tubers become up to 40 cm long and are very difficult to dig out as they break easily. In 1999 the leaf-feeding beetle *Charidotis auroguttata* was released as a biocontrol agent, but success is not yet significant.

Macfadyena unguis-cati

CACTACEAE

Pereskia aculeata
Barbados gooseberry * Barbados-stekelbessie

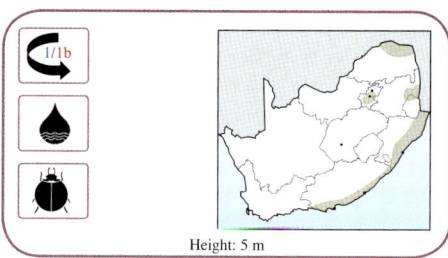

Height: 5 m

Other common names
blade apple * leafy cactus * lemon vine * primitive cactus * Spanish gooseberry * uqwaningi (Z)

Origin and description: This succulent, perennial plant is from tropical South America and was originally planted as an ornamental and for security hedging. Early records suggest that it was already present in present-day KwaZulu-Natal by 1881. The fruits of *P. aculeata* are edible, with the seeds being spread by birds and other animals. It can form an impenetrable hedge and is planted for this purpose. The spines on the young growth are short and hooked and occur in the leaf axils. On old wood they are straight and occur in bunches. The Zulu people plant it over graves as protection against vandalism. Despite being from the cactus family, *P. aculeata* resembles bougainvillea in habit.

Impact: It has established itself in the wild, especially in the KwaZulu-Natal coastal region, but is also found in other tropical areas where it invades natural vegetation, smothering indigenous species and restricting access on account of its unpleasant spines.

Control: Chemical control is partially successful if the cut stumps are treated with a registered herbicide. Triclopyr is registered as a foliar application. All the cut pieces must be collected and burned as the succulent stems can take root if they fall onto the ground. A leaf-chewing biocontrol agent was released in the early 1990s, but success to date has been minimal.

Pereskia aculeata

CAPRIFOLIACEAE

Lonicera japonica
Japanese honeysuckle * Japanse kanferfoelie

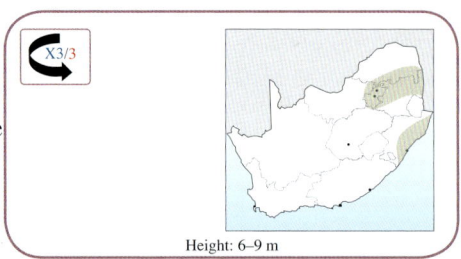

Height: 6–9 m

Origin and description: Honeysuckles are native to the northern hemisphere, with this one coming from Asia. *L. japonica* was planted throughout the world as an ornamental climber, for erosion control and for wildlife forage and cover. It is now classified as a noxious weed in parts of the USA as well as in South Africa, where it has become naturalised. There are many cultivars and subspecies of *Lonicera*.

Impact: Honeysuckle climbs by twisting its stems around vertical structures, including trunks of shrubs and trees. In South Africa it has few, if any, natural enemies, allowing it to spread widely and out-compete native plant species. It invades woodland and riverbanks, usually close to urban areas.

Control: For small patches, repeated pulling of entire plants and root systems may be effective. Systemic herbicides can be used if the leaves are green and healthy, but there are no registrations at present. Unintended plants should be removed.

Lonicera japonica

CONVOLVULACEAE

Convolvulus arvensis
field bindweed * akkerwinde

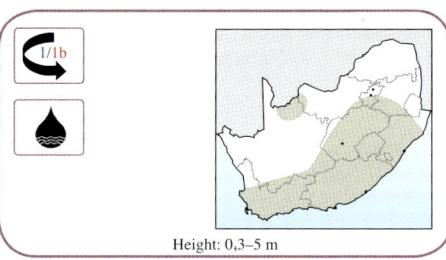

Height: 0,3–5 m

Other common names
wild morning glory * klimop * monnikbaard * warkruid

Origin and description: *C. arvensis* is from Europe and Asia and is distributed throughout South Africa and most other temperate countries. It can be differentiated from the morning glory (*Ipomoea purpurea*, see page 363) by its arrow-shaped leaves as opposed to the latter's more heart-shaped leaves.

Impact: It is a troublesome weed in cultivated lands.

Control: *C. arvensis* is difficult to control on account of its long underground runners. The runners can survive herbicide applications and even the most severe frosts. Glyphosate is registered as a foliar spray to be applied at the onset of flowering.

Convolvulus arvensis

CONVOLVULACEAE

Convolvulus farinosus
wild bindweed * klimop

Convolvulus sagittatus **subsp.** *sagittatus*
weed * bobbejaantou

Convolvulus sagittatus **var.** *ulosepalus*
wild bindweed * wilde-akkerwinde

Origin and description: *C. farinosus* and the many subspecies and variants of *C. sagittatus* are indigenous to South Africa.

Impact: All these species occur as weeds and are difficult to control in cultivated lands, gardens and waste places. *C. sagittatus* var. *ulosepalus* is a particularly serious problem in wheat fields and vineyards along the Orange River.

Control: Chemical control of these plants is best achieved with systemic chemicals applied at the end of the growing season when nutrient movement is downwards to the roots and rhizomes. However, various cultural techniques can reduce the problem. Sheep will readily graze the foliage, for example, and pigs are fond of the underground parts. Lucerne (*Medicago sativa*) can compete successfully with bindweed and to a certain extent can crowd out the roots. Repeated cutting of the lucerne also keeps the foliage of the weeds in check.

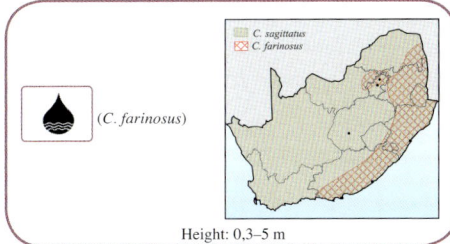

Height: 0,3–5 m

Other common names
C. farinosus: inabulele (X) * umkoka (Z)

Convolvulus sagittatus

Convolvulus sagittatus var. *ulosepalus*

Convolvulus farinosus

CONVOLVULACEAE

Cuscuta campestris
dodder * dodder

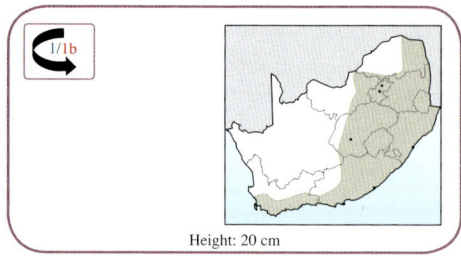
Height: 20 cm

Origin and description: This particular species of *Cuscuta*, which is the most common in South Africa, is in fact a native of North America but is now a widespread weed in many countries in Europe, Africa and Asia as well as Australia and Polynesia. It has been known in South Africa for over 100 years. It has yellow, thread-like stems and flowers in dense globe-shaped clusters. There are also 17 indigenous species of dodder in South Africa.

Impact: All dodders are parasitic, attacking many crops. *C. campestris* has no leaves or chlorophyll and is parasitic on a wide range of plants but is most important as a pest of lucerne and other legumes. Being parasitic, it draws nutrients from the host plant, which can be seriously debilitating, causing substantial yield losses and it can even, in densely infested patches, kill the host plant. It is dispersed by seeds and pieces carried by water or when the crop is harvested. The seeds germinate in soil but are without roots and will die if a host is not found. Germination of seeds is stimulated by the germination of the seeds of host plants in the soil nearby.

Control: *C. campestris* is not controlled by the usual herbicides and the only successful method is to cut out and burn infected plants before the dodder can produce seeds.

Other common names
umankunkunku (Z) * unyendeny-ende (Z)
Similar species
C. suaveolens ⊂ 1b (North America)
lucerne dodder * fringed dodder

Cuscuta campestris

CONVOLVULACEAE

Ipomoea alba
moonflower * maanblom

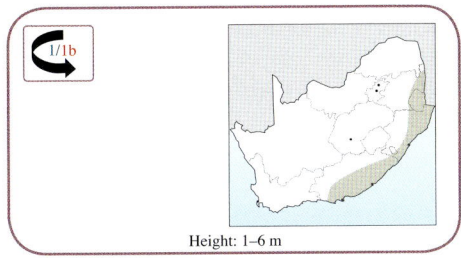

Height: 1–6 m

Origin and description: There are many indigenous species of *Ipomoea*, which are all relatively minor weeds. This species is from tropical America and having escaped cultivation as an ornamental, is a very aggressive invader. It is called 'moonflower' because it blooms at night and the large, white flower looks like the moon.

Impact: The moonflower can now be found scrambling over trees, hedges and fences in urban areas and nearby forests, riverbanks and coastal bush. It can quickly smother and kill indigenous species.

Control: This plant is a Category 1 species in the whole of South Africa and its use as a garden plant is to be discouraged. Unwanted plants should be physically removed; no specific herbicides are registered.

Ipomoea alba

CONVOLVULACEAE

Ipomoea hederifolia
scarlet morning glory

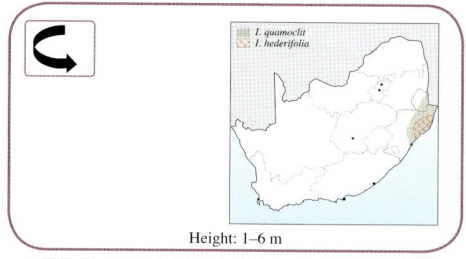

Height: 1–6 m

Origin and description: *I. hederifolia* is native to the Americas but is now a common weed in many tropical and subtropical countries, where it is often reported as a weed of sugarcane. It has been recorded for some time in central Africa, but has only recently been found in South Africa. It was probably brought here as an ornamental and can now be found naturalised in the warmer areas like the KwaZulu-Natal north coast.

Impact: This plant clearly has the potential to be a troublesome weed since it flourishes in other tropical regions. It can be found invading roadsides and waste areas where it climbs over other vegetation. It may not be long before it invades local crops such as sugarcane.

Similar species
I. quamoclit (cardinal creeper), also from the Americas, has very similar flowers but very fine, needle-like leaves. It is already a recognised weed of sugarcane in South Africa

Control: This plant should be removed or destroyed wherever it is found.

Ipomoea hederifolia

CONVOLVULACEAE

Ipomoea purpurea
common morning glory * purperwind

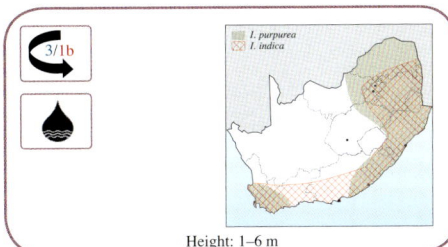

Height: 1–6 m

Origin and description: There are many indigenous species of *Ipomoea*, which are all relatively minor weeds, but *I. purpurea* is native to tropical and subtropical America and is an extremely troublesome weed in crops and gardens in most areas. It was recorded in the Durban area as far back as 1830. The flowers are usually purple or white, but occasionally have stripes of contrasting colours. The main leaves are heart-shaped, but the first two cotyledonous leaves are lobed. In frost-free areas, individual plants of *I. purpurea* can survive from year to year and can reach enormous heights. In other areas, however, the weed is an annual.

Impact: The vigorous climbing habit of *I. purpurea* can rapidly smother infested crops and interfere with harvesting by stringing all the plants together. The seed is said to contain a powerful hallucinogen. Maize grain is downgraded with only a small number of seeds present.

Control: Germination of *I. purpurea* takes place over an extended period and from great depths. It is therefore not unusual for a maize crop in autumn to still have freshly emerging seedlings present, capable of climbing up the drying plants. For this reason it is an extremely difficult weed to control as it escapes most pre- and even post-emergence herbicide treatments. In some crops, such as tomatoes, there are no selective herbicides that will control it. However, *I. purpurea* is very sensitive to the hormone-type herbicides.

Other common names
clock plant * ibhoqo (Z) * ijalamu (Z) * imotyikatsana (X) * ubatata wentaba (Z)

Similar weedy species
I. carnea subsp. *fistulosa* 1b (=*I. fistulosa*) (morning glory bush), from tropical America, problem in northern KwaZulu-Natal
I. cairica (coastal morning glory * Messina creeper * ihlambe (Z)), uncertain origin, probably indigenous, problem on the KwaZulu-Natal coast
I. coscinosperma, indigenous and occasional problem on the Springbok Flats
I. indica 1/1b (blue morning glory), from Hawaii and New World tropics
Convolvulus spp. (see page 368)

Ipomoea purpurea

Ipomoea indica

Creepers and climbers

Ipomoea purpurea

FABACEAE

Pueraria lobata* var. *lobata *(= P. montana var. lobata)*
kudzu vine * kudzuranker

Origin and description: A native of southern Japan and southeastern China, the kudzu vine was probably introduced into South Africa as an ornamental and for erosion control. Although the plant produces seeds, it is spread more frequently by vegetative expansion using its stolons (runners) that root at the nodes and by rhizomes.

Impact: The kudzu vine invades forest margins, riverbanks and similar niches in the warmer, moister eastern parts. It smothers desirable plants and causes them to collapse under the extra weight. Being a legume, this vine can fix nitrogen and enhance the fertility of the soil, but this can also upset indigenous species.

Control: Triclopyr is registered for chemical control, but this plant can be killed if the root crown and all rooting runners are destroyed. This can be done by cutting off the crowns from the roots, usually just below ground level. This immediately kills the plant. The removed crown must be destroyed. Repeated mowing or heavy grazing can exhaust plants and will assist with control measures.

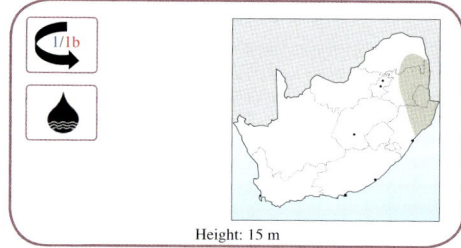

Height: 15 m

Pueraria montana

FABACEAE

Vicia benghalensis
narrow-leaved purple vetch * smalblaarperswieke

Vicia sativa subsp. *sativa*
broad-leaved purple vetch * breëblaarperswieke

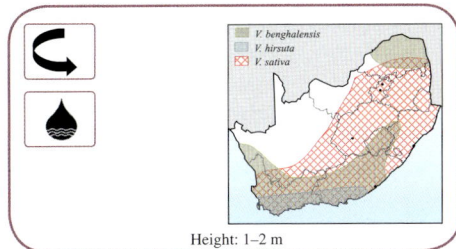

Height: 1–2 m

Similar species
V. hirsuta (hairy tare * small purple vetch), from Eurasia to North Africa
V. villosa (Russian vetch), from Europe to Afghanistan

Origin and description: Six species of *Vicia* or vetch have been introduced from Eurasia as fodder crops and are now naturalised. These two in particular have become widespread and potentially serious weeds. These plants are found throughout the country, but especially in the winter-rainfall region and in irrigated winter crops of the summer-rainfall regions. They are primarily annual plants, but can survive for more than one season. *V. benghalensis* has a long group of flowers on a stalk, whereas in *V. sativa* the flowers are solitary in the axils of the leaves. Despite their names, the leaves look similar.

Impact: Because of their scrambling growth habit, these weeds can smother and destroy crops. These plants were very important weeds of the wheat fields in the southwestern Cape during the 1940s and 1950s.

Control: Due to an extended germination period, these weeds can often escape pre-emergence herbicides. They are, however, very susceptible to the hormone-type herbicides, and continued use of these chemicals has substantially reduced the status of these plants as weeds.

Vicia benghalensis

Vicia sativa

Creepers and climbers

PASSIFLORACEAE

Passiflora subpeltata
wild grenadilla * wildegrenadella

Passiflora suberosa
devil's pumpkin

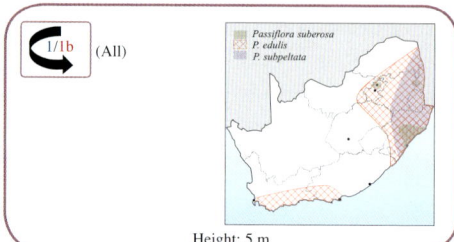

Height: 5 m

Origin and description: There are 14 species of *Passiflora* that have been introduced into southern Africa, of which seven have become naturalised, and of these, four are Category 1. Some of these species were introduced as climbing ornamentals and others for their fruit. They have now spread into various niches in most areas of the country. These two are from South America and are now invasive weeds.

Impact: Some of these species are common in the subtropical forest areas of the summer-rainfall regions. *P. subpeltata* and *P. suberosa*, for instance, are common in the forests of the KwaZulu-Natal midlands, where they clamber over trees and saplings, interfering with growth and harvesting. *P. suberosa* is also commonly seen climbing fences in Durban and Pietermaritzburg.

Control: These are perennial plants and as such they have to be controlled effectively. Systemic chemicals or total physical removal are the best methods.

Other common names
P. subpeltata: granadina
P. suberosa: indigo berry

Similar species
P. caerulea 1/1b (also spelt *P. coerulea*) (blue passion flower * siergrenadella), from South America
P. edulis X2/2 (passion fruit * grenadella), from Brazil, cultivated for its fruit
P. mollissima 1/1b (banana poka * bananadilla * piesangdilla)
P. foetida (wild water lemon), from Central and South America

Passiflora suberosa

Passiflora subpeltata

Passiflora edulis

Creepers and climbers

Passiflora suberosa

Passiflora caerulea

Passiflora molissima

POLYGONACEAE

Fallopia convolvulus *(=Polygonum convolvulus) (=Bilderdykia convolvulus)*
climbing knotweed * slingerduisendknoop

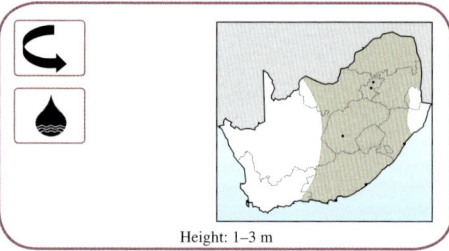

Height: 1–3 m

Origin and description: A native of Europe, *F. convolvulus* is now a widespread and troublesome weed in South Africa, particularly in winter wheat and vegetables. It is especially a problem in the eastern Free State and at the Loskop Irrigation Scheme. In KwaZulu-Natal, it is a minor weed in Weenen and Winterton. It has probably been spread by contaminated wheat seed.

F. convolvulus is a climbing annual with a deep taproot. The flowers, unlike those of *Ipomoea purpurea* (see page 363) with which it is sometimes confused, are small and green or white. The leaves are narrow and arrow-like and the fruits distinctly triangular. It is easily identifiable in its seedling stage.

Impact: The climbing habit of this weed means that it can grow above the crop canopy for sunlight and can easily smother the crop, thereby affecting yields and interfering with harvesting. The fact that it is usually still green when wheat is ripe exacerbates the problem.

Control: *F. convolvulus* is relatively tolerant to some herbicides, especially the hormone-type herbicides, so care must be taken with a suitable choice of chemical. Particular attention must be paid to correct and efficient application.

Other common names
bearbind * black knotweed * wild buckwheat * wildebokwiet * morarana (S)

Fallopia convolvulus

SAPINDACEAE

Cardiospermum grandiflorum
balloon vine * blaasklimop

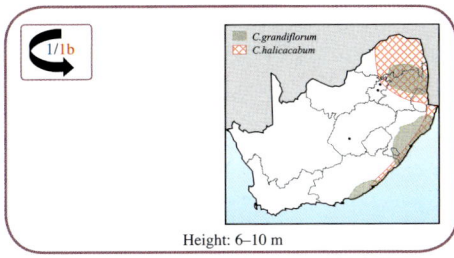

Height: 6–10 m

Origin and description: A native of South America, *C. grandiflorum* is now a serious alien invader in many parts of South Africa, especially the eastern regions. It was probably originally introduced as an ornamental and was first recorded in Eshowe in 1937.

Impact: It is a creeping annual or perennial plant (reproducing only by seed), and can be found sprawling over trees, fences, embankments and similar places. It can grow to enormous lengths and is capable of smothering a tree of up to 10 m tall, thereby completely destroying native woodland. The capsules are blown by the wind and can float, meaning they are readily distributed by water and easily reach their preferred habitat, which is moist yet sunny.

Control: *C. grandiflorum* is relatively easy to control. As long as the root is destroyed, the rest of the plant can be left to wither and die. There are no specific herbicides registered for this weed, so the best method of eradication is to dig up the root.

Other common names
heart pea * heart seed * opblaasboontjie * intandela (Z) * uzipho (Z)

Similar species
C. halicacabum: X3/3 The most easily distinguishable difference between these two species is that the fruit capsules of *C. halicacabum* are rounded and not pointed like those of *C. grandiflorum* X3/3

Cardiospermum grandiflorum

Cardiospermum grandiflorum

SOLANACEAE

Solanum seaforthianum
potato creeper * aartappelranker

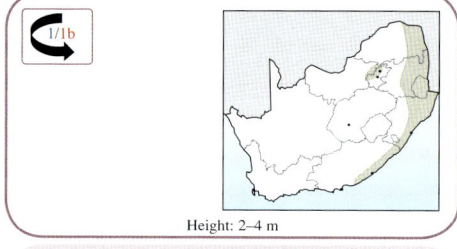

Height: 2–4 m

Other common names
Brazilian nightshade *climbing nightshade * ijalamu (Z)

Origin and description: *S. seaforthianum* was introduced from tropical America as an ornamental, since it survives under adverse conditions and produces large clusters of attractive, fragrant, violet-blue flowers. It is now a common invader in the warmer eastern areas. Easily identified by the striking purple flowers and the berries that go from green to bright red as they mature. The plant reproduces only by means of these berries.
Impact: It is usually found invading riverine bush, woodland and bush. It scrambles over indigenous bush and can replace desirable species. Although the berries are toxic to mammals, birds are very fond of them, and this could potentially result in the neglect of indigenous fruit and thereby cause the failure of adequate seed dispersal.
Control: There are no recommendations for the control of this plant. It should be removed by hand wherever it is encountered.

Solanum seaforthianum

SUCCULENTS

Within the section, the plants are arranged alphabetically according to family and genus.

SPECIES	Cactus-like	Flat or decumbent	Red flowers	Yellow flowers	White flowers
Opuntia spp., p. 376–381	■		■	■	■
Opuntia aurantiaca, p. 376	■		■		
Opuntia imbricata, p. 376	■		■		
Opuntia ficus-indica, p. 377	■		■	■	
Opuntia humifusa, p. 378	■		■	■	
Opuntia monocantha, p. 378	■			■	
Opuntia lindheimeri, p. 380	■			■	
Opuntia robusta, p. 380	■			■	
Cereus jamacara, p. 372	■				■
Harrisia martinii, p. 375	■			■	
Echinopsis spachiana, p. 374	■				■
Bryophyllum delagoense, p. 381			■		
Agave sisalana, p. 372				■	
Agave americana, p. 372				■	

AGAVACEAE

Agave sisalana
sisal * garingboom

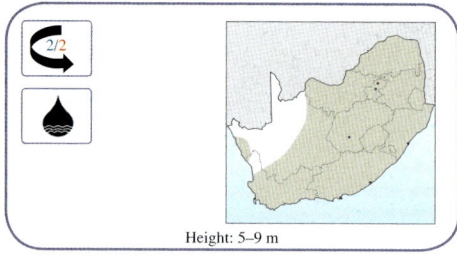

Height: 5–9 m

Origin and description: Introduced from Central and North America for various uses, which include as an ornamental, for fodder, for security hedging and as a source of fibre. Several species of *Agave* have become naturalised in South Africa and relic plantings remain throughout the country. *A. americana* is still harvested in the Karoo for the production of 'agava', which is an alcoholic drink similar to tequila.

Impact: *A. sisalana* forms an impenetrable barrier as the leaves are tipped with a spine. In fact, the

Other common names
hemp plant * sisal hemp
Similar species
A. americana X2/1b and various subspecies and varieties (American aloe * century plant * blou-aalwee * lekhala (S)), also from America
A. vivipara (kleingaringboom), from Mexico

plant is still used as a barrier around kraals and at some international borders. Large areas of Zululand that are infested with this plant as well as encroaching bushes have become inaccessible to humans, large game and livestock.

Control: *A. sisalana* can be controlled with the direct injection of concentrated MSMA into the sisal bole. When the plants have died and dried out, the area can be cleared by fire. Physical removal is almost restricted to the use of bulldozers.

Agave sisalana

Agave americana

CACTACEAE

Cereus jamacara
queen of the night * nagblom

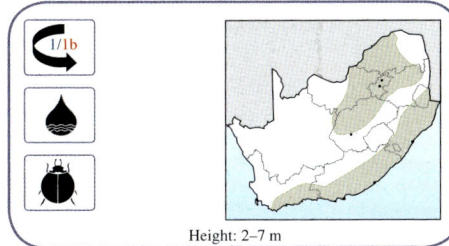

Height: 2–7 m

Other common names
Peruvian apple cactus * bobbejaanpaal

Origin and description: Introduced from South America as an ornamental and as a barrier plant, this cactus is now a serious alien invader in some parts of South Africa, particularly Mpumalanga. It is especially prevalent north of Pretoria around Kameelpoort and towards Rustenburg and Groblersdal and also in Namibia. In spring it produces attractive white flowers that open mainly at night and, unless it is cool and cloudy, will close again the next morning. The seeds are spread by birds and monkeys that feed on the fruit. The seeds fall under the trees where the birds and monkeys sit, and land on a shaded seedbed, which is ideal for the germination and growth of the cactus.

Impact: *C. jamacara* is found mainly in the veld, where it grows under and among trees. It replaces indigenous vegetation and prevents animals from finding food and shade. Chopped or broken branches are capable of taking root and forming new plants, so they should be buried deeply or burned. Pieces should not be carted away and discarded, as this is one of the most common ways in which new infestations begin. *C. jamacara* was proclaimed a weed in 1982.

Control: Small plants can be sprayed and larger ones injected with MSMA. They can also be chopped down, but the stem base must be dug up. It should then be deeply buried or burned. Some success is being achieved with the stem borer *Alcidion cereicola*, which was released in 1990. Every effort should be made to eradicate this plant, both on farms and in gardens, while serious infestations are still relatively uncommon.

Cereus jamacara

CACTACEAE

Echinopsis spachiana
torch cactus * orrelkaktus

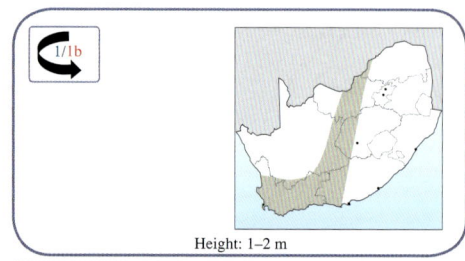
Height: 1–2 m

Origin and description: Introduced from Argentina as an ornamental and for hedging, this cactus has now invaded arid areas such as the Karoo. It grows in sandy, gravelly soils and is covered in tufts of sharp spines, with the central spine of each tuft always being much longer than the others. There are 128 species of *Echinopsis* in South America, but only this one has found a home in South Africa.

Impact: *E. spachiana* competes for precious moisture in savanna and arid areas. It will grow under trees, removing moisture and preventing wild and domestic animals from accessing the shade. In this way it can reduce the carrying capacity of the veld and the spines can cause injuries to the animals. This cactus is spreading rapidly and has the potential to become a serious problem.

Control: No herbicides are registered for this particular cactus, but it is likely to be susceptible to those methods used on other cacti.

Echinopsis spachiana

CACTACEAE

Harrisia martinii
moon cactus* toukaktus

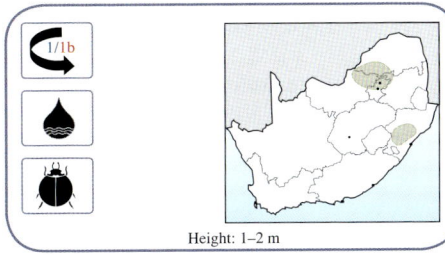
Height: 1–2 m

Other common names
torch cactus * kabelkaktus

Origin and description: Like the other cactus species (*Opuntia* spp.), the moon cactus also comes from South America, specifically Argentina, and was probably introduced around the turn of the century. It is now found in KwaZulu-Natal around Pietermaritzburg and in Mpumalanga and Gauteng, especially in the thornveld around Pretoria. *H. martinii* propagates vegetatively and by means of seeds, which are much favoured by birds, especially the black-eyed bulbul. However, the principal methods of dispersal are by birds and succulent collectors. It was declared a weed in 1968 but is still being cultivated (illegally) as an ornamental in gardens.

Impact: It sprawls over valuable grazing in the thornveld and can climb over small trees, completely smothering them.

Control: As with all cacti, *H. martinii* can be sprayed or injected with MSMA, or alternatively actively growing plants can be sprayed with triclopyr. Physical removal must be total as small stem sections can root and form new plants. The plant can be kept in check by biocontrol agents, but as usual with a plant under biological control, there are periodic flare-ups. *H. martinii* has been the subject of an intensive and successful control campaign that virtually eliminated it. However, in recent years it has reappeared as a problem plant and control efforts are being renewed.

Harrisia martinii

CACTACEAE

Opuntia aurantiaca
jointed prickly pear * litjiesturksvy
Opuntia imbricata
imbricate prickly pear * kabelturksvy

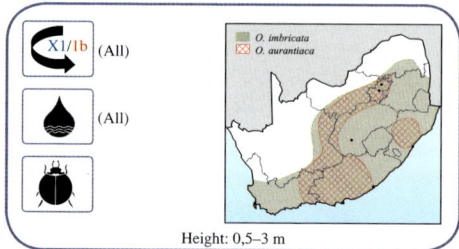

Height: 0,5–3 m

Origin and description: These 'prickly pears' are the more common ones of approximately 11 species that are declared weeds in South Africa. All of them are natives of the Americas. There is only one indigenous species of cactus of any kind in Africa, namely *Rhipsalis baccifera*, which is a small plant growing as an epiphyte (on another plant, but not parasitic) in evergreen forest. *O. aurantiaca* was originally introduced as a hybrid in the middle of the 19th century. It was planted for ornamental purposes in a garden in the Stockenstroom district of the Eastern Cape and was spread to remote mission stations by missionaries. It has escaped into the wild, becoming one of South Africa's most noxious and costly weeds.

O. imbricata is an inconspicuous perennial seldom exceeding a height of 50 cm. The length of its joints and spines depends on habitat and climate. Each piece that breaks off the main plant

Other common names
O. aurantiaca: jointed cactus * kaatjie * litjieskaktus * platturksvy

is capable of rooting and producing a new plant. If the plant is cut down, all the pieces must be collected and destroyed.

Impact: The ease of spread, rapid growth and unpleasant spines of these cacti result in infested areas rapidly becoming inaccessible, even to livestock. All invasive cacti seriously degrade veld, injure stock and damage their hair or wool.

Control: Chemical control is restricted to the spraying or injecting of MSMA or glyphosate, but it is time-consuming and costly. The introduction of cactoblastis as a biocontrol agent has greatly reduced the problem.

Opuntia imbricata

CACTACEAE

Opuntia ficus-indica
sweet prickly pear * boereturksvy

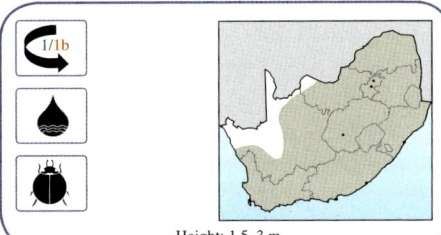

Height: 1,5–3 m

Origin and description: *O. ficus-indica* is one of several species of *Opuntia* that were introduced from Central America and was intended for use as a hedge, as fodder and for its succulent fruit. The date of its arrival is not certain, but it was probably during the 19th century. Although widespread, the main areas of infestation were in the Eastern Cape from Humansdorp to Aliwal North. However, serious problems also existed around Port St Johns, Dundee in KwaZulu-Natal and several other places further north. This cactus propagates easily from the leaf-pads or cladodes. Even a small piece lying on the ground can produce roots and flourish.

Impact: *O. ficus-indica* can be an aggressive invader, rendering heavily infested land virtually useless. Commercial varieties have been de-

Other common names
Indian fig * mission prickly pear * grootdoringturksvy * umthelekisi (Z)
Similar species
O. microdasys 1b (bunny ears * dwarf cactus * garden cactus * yellow bunny-ears)
O. spinulifera 1/1b (large round-leaved cactus * saucepan cactus * grootrondeblaarturksvy)
O. stricta 1/1b (Australian pest pear * suurturksvy)

Opuntia ficus-indica

veloped that are spineless and are cultivated in the warmer, drier areas of South Africa. The fruit is a delicacy and stock and game readily browse the cladodes ('leaves') but avoid the fruit. The small spines on the fruit of wild plants are highly irritating.
Control: Chemical control is possible with several herbicides such as MSMA and glyphosate. However, biological control with cactoblastis and cochineal (see page 25) has been so successful that special control measures are rarely required. Dense infestations are now rare and sporadic in nature.

Opuntia ficus-indica

CACTACEAE

Opuntia humifusa
large-flowered prickly pear

Opuntia monocantha
drooping prickly pear * suurturksvy

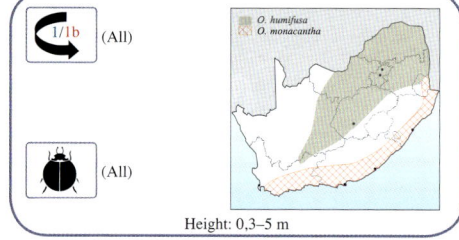
Height: 0,3–5 m

Origin and description: Two more species of *Opuntia* that are declared weeds in South Africa. *O. humifusa* was introduced from North America and was intended as an ornamental. *O. monocantha* comes from South America. These cacti propagate easily from the leaf-pads or cladodes. Even a small piece lying on the ground can produce roots and flourish.

Other common names
O. humifusa: creeping prickly pear * devil's tongue
O. monocantha: barbary fig * cochineal prickly pear * luisiesturksvy

Impact: Sometimes cultivated for the tasty fruit, or as a barrier hedge. The small spines on the fruit of wild plants are highly irritating. *O. humifusa* is found in the dry highveld areas, possibly where it has escaped cultivation, whereas *O. monocantha* favours the subtropical, moist coastal areas.
Control: Chemical control is possible with several herbicides such as MSMA and glyphosate.

Opuntia humifusa

Opuntia humifusa

Opuntia monocantha

CACTACEAE

Opuntia lindheimeri
small round-leaved prickly pear * klein-rondeblaarturksvy
Opuntia robusta
blue-leaf cactus * robusta-turksvy

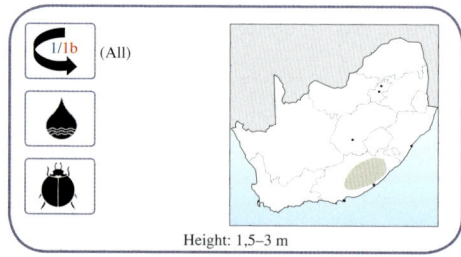

Height: 1,5–3 m

Origin and description: *O. lindheimeri* and *O. robusta* are two species of *Opuntia* that were introduced from Central America and were intended for use as a hedge, as fodder and for their succulent fruit. *O. lindheimeri* has a comparatively limited distribution and is found mainly around Somerset East in the Eastern Cape, whereas *O. robusta* is more widespread. These cacti propagate easily from the leaf-pads or cladodes. Even a small piece lying on the ground can produce roots and flourish.

Impact: These cacti are less serious invaders than some other species, but can still interfere significantly with access to and the use of land.

Control: Chemical control is possible with several herbicides such as MSMA and glyphosate.

Opuntia robusta

Opuntia lindheimeri

CRASSULACEAE

Bryophyllum delagoense *(=Kalanchoe delagoensis)*
chandelier plant * kandelaarplant

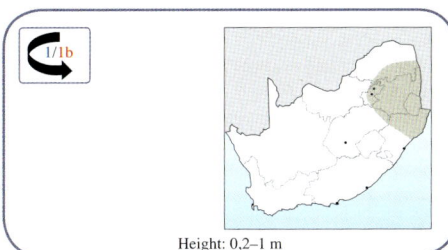
Height: 0,2–1 m

Origin and description: Originally from Madagascar, and cultivated in South Africa as ornamentals, several species of *Bryophyllum* have established themselves in certain parts of the eastern highveld and lowveld, coastal KwaZulu-Natal as well as in other parts of the world such as Asia and Australia. They are still popular house and garden plants and add interest to rockeries, but they can reproduce rapidly both from plantlets produced at the leaf tips that drop off and take root, as well as from seed.

Impact: These plants invade rocky places in open areas and in urban spaces. All three of these species are very poisonous to animals, both wild and domestic. *B. delagoense* is an invader in the Kruger National Park having escaped from gardens in the tourist camps.

Control: Unwanted plants should be uprooted and totally removed.

Other common names
mother of millions * mother of thousands
Similar species
B. pinnatum 1b *(=Kalanchoe pinnata)* (cathedral bells * life plant * miracle leaf)
B. proliferum 1b *(=Kalanchoe proliferum)* (green mother of millions)

Bryophyllum pinnatum

Bryophyllum delagoense

WATER WEEDS

Within the section, the plants are arranged alphabetically according to family and genus.

SPECIES	Yellow flowers	Blue flowers	White flowers	No visible flowers	Finely divided leaves
Nymphaea mexicana, p. 389	■				
Hydrocleys nymphoides, p. 389	■				
Eichhornia crassipes, p. 390		■			
Pontederia cordata, p. 392		■			
Nasturtium officinale, p. 385			■		
Pistia stratiotes, p. 383				■	
Azolla filiculoides, p. 384				■	
Myriophyllum aquaticum, p. 386				■	■
Hydrilla verticillata, p. 386				■	■
Lemna gibba, p. 388				■	
Salvinia molesta, p. 393				■	

ARACEAE

Pistia stratiotes
water lettuce * waterslaai

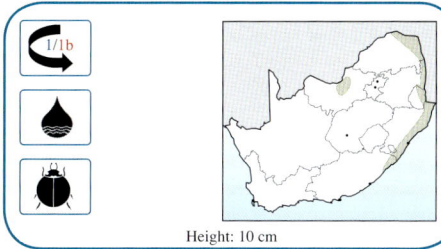

Height: 10 cm

Origin and description: *P. stratiotes* is a weed of uncertain origin. Although it was first described in Egypt in AD 77, it is now found in tropical areas throughout the world. In South Africa it was first seen in northern KwaZulu-Natal but it has subsequently been found in eastern Mpumalanga and the Eastern Cape. It is now a declared weed and a major problem in certain water systems. Resembling a lettuce floating on the water, this plant has a mass of roots and short rhizomes suspended below. In the centre there is a large number of inconspicuous unisexual flowers, which are capable of producing limited numbers of seed. The principal means of propagation, however, is vegetative. It is commonly cultivated in water gardens and aquaria.

Impact: *P. stratiotes* can block watercourses and provides a habitat for mosquitoes and bilharzia-carrying snails. It can reduce the biodiversity of a waterway by blocking light, and killing native, submerged plants. The plant can survive periodical drying up of its habitat, but experimental biological control agents cannot, and their effect is therefore eliminated.

Other common names
Nile cabbage * Nile lettuce * shell-flower * water cabbage * water fern

Control: Mechanical or hand removal of the plants is usually effective, but herbicides such as terbutryn have been registered for aerial application. If large areas are sprayed at one time, the massive volume of decomposing vegetable matter can cause severe oxygen depletion. Considerable success is being achieved with biological control using the South American weevil *Neohydronomus affinis*, which was first released in 1985. In any chemical control programme, surviving plants should not be sprayed as they act as a refuge for the biocontrol agents, which will continue to keep the weed in check.

Pistia stratiotes

AZOLLACEAE

Azolla filiculoides
water fern * watervaring

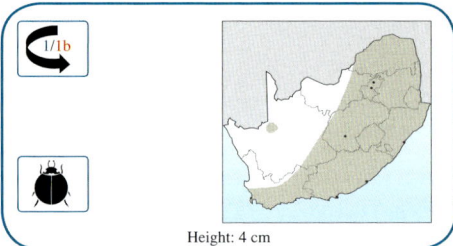
Height: 4 cm

Origin and description: There are three species of *Azolla* in Africa. This one is redder but it would require a microscope for proper identification. *A. filiculoides* is a native of South America and was brought over in the 1950s, probably as an aquarium plant. It has now established itself over a wide area of South Africa, from the Northern Cape (Gariep Dam) to central Mpumalanga and northern Limpopo. It has recently been found in Durban. *A. pinnata* is indigenous and *A. nilotica*, also from South America, has only been found as far south as Zimbabwe.

Azolla is a true fern and reproduces by means of simple vegetative methods (division) or can produce spores in the process of sexual reproduction. These spores can be transported over long distances by floodwaters, birds and other animals. *Azolla* has a symbiotic relationship with a blue-green alga, *Anabaena azollae*, which is present in the upper lobe of each leaf and is capable of fixing sufficient atmospheric nitrogen to supply all the nitrogen needs of the plant. In parts of Asia, the water fern is used as green manure and as a pig and duck food.

Impact: *A. filiculoides* can be found in small dams and rivers and sheltered areas of large ones. It is often found in the small dams on golf courses and can form a mat up to 0,3 m thick. It clogs waterways, provides shelter for mosquitoes and bilharzia-carrying snails and prevents birds, other animals and people from making normal use of the water. Furthermore, it reduces the

> **Similar species**
> *A. filiculoides*, not to be confused with *Salvinia molesta*, which is also called 'water fern' (see page 393)

Azolla filiculoides

quality of drinking water and severely affects aquatic biodiversity.

Control: Herbicides are probably not an option, but filtration or physical removal is. Fortunately a weevil (*Stenopelmus rufinasus*), that was brought over from South America, is proving a major success. In fact, it has on occasions given complete control on a water body where not a single plant remains, which is remarkable for biocontrol. Although there has been a dramatic collapse of infestations in South Africa as a result, it is likely that there will still be sporadic outbreaks of the weed.

Azolla filiculoides

BRASSICACEAE

Nasturtium officinale *(= Rorippa nasturtium-aquaticum)*
watercress * sterkkos

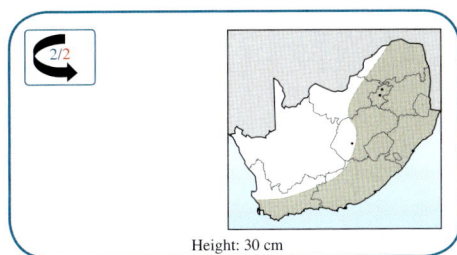
Height: 30 cm

Origin and description: *N. officinale* is from Europe and Asia and was introduced as a horticultural crop during the early settler days. It has escaped into the wild and is now a widespread weed. *N. officinale* or watercress is well-known throughout the world as a salad plant.

Impact: It is a perennial plant that usually lives in water but is rooted in the ground. It can clog waterways, increase water loss and act as shelter and breeding site for bilharzia-carrying snails and mosquito larvae. It is common in ditches, at the edge of dams and on riverbanks where it competes aggressively with indigenous species.

Control: There are no herbicides registered for this weed and it is best removed by hand.

Other common names
brongras * bronkhorstslaai * bronkors * bronslaai * stercors * kerese (S)

Nasturtium officinale

HALORAGACEAE

Myriophyllum aquaticum
parrot's feather * waterduisendblaar

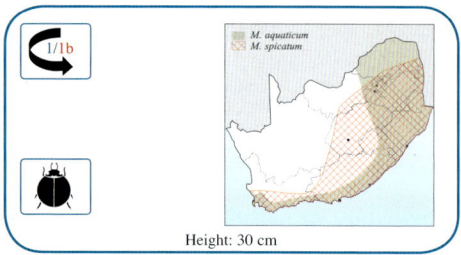

Height: 30 cm

Origin and description: A native of South America, *M. aquaticum* is now found throughout the world. It was introduced into South Africa in about 1919, being first recorded in the Noorder-Paarl in the Western Cape. It is now a declared weed and is widespread in South Africa, especially in the southwestern Cape, KwaZulu-Natal, and southern and eastern Mpumalanga. It is a rooted water plant that can grow into dense mats in clear, polluted or brackish water. Although it produces flowers, only female plants are known to occur in southern Africa. Reproduction is entirely vegetative.

Impact: Parrot's feather clogs water bodies, interfering with fishing, fish culture and navigation. It can cause drowning of animals and humans. Apart from blocking watercourses, the weed provides a breeding place for mosquitoes and bilharzia-carrying snails. In frost areas the above-water parts die back, producing a mass of rotting vegetation. Infested waterholes that dry up will rapidly become re-infested when the rains return.

Control: There are no chemicals registered for use on this weed at present and physical removal is the only alternative. Any small piece that remains will take root and grow, so removal must be thorough and cleared areas regularly checked for regrowth. Biological agents (*Lysathia* sp.) that were released in 1994 are proving to be significant.

Other common names
water feather *
water milfoil * uphaphe (Z)
Similar species
M. spicatum (spiked water milfoil), from Eurasia and is more submerged
Ceratophyllum demersum, indigenous, submerged and can clog water bodies in the eastern parts

Myriophyllum aquaticum

HYDROCHARITACEAE

Hydrilla verticillata
hydrilla

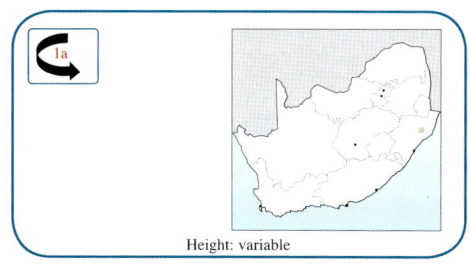

Height: variable

Origin and description: A water plant native to the warmer parts of Asia that was probably introduced by accident. This is because it is so similar to *Elodea canadensis* and *Egeria densa*, which are two species of alien water plants that are already well established in South Africa. In fact, the only significant difference to these two is the possession of leaflets with serrated edges and the number of these leaflets in a whorl. This is the only species of *Hydrilla* in the world but there are also some very similar indigenous species. It was first recorded in the region from the Komati River in Mozambique in 1961 but so far, hydrilla has only been found in the Pongolapoort Dam in northern KwaZulu-Natal. That does not mean it has not invaded elsewhere. It is a resilient waterweed with a high resistance to salinity and can tolerate a very wide range of conditions.

Impact: Hydrilla can form dense underwater masses, which can replace all other plants. It can grow in water up to 12 m deep and can hinder commercial and recreational use of the water, even making swimming dangerous. Major infestations can limit the size and weight of sport fish.

Control: Hydrilla can be controlled by aquatic herbicides, although no specific registration exists in South Africa. It is eaten by grass carp and is the subject of biological control studies. Any control or removal programme must be followed up regularly since the tubers can lie dormant in the sediment for a number of years. Any visitors to this dam must be careful not to take fragments of the weed away with them, such as on their boat or trailer.

Hydrilla verticillata

LEMNACEAE

Lemna gibba
duckweed * damslyk

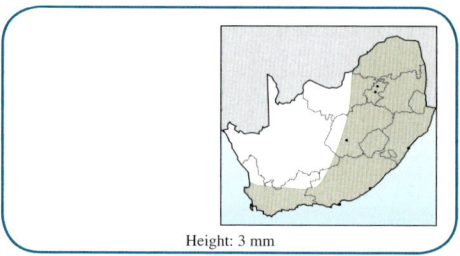
Height: 3 mm

Origin and description: *L. gibba* is an indigenous weed of pools, dams and river backwaters throughout southern Africa. It is a very small, free-floating plant with each leaf-like structure bearing a single root up to 10 cm long. The minute flowers are seldom seen and are contained in small pockets. Duckweed is capable of producing seeds but it reproduces mainly by budding. It favours nutrient-rich water and can quickly cover a wide expanse of water. There are three species of *Lemna* in South Africa, two species of the very similar *Spirodela*, and two species of *Wolffia*. They are all referred to as duckweed or 'damslyk'.

Other common names
L.gibba: duck's meat * frog-buttons * eendekroos
S.spirodela: great duckweed * water flaxseed

Impact: Young plants are eaten by ducks, but as the plants mature and decay they absorb oxygen and give off carbon dioxide and hydrogen sulphide, making the water unpalatable to livestock and unsuitable for fish. The plant itself is very palatable and in some countries, dried duckweed is used as a highly nutritious cattle feed. A duckweed farm can produce up to 30 tons of dry matter per hectare per year.

Control: Although registrations exist elsewhere, in South Africa there are no chemicals registered for the control of *L. gibba*. In order to minimise the risk of invasion it is important to prevent the inflow of nutrients (e.g. run-off from fertilised fields) and to keep the water agitated.

Lemna gibba

388

NYMPHAEACEAE

Nymphaea mexicana
yellow water lily * geelwaterlelie

Hydrocleys nymphoides
water poppy

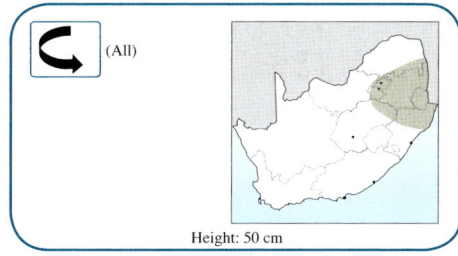

Height: 50 cm

Origin and description: Water poppies and water lilies were introduced from South America as ornamentals. *N. mexicana* has invaded rivers and dams from the Vaal River to the Lowveld and it has also been found in KwaZulu-Natal. *H. nymphoides* can be seen in botanical gardens but has yet to be found in the wild; it has the potential to become seriously invasive. There are hybrids of both these species and several related indigenous species of both, which can all become problematic under certain circumstances. *N. mexicana* has long, spongy, creeping stolons, that look like small bananas and which can produce fresh plants.

Impact: Water lilies and poppies can carpet the surface of the water with their large, flat leaves. These dense patches can exclude other species and create stagnant areas with low oxygen levels under the floating mat. They can also interfere with recreational activities such as boating, fishing or swimming.

Control: Herbicides are generally not recommended for larger areas because the dying and decomposing plant material removes oxygen from the water. It is not thought that grass carp eat lilies, so the best method of removal is physical. In ornamental ponds, the bottom can be covered with a barrier material to prevent rooting.

Similar species
Nymphaea lotus (white water lily), indigenous
Nymphoides indica (yellow pond lily), indigenous
Nymphoides peltata (fringed water lily * yellow floating heart), Eurasia

Hydrocleys nymphoides

Nymphaea mexicana

PONTEDERIACEAE

Eichhornia crassipes
water hyacinth * waterhiasint

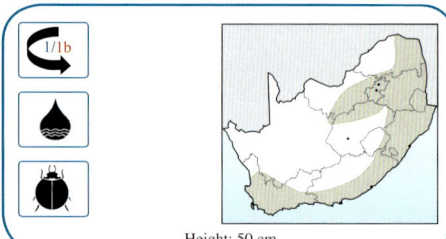

Height: 50 cm

Origin and description: Originally from the Amazon River in South America. People started using this attractive plant as a feature in fishponds as far back as 1884. It was introduced into South Africa for this purpose shortly before 1910. It immediately escaped and is now well established in all provinces and indeed, throughout Africa. *E. crassipes* is now considered the world's worst waterweed.

Generally this weed is free-floating but it can become anchored in shallow water. Only when it is anchored does it produce flowers and seeds, otherwise it reproduces by runners. The seeds have a recorded longevity of 15 years. The roots are long and feathery, helping to keep the plant balanced; the petiole (leaf stalk) is swollen and filled with air to keep the plant afloat.

Impact: Infested lakes, dams and rivers become blocked, preventing navigation, harbour access and fishing and increasing evaporation; biodiversity is destroyed and the entire ecosystem is threatened. Under ideal conditions the number of plants can double every five to 15 days and the floating masses can block irrigation canals and interfere with pumps and hydroelectric schemes. They also harbour mosquito larvae and bilharzia-carrying snails. Cattle have drowned by stepping onto a seemingly solid surface of *E. crassipes*. Many infestations in Africa that were coming under control in the 1990s are now flourishing again on account of pollution and an increased nutrient load of the water.

Control: The plant is easy to pull from the water and can be killed by spreading it on the ground in the sun to dry and then burning it. Chemical control is also possible, but this must be done under strict supervision to prevent contamination of the water and is usually for maintenance treatments or for small patches that pose a threat. Rapid decomposition of large weed masses in the water may cause an oxygen shortage, resulting in the deaths of aquatic flora and fauna. Several natural enemies, including weevils, moths and fungi, have been introduced onto major infestations over the years with varying

Other common names
*Nile lily * Nyllelie * snotterbel *amazibo (Z)*

Similar species
E. azurea (anchored water hyacinth), from Central and South America. It has not yet been recorded in South Africa, so its introduction should be seriously avoided. It is very similar, but is always rooted, does not have swollen petioles and the petals of the flowers are serated

Eichhornia crassipes

degrees of success. These agents take two to six years to become established but can be very effective in the long term. Using the plant to make paper, fibre-board, yarn, charcoal briquettes or even biogas and fish food could greatly assist control programmes.

Water weeds

Eichhornia crassipes

PONTEDERIACEAE

Pontederia cordata
pickerel weed * jongsnoekkruid

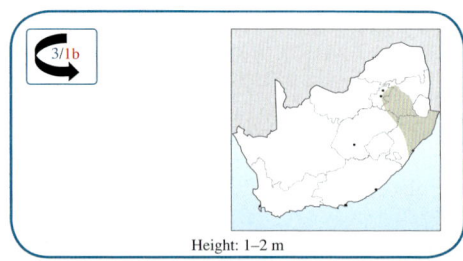
Height: 1–2 m

Origin and description: Indigenous to North, Central and South America, pickerel weed has been introduced to South Africa as an ornamental aquatic plant. It is now a weed of dams and waterways, especially in the Kwazulu-Natal Midlands, but also in Gauteng, Mpumalanga and in the Kruger Park. The South African form, however, does not appear to produce fruit and it only spreads vegetatively from fragmented rhizomes.
Impact: This plant can be a prolific grower and under the right conditions can totally fill a dam, rendering it useless and destroying biodiversity.
Control: This plant is best removed physically, because herbicide-treated, decaying plant masses can seriously de-oxygenate and degrade the water. Care should be taken to remove the entire root system.

Pontederia cordata

SALVINIACEAE

Salvinia molesta
salvinia* watervaring

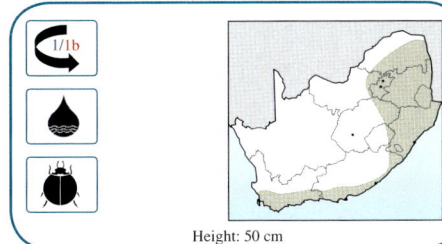
Height: 50 cm

Origin and description: Introduced from tropical America as an ornamental pond plant, this free-floating aquatic fern is now a widespread perennial waterweed in parts of Asia and many countries in Africa, including South Africa. *S. molesta* is frequently found in aquaria and small dams from which infestations can spread to nearby rivers. Under favourable conditions the plant multiplies rapidly, such as in Lake Kariba, where it was first recorded in 1959. Three years later the infestation covered 1 000 km² of the lake margin. The leaves of the weed are covered in coarse hairs; these hairs trap air bubbles and then keep the plant afloat. It is a sterile hybrid that propagates only vegetatively and can easily regenerate from any small fragment as long as it includes a growing point.

Impact: *S. molesta* forms mats up to 50 cm thick that can choke waterways, increase water loss, interfere with recreational use of the water and provide a haven for mosquitoes and bilharzia-carrying snails. Furthermore, it reduces the quality of the water and severely affects the aquatic biodiversity.

Control: Snout-beetles (*Cyrtobagous salviniae*) were released as biological control agents in 1985 and are proving very effective. Diquat was registered for both ground and aerial application. In small areas the weed can be controlled mechanically, but great care must be taken to remove every small piece that could break off and form a new plant.

Other common names
African pyle * Kariba weed * water fern

Similar species
S. hastate (indigenous)
S. molesta should not be confused with *Azolla* spp., which is also called 'water fern' and 'waterva-ring' (see page 384)

Salvinia molesta

GLOSSARY

adventive	A recent introduction which is now established but still spreading.
aggregate	A non-uniform group of forms or variants within a species.
alien	A plant introduced by man from elsewhere.
allelopathy	The production and release of chemical agents from one plant that can influence the growth of another.
annual	A plant completing its life cycle, from germination to ripening of the seed, in less than one year.
application	Method of applying a plant protection product such as a herbicide.
aquatic	Growing in water.
aromatic	Having a strong smell, especially when crushed.
awn	A bristle-like projection borne at the end or from the side of an organ; frequently present in the spikelets of grasses.
axil	The angle between a leaf and the stem to which it is attached.
basal stem	Referring to the application of a herbicide to the root crown, exposed roots and stem up to a height of about 250 mm of a growing tree or shrub.
berry	Juicy fruit with seeds immersed in pulp.
biennial	A plant that requires two summers for development, with a winter in between. The vegetative phase is completed in the first year and the reproductive phase in the second year.
biological control	The use of living organisms to control pests.
blade	The flattened and often broad part of a leaf.
bulb	An underground storage organ made up of fleshy leaf bases.
bulbil	A small bulb arising from the base of a larger bulb.
bur	A prickly or spiny fruit.
capsule	A dry fruit which opens at maturity to release the seeds.
chlorophyll	The green colouring matter in plants.
common	Occurring frequently and in abundance within a specified range.
competitive	Referring to a plant that is able to flourish in the presence of and often to the detriment of other plants.
conservation tillage	Reduced tillage practices with the primary aim of preventing soil erosion.
coppice	Regrowth from a stump or stem that has been cut down.
cosmopolitan	Occurring in all parts of the world (under suitable conditions).
cotyledonous leaves	In dicotyledons, the first two leaves that appear above the ground.
creeper	A trailing shoot that takes root along most of its length or at the nodes, thereby assisting in the gradual spread of a plant.
cultivar	A horticultural variety, strain or race that was produced by cross-breeding.
cultivation	The act of growing plants in general, tilling the soil or disturbing the soil mechanically with the specific aim of destroying weeds.

GLOSSARY

cut-stump	Referring to the application of a herbicide to the cut surface of freshly cut stumps.
deciduous	Shedding leaves at the end of the growing season.
decumbent	With a stem that reclines, at least at first, even if it rises later.
dicotyledons	Plants with two cotyledons or halves in the seed; otherwise known as broadleaf plants.
donga	An eroded gully with steep sides.
ecology	The study of the interrelationships between organisms, and between them and their environment.
eelworm	See nematode.
evergreen	Retaining leaves throughout the year, even during winter.
fallow land	Cultivated land that is unused or is being rested for one or more years.
fodder	Animal food.
foliar application	Applying herbicides to the leaves.
frill	Referring to the application of a herbicide to a frill which is cut around the total circumference of a tree, taking care to cut through the bark to the cambium.
fruit	The dry or fleshy structure containing the seeds.
fynbos	Veld in which the main vegetation is indigenous non-grasses. The plants are usually adapted for arid conditions and have fine ('fyn'), waxy leaves; typically occurs in the southwestern Cape.
gland	A small globular structure containing liquid, either sunk into the leaf or borne at the tip of a slender stalk (glandular hair).
habit	Appearance, outer form.
herbaceous	Not woody; soft and leafy.
herbicide	A substance for controlling weeds.
hybrid	The offspring produced by mating two genetically different plants.
hygroscopic	Having the ability to absorb moisture from the air.
indigenous	Native or originating from a place.
inflorescence	Any arrangement of more than one flower; the flowering portion of a plant.
internode	Part of the stem between two nodes.
introduced	See 'alien'.
invader	A species that is not indigenous to a particular area.
invading	The process of an alien plant establishing itself in a new area.
latex	Milky juice in some plants.
lands	In the South African sense, cultivated fields.
minimum tillage	Same as conservation tillage, except that the ultimate aim, for example, may be to reduce costs and not necessarily to prevent erosion.
monocotyledons	Plants with one embryonic leaf in the seed, e.g. the grasses.

GLOSSARY

Native	See 'indigenous'.
naturalised	Introduced, but now established and able to reproduce independently in a new environment.
nematodes	Minute, worm-like organisms that usually live in the soil and can attack and damage plant roots; also called eelworms.
node	The point on a main stem or branch where leaves or buds arise.
noxious weed	A weed proclaimed in the Weeds Act, 1937 (Act No. 42 of 1937).
parasite	A plant that obtains its sustenance from another living plant (the host) to which it is attached.
perennial	A plant living for two or more years. It reproduces vegetatively as well as through seeds.
perennial crop	A crop that grows for more than one year, e.g. sugarcane, pineapples, etc.
pioneer	A plant capable of invading bare or disturbed sites and persisting there until replaced by other species.
pod	A type of fruit that splits longitudinally into two valves.
post-emergence	Referring to the use of a herbicide after the appearance of the weed (sometimes used to describe the use of a herbicide after the emergence of the crop).
pre-emergence	Referring to the use of a herbicide before the emergence of the weeds (or crop).
prostrate	With stems lying flat on the ground.
rhizome	A stem that is usually horizontal and underground, and that produces rootlets and aerial shoots.
ringbarking	The removal of the bark of a shrub or tree in a complete ring around the trunk.
rootstock	Strictly a short, erect underground stem. The term is used loosely, however, to describe perennial underground roots or other organs.
runner	An elongated stem growing horizontally above the ground and rooting at the nodes to form new plants.
seedling	An imprecise term, but it generally refers to a plant before it reaches the 6-8-leaf stage.
shrub	A perennial woody plant with usually one or more stems arising from or near the ground.
siliqua(e)	Fruit of the Brassicaceae resembling a 'pod', more than three times as long as it is broad. The two valves break away from the central portion from base to apex.
selective	Referring to a herbicide that can be sprayed over a crop and which selectively controls the weeds without harming the crop.
spikelet	The unit of the grass inflorescence.
stolon	See 'runner'.
stool	A clump of shoots.

GLOSSARY

succulent	A plant with fleshy and juicy stems and leaves that contain reserves of moisture.
sucker	A shoot arising from the roots of a woody plant, often some distance away from the main stem.
symbiotic	Referring to two or more species of organisms living in a close relationship that is to their mutual benefit.
sward	A grass turf or sod.
tap root	An unbranched, vertically descending root.
tiller	A side shoot. Once grasses, in particular, have become established, they go through a tillering growth phase before vertical growth commences.
translocation	The movement of nutrients or herbicides in the sap of a plant from one place to another, e.g. from the leaves to the roots or vice versa.
tree	A large woody plant with a single stem or trunk.
tuber	A short, thickened portion of an underground stem bearing dormant buds, e.g. a potato.
variety	A subdivision of a species officially ranking between subspecies and forma.
vegetative reproduction	Asexual reproduction through the formation of growth structures that are an extension of the original plant.
veld	A general term for countryside and natural vegetation, often implying a strong grassy component.
volunteer	A crop plant that is regenerating as a weed, often in a crop following on the previous crop.
widespread	Occurring in most parts of the country.

BIBLIOGRAPHY

BOTHA, C. 2001. *Common weeds of crops and gardens in southern Africa.* Grain Crops Institute, Agricultural Research Council, Pretoria.

BROMILOW, C. 2001. *Problem plants of South Africa.* Briza Publications, Arcadia.

CIBA-GEIGY (PTY) LTD. 1985. *Weeds of crops and gardens in southern Africa.* Seal Publishing, Johannesburg.

DEPARTMENT OF AGRICULTURE. 2001. Conservation of Agricultural Resources Act, 1983 (Act No. 43 of 1983), Regulations: Amendment. *Government Gazette,* 30 March 2001.

HENDERSON, L. 2001. *Alien weeds and invasive plants.* Agricultural Research Council, Pretoria.

HENDERSON, L. 2007. Invasive, naturalised and casual alien plants in southern Africa: a summary based on the Southern African Plant Invaders Atlas (SAPIA). *Bothalia* 37,2: 215–248.

HENDERSON, L., CILLIERS, C.J. 2002. *Invasive aquatic plants.* Plant Protection Research Institute, Agricultural Research Council, Pretoria.

HENDERSON, M., FOURIE, D.M.C., WELLS, M.J., HENDERSON, L. 1987. *Declared weeds and alien invader plants in South Africa.* Department of Agriculture and Water Supply, Pretoria.

HENDERSON, M.D., ANDERSON, J.G. 1966. Common weeds in South Africa. *Memoirs of the Botanical Survey of South Africa*, No. 37.

PIENAAR, K. 1992. *What flower is that?* Struik, Cape Town.

POOLEY, E. 1998. *A field guide to wild flowers in KwaZulu-Natal and the eastern region.* Natal Flora Publications Trust, Durban

VAN OUDTSHOORN, F.P. 1992. *Guide to grasses of South Africa.* Briza Publications, Arcadia.

VAN WYK, A.E., MALAN, S.J. 1988. *Field guide to the wild flowers of the Witwatersrand and Pretoria region.* Struik, Cape Town.

VAN WYK, B-E., VAN HEERDEN, F.R., VAN OUDTSHOORN, B. 2002. *Poisonous plants of South Africa.* Briza Publications, Arcadia.

WELLS, M.J., BALSINHAS, A.A., JOFFE, H., ENGELBRECHT, V.M., HARDING, G., STIRTON, C.H. 1986. Catalogue of problem plants in southern Africa. *Memoirs of the Botanical Survey of South Africa*, No. 53.

WEBSITES
www.sibis.sanbi.org
www.biodiversityexplorer.org
www.agis.agric.za
www.issg.org
www.plantzafrica.com
www.hear.org/pier/
www.weeds.org.au
www.fernkloof.com

INDEX

A
aarbeikoejawel 55, 144
aartappelranker 57, 370
abele 155
Abessiniese coleus 54, 186
abiekwasgeelhout 162
Acacia ataxacantha 17
Acacia baileyana 44, 121
Acacia borleae 17
Acacia burkei 17
Acacia caffra 17
Acacia cyanophylla 126
Acacia cyclops 27, 44, 122
Acacia dealbata 44, 123, 124
Acacia decurrens 44, 123
Acacia elata 44, 125, 126
Acacia erioloba 17
Acacia erubescens 17
Acacia exuvialis 17
Acacia fleckii 17
Acacia galpinii 17
Acacia gerrardii var. *gerrardii* 17
Acacia grandicornuta 17
Acacia hebeclada 17
Acacia hereroensis 17
Acacia implexa 44
Acacia karroo 17
Acacia longifolia 26, 27, 44, 125
Acacia luederitzii var. *luderitzii* 17
Acacia luederitzii var. *retinens* 17
Acacia mearnsii 27, 44, 123, 124
Acacia melanoxylon 27, 44, 122
Acacia mellifera subsp. *detinens* 17
Acacia nebrownii 17
Acacia nigrescens 17
Acacia nilotica subsp. *kraussiana* 17
Acacia paradoxa 44
Acacia pendula 44
Acacia permixta 17
Acacia podalyriifolia 44, 121, 140
Acacia pycnantha 27, 44, 125, 126
Acacia reficiens 17
Acacia robusta 17
Acacia saligna 28, 44, 126, 127, 143
Acacia senegal var. *rostrata* 17
Acacia sieberiana var. *woodii* 17
Acacia spp. 11, 15, 33, 112
Acacia stricta 44
Acacia tenuispina 17
Acacia terminalis 125
Acacia tortilis subsp. *heteracantha* 18

ACANTHACEAE 310
Acanthospermum australe 316, 317
Acanthospermum brasilium 316, 317
Acanthospermum hispidum 316
Acanthospermum spp. 197, 308
Acer buergerianum 44, 114
Acer negundo 44, 114
Acer spp. 112
ACERACEAE 114
Achillea millefolium 200, 210
Achyranthes aspera var. *aspera* 44, 197, 203
Achyranthes repens 203
adelaarsvaring 265
African flame tree 57
African goosegrass 82
African purslane 310
African pyle 393
African weeping wattle 19
African wild clover 333
Afrikaanse vlamboom 57
Agave vivipara 372
AGAVACEAE 372
Agave americana 44, 371, 372
Agave sisalana 44, 371, 372
Ageratina adenophora 44, 168, 200, 210, 211
Ageratina riparia 44, 211
Ageratum conyzoides 44, 212
Ageratum houstonianum 44, 212
Ageratum spp. 199
Agrimonia odorata 297
Agrimonia procera 44, 197, 297
agrimony 44, 297
Agrostis montevidensis 63
agurkie 330
Ailanthus altissima 11, 44, 112, 158
AIZOACEAE 310
akaroa 77
akker-ereprys 346
akkermonie 297
akkerwinde 47, 358
Albizia julibrissin 44, 127, 128
Albizia lebbeck 45, 127, 128
Albizia lophanta 131
Albizia procera 45, 128
Albizia spp. 11, 112, 128
aleppo grass 106
aleppo pine 54, 147
aleppoden 54, 147
algaroba 132, 176
Algerian ash 49

Algeriese esseboom 49
Alhagi camelorum 175
Alhagi maurorum 45, 163, 175
Alisma plantago-aquatica 45
ALLIACEAE 202
Allium gracile 202
Alnus glutinosa 45
Alpinia zerumbet 45, 307
Alston's saltbush 173
Alternanthera pungens 227, 236, 308, 311
Alternanthera repens 311
amafusine 74
amahashe 290
amalenjane 217
AMARANTHACEAE 203, 204, 205, 206, 311, 312, 313
Amaranthus 36, 37, 261, 341
Amaranthus deflexus 204, 206
Amaranthus hybridus subsp. *hybridus* 205
Amaranthus spp. 37, 201, 205, 261
Amaranthus spinosus 197, 206
Amaranthus thunbergii 204
Amaranthus viridis 206
amate 117
amazibo 390
Amazon sword plant 48
Ambrosia artemisiifolia 201, 213
amelanyane 342
American aloe 372
American ash 49
American bramble 56, 192
Amerikaanse braambos 56, 192
Amerikaanse driedoring 49
Amerikaanse esseboom 49
Ammi majus 200, 207
Ammophila arenaria 45
amourette 178
Amsinckia menziesii 199, 200, 245, 246
ANACARDIACEAE 115, 116
Anagallis arvensis 267, 308, 323, 343, 344
anchored water hyacinth 390
annual bluegrass 100
annual ragweed 213
annual ryegrass 87
annual scleranthus 259
annual tree mallow 277
annual veld grass 82
annual yellow sweet clover 271
Anredera basalloides 354
Anredera cordifolia 45, 350, 354, 355

ant tree 58, 150
Anthemis arvensis 214
Anthemis cotula 214
Anthemis spp. 199
Anthospermum aethiopicum 19
Antigonon leptopus 45
APIACEAE 207, 208, 209, 313
apiesdoring 17
Apium leptophyllum 208
APOCYNACEAE 165, 166, 168, 315
Appelblaar 18
appelliefie 300, 301
apple-leaf 18
apple-of-Peru 52, 300
ARACEAE 351, 383
ARALIACEAE 117, 352
Araujia sericifera 45, 350, 353, 354
Arctotheca calendula 199, 214
Arctotis leiocarpa 215
Arctotis venusta 200, 215
Ardisia crenata 45, 163, 187
Ardisia elliptica 45
Argemone mexicana 45, 286, 287
Argemone mexicana forma *mexicana* 286
Argemone ochroleuca subsp. *ochroleuca* 45, 286, 287
Argemone spp. 197
Argemone subfusiformis 286
Argyle apple 140
Aristea africana 38
Aristida spp. 20
Aristolochia elegans 45, 350, 352, 353
ARISTOLOCHIACEAE 352
armmanslusern 341
armoedskruid 224
Armstrong's weed 168
arrow-head philodendron 351
arrow-head vine 351
arrow-head vines 57
arrow-leaf sida 279
arsenic bush 180
Arundo donax 45, 64, 65, 98
asbossie 170
ASCLEPIADACEAE 353
ash-leaved maple 114
ASPARAGACEAE 167
Asparagus laricinus 163, 167
Asparagus spp. 18
aster 319
ASTERACEAE 169, 170, 171, 210, 212, 213, 214, 215, 216, 217, 219, 220, 221, 222, 223, 224, 225, 226, 227, 228, 229, 230, 231, 232, 234, 235, 236, 238, 239, 241, 242, 243, 244, 245, 316, 317, 318, 319, 320
asthma plant 330
Athanasia crithmifolia 19
Athel tamarisk 162
Athel tree 57, 162
Atriplex inflata 45
Atriplex lindleyi subsp. *inflata* 45, 201, 260
Atriplex nummularia subsp. *nummularia* 45, 163, 173
Atriplex semibaccata 309, 324
Augustinuskweek 107
Australian albizia 131
Australian blackwood 44
Australian cabbage tree 56, 117
Australian cheesewood 54, 150
Australian crimson oak 49
Australian myrtle 50, 143
Australian pest pear 377
Australian salt bush 173, 324
Australian silky oak 49, 151
Australian tree fern 41
Australian water pear 57, 145
Australian wattle 123
Australiese basboom 123
Australiese boomvaring 41
Australiese brakbos 173, 324
Australiese gras 77
Australiese kasuur 54, 150
Australiese kiepersol 56, 117
Australiese mirteboom 27, 50, 143
Australiese rooi-eik 49
Australiese silwereik 49, 151
Australiese swarthout 44
Australiese waterpeer 57, 145
Avena barbata 65
Avena byzantina 65
Avena fatua 65, 66
Avena sativa 65
Avena sterilis 65
Azima tetracantha 18
azolla 45
Azolla filiculoides 382, 384
Azolla nilotica 384
Azolla pinnata 45, 324
Azolla sp. 393
AZOLLACEAE 384

B
baaibos 122
Babiana stricta 38
baboonflower 35
baby pepper 291
baby's tears 325
Babylon willow 157
bachelor's button 220, 312
bahia paspalum 93
bahiagras 93
Bailey's wattle 44, 121
Bailey-se-wattel 44, 121
balloon vine 46, 369
bamboes 66
bamboo reed 64
Bambusa balcooa 66, 67
Bambusa glaucescens 66, 67
banana poka 366
bananadilla 53, 366
band plant 315
bankrotbossie 20, 170, 236
bankrupt bush 20, 170, 236
Barbados gooseberry 54, 356
Barbados-stekelbessie 356
barbary fig 378
barnyard grass 80
barnyard millet 80
Bartlettina 45
Bartlettina sordida 45
basden 54, 148
BASELLACEAE 354
bastard mustard 258
bastard raisin 18
baster haak-en-steek 17
baster kakiebos 217
basterappelliefie 52, 300
basterknoffel 202
basterlebbeck 45, 127
basterrosyntjie 18
bastersinjaalgras 68
Bathurstbush 243
Bauhinia galpinii 129
Bauhinia petersiana subsp. *macrantha* 129
Bauhinia purpurea 45, 128
Bauhinia spp. 112
Bauhinia variegata 45, 129
bead tree 137
bearbind 368
beard grass 100
beauty-of-the-night 281
beddinggras 85
bedstraw 298
beefwood 46
beesgras 70, 110
beggar tick 217
beggarsticks 217
begraafplaasblom 46, 315
belambra boom 146
belhambra 54, 146
belly thorn 17
Benghal wandering Jew 326
Berberis thunbergii 45
bergdoring 17
bergkatdoring 167
bergrooigras 88

Berkheya erysithales 216
Berkheya rigida 216
Berkheya spp. 197
Bermuda buttercup 338
Bermuda grass 76
berry saltbush 324
berry tree 137
bessieboom 137
bewertjies 69
Bidens bipinnata 217, 218
Bidens biternata 217
Bidens formosa 226
Bidens pilosa 217, 218
Bidens spp. 197
biesie 38
BIGNONIACEAE 118, 172, 355
Bilderdykia convolvulus 368
Billardiera heterophylla 45
billy-goat weed 212
bima 316
bindweed 47, 359
bird knotgrass 341
bird's eye 343
bird's eye speedwell 346
bird-of-paradise flower 46, 176
birds's brandy 196
birdseed 251
birdseed grass 98
bishop's weed 207
bisi 285
bitter apple 57, 193, 302, 330
bitter Karoo bush 20, 38
bitter melon 329
bitterboela 330
bitterbos 19
bitterbossie 236
bitterbush 19
bitterkaroobossie 20, 38
bitterweed 213
blaasklimop 46, 369
black alder 45
black cherry 55
black elder 120
black fellows 217
black iron bark 48, 142
black jack 217
black knotweed 368
black locust 55, 133
black medic 333
black monkey thorn 17
black mulberry 52, 138
black nightshade 303
black poplar 156
black seed wild sorghum 106
black thorn 17
black wattle 44, 123
blackberry 192
blackberry bramble 192

blackbutt peppermint 142
Blackiella inflata 260
blackwood 44, 122
bladder flower 353
bladder hibiscus 276
bladder weed 276
bladdoring 17
blade apple 356
blade thorn 17
blanket-stabbers 217
blasie-soutbos 45, 260
bleeding heart tree 49
blinkblaar-wag-'n-bietjie 19
blinkgras 88
bloedbessie 55, 291
blombos 20
blompeer 18
bloodberry 55, 291
blou lupine 269, 270
blou-aalwee 372
bloubasboom 123
bloubekruip 330
bloubitter 300
bloubos 18, 302
bloubossie 38, 246, 261
bloubottelboom 174
bloudisseldoring 246
bloudisselblommetjie 326
blouduiwel 53
bloudwergmispel 47
blou-echium 48, 246
bloughwarrie 18
blouhaak 17
bloukom 142
bloukoringblom 220
blousaadbuffelsgras 90
blousaadgras 21, 80, 90, 91
blousaadsoetgras 90
blouselblommetjie 326, 343
bloutamboekiegras 85
blouwaterbossie 58, 219, 306
blouwattelboom 123
blue bottle 220
blue Brazilian 118
blue broomrape 53
blue bush 18
blue buttons 315
blue devil 246
blue echium 48, 246
blue guarri 18
blue jacaranda 118
blue lupin 269
blue morning glory 363
blue panic 91
blue passion flower 53, 366
blue porter weed 305
blue rat's tail 305
blue sailors 222

blue seed grass 21
blue thatching grass 85
blue thorn 17
blue wattle 123
blue weed 212
bluebell creeper 45
bluegum 142
blue-leaf cactus 53, 380
blueweed 246, 331
bobatsi 304
bobbejaandraad 341
bobbejaandruif 290
bobbejaandruifboom 54, 146
bobbejaanpaal 373
bobbejaantjie 38
bobbejaantou 359
bobbejaanvinkel 209
bobbin weed 38
boer love grass 83
boereturksvy 53, 377
Boerhavia diffusa 337
Boerhavia erecta 337
Boerhavia spp. 309
boesmansgras 85
boetebessie 19
boetebos 58
boetebossie 243
bohomane 203
bohome 297
bohomenyana 332
bokhaarklawer 271
bokhara clover 271
bokkruid 212
bokrambossie 236
Bolandse-kakiebos 227
bolepo-ba-seokha-sa-merung 322
bolila 266, 291, 338
bolilanyana 291
bono-sa-lekhoaba 239
boobialla 138
boommalva 277
boomtamatie 57
BORAGINACEAE 245, 246, 248, 249
bore-ba-ntja 108, 110
bosbraam 192
bosgifappel 304
bosluisboom 174
bosrooibloekom 48
Boston fern 280
bosveldbeesgras 21, 110
botterblom 214
bottle tree 45
Bougainvillea sp. 41
Boussingaultia baselloides 354
box elder 44, 114
box thorn 20
braam 56, 192

Brabantse klawer 333, 334
Brachiaria deflexa 68
Brachiaria eruciformis 68
Brachychiton populneus 45
bracken 265
brak 324
brakbaardgras 100
brakgras 100
bramble 56, 191, 192
brandmelkkruid 266
brandneuker 304
Brasiliaanse glorie-ertjie 182
Brasiliaanse koejawel 55, 144
Brasiliaanse peperboom 56, 116
Brassaia actinophylla 117
Brassica napus 14
BRASSICACEAE 250, 251, 252, 253, 254, 321, 385
Brayulinea densa 313
Brazilian glory pea 182
Brazilian guava 55, 144
Brazilian holly 116
Brazilian nightshade 370
Brazilian pepper tree 56, 116
Brazilian rosewood 118
Brazilian verbena 58
bread-and-cheese 335
breëblaarborselgras 102
breëblaarbuffelsgras 90
breëblaarperswieke 365
breëblaarsetaria 102
bridal wreath 354
bristle mallow 336
bristly foxtail 104
bristly oxtongue 234
brittle willow 56
Briza maxima 69
Briza minor 69
broadleaf ribwort 291
broadleaf wattle 125
broad-leaved purple vetch 365
broad-leaved setaria 102
brome grass 70
Bromus catharticus 70
Bromus diandrus 36, 71
Bromus pectinatus 71
Bromus unioloides 70
Bromus willdenowii 70
broncho grass 70, 71
brongras 385
bronkhorstgras 71
bronkhorstslaai 385
bronkors 52, 346, 347, 385
bronslaai 385
brood-en-botter 335
brother berry 19
bruinvingergras 73
brush cherry 145

Bryophyllum delagoense 45, 371, 381
Bryophyllum pinnatum 46, 381
Bryophyllum proliferum 46, 381
bubati 304
buckhorn plantain 291
Buddleja davidii 46, 163, 193
Buddleja madagascariensis 46, 193
Buddleja salviifolia 193
buffalo grass 89, 91, 102, 108
buffalo quick paspalum 93
buffalo thorn 19
buffelsgras 91
buffelskweek 107, 108
buffelskweekpaspalum 93
bugtree 161
bugweed 57, 161
bull thistle 223
bunnie's tails 86
bunny ears 53, 377
bur bristle grass 104
bur clover 332
bur marigold 217
Burchell-senecio 238
burgan 50
burgrass 72, 108
Burkea africana 18
burnut 349
burs 38
burweed 44, 203, 220, 243
bush grass 102
bush stinging nettle 304
bushveld herringbone 110
bushveld signal grass 21
bushy yate 140
buttercup bush 180
butterfly bush 193
butterfly orchid tree 45, 129
butterfly tree 129

C

Cabomba 46
Cabomba caroliniana 46
CACTACEAE 356, 373, 374, 375, 376, 377, 378, 380
Cactoblastis cactorum 25
cactus 25, 53, 356, 371, 373, 374, 375, 376, 377
Caesalpinia decapetala 46, 163, 175, 176
Caesalpinia gilliesii 46, 176
Caesalpinia spinosa 176
Calabrian pine 147
Caledonbos 227
calico flower 352
Californian privet 51, 187

Callisia repens 46, 309, 325
Callistemon citrinus 46, 139
Callistemon rigidus 46, 139, 140
Callistemon spp. 11, 112
Callistemon viminalis 139, 140
calotropis 46
Calotropis procera 46
Camden woollybutt 142
camel thorn 17
camel thorn bush 45, 175
camel's foot 129
camphor bush 19
camphor inula 227
camphor laurel 135
camphor tree 47, 135
camphorwood 135
Campuloclinium macrocephalum 46, 199, 219, 306
Canadian elder 56, 120
Canadian water weed 48
canary pine 54, 148
canary ivy 49
canary weed 236, 238
canarybird bush 47, 269
candle thorn 17
Canna glauca 255
Canna indica 46, 255, 256
Canna spp. 197, 256
Canna x generalis 255, 256
CANNABACEAE 257
Cannabis sativa 201, 257, 276
CANNACEAE 255
Cape dandelion 214
Cape khakiweed 227
Cape kweek 107
Cape lilac 137
Cape marigold 214
Cape pepper cress 251
Cape pigweed 204, 205
Cape wattle 131
Cape weed 214
Cape wild mustard 254
Cape willow 157
CAPPARACEAE 258
CAPRIFOLIACEAE 120, 358
Capsella bursa-pastoris 198, 250
cardinal creeper 362
Cardiospermum grandiflorum 46, 350, 368, 369, 370
Cardiospermum halicacabum 46, 369
Carolina fanwort 46
carpet grass 107
carrizo 98
carrot weed 313, 321
cart-track plant 291
CARYOPHYLLACEAE 259, 260, 322, 323

cascade creeper 354
case weed 250
casey paspalum 94
caspian manna 175
Cassia bicapsularis 180
Cassia coluteoides 180
Cassia corymbosa 180
Cassia didymobotrya 178
Cassia floribunda 180
Cassia hirsuta 179
Cassia occidentalis 180
castorbean 174
castor-oil plant 55, 174
Casuarina cunninghamiana 46
Casuarina equisetifolia 46
Casuarina pp. 11
cat's claw 176
cat's claw creeper 51, 355
cat's ears 320
cat's head 340
cat's milk 266
cat's tail 20, 98, 103, 104
cat's tail dropseed 20
catchweed 298
catclaw mimosa 178
caterpillar weed 248
Catharanthus roseus 46, 315, 316
cathedral bells 46, 381
Catophractes alexandri 18
Caucasian wingnut 42
cedar wattle 125
Celtis australis 46
Celtis occidentalis 46
Celtis sinensis 46
Cenchrus brownii 72
Cenchrus incertus 72
Cenia turbinata 317
Centaurea calcitrapa 220
Centaurea cyanus 200, 220
Centaurea melitensis 197, 220, 221
Centaurea salmantica 220
Centaurea solstitialis 220
Centella asiatica 313, 314
century plant 372
Ceratophyllum demersum 386
Ceratotheca triloba 289
Cereus jamacara 47, 371, 373
cestrum 47
Cestrum aurantiacum 47, 158, 159
Cestrum diurnum 159
Cestrum elegans 47, 158, 159
Cestrum laevigatum 47, 113, 158, 159, 290
Cestrum parqui 47, 158
Cestrum spp. 47
Ceylon raspberry 56, 192
Ceylon rose 165

chaff flower 203
Chamaesyce hirta 330
Chamaesyce inaequilatera 331
Chamaesyce prostrata 331
chameleon plant 49
chamomile 214
chandelier plant 45, 381
changeable velvetberry 305
Charleston grass 107
charlock 252
cheeseweed 335
CHENOPODIACEAE 173, 260, 261, 263, 264, 324
Chenopodium album 261, 262, 263
Chenopodium ambrosioides 263
Chenopodium bontei 324
Chenopodium carinatum 309, 324, 325
Chenopodium murale var. *murale* 263
Chenopodium spp. 201
cherry guava 144
cherry pie 196
cherry plum 55
chickweed 267, 323
chicory 47, 222
Chilean cestrum 47, 158
Chilean flame creeper 58
Chileense inkbessie 47
China tree 137
China berry 137
Chinese ahorn 44, 114
Chinese elm 43, 58
Chinese esdoring 114
Chinese firethorn 55, 190
Chinese iep 43, 58
Chinese liguster 51, 188
Chinese mallow 335
Chinese maple 44, 114
Chinese netelboom 46
Chinese nettle tree 46
Chinese privet 51, 188
Chinese sagewood 46, 193
Chinese saliehout 46, 193
Chinese sumac 158
Chinese tamarisk 57
Chinese wax-leaved privet 51, 188
chir pine 54, 148
Chloris abyssinica 72
Chloris gayana 72, 73
Chloris pycnothrix 72, 73
Chloris virgata 74
Chondrilla juncea 47
Christmas berry 187
Christmas berry tree 116
chromolaena 47
Chromolaena odorata 24, 28, 47, 163, 168, 211

Chrysanthemoides monilifera 19
Chrysanthemum segetum 200, 221
Chrysocoma tenuifolia 19
chuana soap 263
chufa 60
Cichorium intybus sunsp. *intybus* 47, 197, 222
Ciclospermum leptophyllum 199, 208
Cineraria lobata 38
Cinnamomum camphora 47, 112, 135
Cirsium arvense 223
Cirsium japonicum 47
Cirsium vulgare 47, 197, 223
Citrullus lanatus 308, 329, 330
Cleome gynandra 198, 258
Cleome monophylla 258
Cleome rubella 258
Cleome spp. 198
Cliffortia ruscifolia 19
climbing knotweed 368
climbing nightshade 370
clock plant 363
clotbur 243
cluster pine 54, 149
coast (duck) grass 78
coast button grass 78
coast tea tree 143
coastal morning glory 363
cobbler's pegs 217
cochineal prickly pear 53, 378
cocksfoot 77
cockspur 220
cockspur grass 80
coco grass 62
Coffea arabica 41, 47
coffee tree 41, 47
coffee weed 182, 222
cogongrass 85
Coleus grandis 186
Colocasia esculenta 40
colocynth 329
Colophospermum mopane 18
Combretum apiculatum 18
commelina 326
Commelina benghalensis 309, 326
COMMELINACEAE 325, 326, 327
Commiphora pyracanthoides 18
common bamboo 66
common blackjack 217
common buffalo grass 89
common bulrush 98
common canary grass 97
common carrot-seed grass 21
common dandelion 231, 320
common dodder 47

common dropseed 107
common dubbeltjie 349
common evening primrose 282
common guarri 18
common hackberry 46
common heliotrope 248
common hook-thorn 17
common lantana 196
common morning glory 363
common paspalum 92
common pigweed 261
common privet 51, 188
common ragweed 213
common reed 64, 98
common stinging nettle 304
common thorn apple 48, 299
common toadflax 51
common vervain 306
common wild oats 65
common wild sorghum 104
common yarrow 210
compass plant 232
coneton 160
CONVOLVULACEAE 328, 358, 359, 360, 361, 362, 363
Convolvulus arvensis 47, 358
Convolvulus farinosus 359
Convolvulus sagittatus subsp. *sagittatus* 359
Convolvulus sagittatus var. *ulosepalus* 359
Convolvulus spp. 350, 363
Conyza albida 224
Conyza bonariensis 36, 224, 225
Conyza canadensis 224
Conyza sumatrensis var. *sumatrensis* 224
Conyza chilensis 224
Conyza spp. 199, 224, 225
Cootamundra wattle 121
copal tree 158
coral berry 291
coral bush 187
coral creeper 45
coralberry tree 45, 187
Corchorus trilocularis 198, 275
coreopsis 47
Coreopsis lanceolata 47, 200, 225
cork tree 18
corn chamomile 214
corn chrysanthemum 221
corn marigold 221
corn spurry 322
cornflower 220
Coronopus didymus 309, 317, 321, 322
Cortaderia jubata 47, 75
Cortaderia selloana 47, 75

cosmos 226
Cosmos bipinnatus 200, 226
cossack asparagus 98
cotoneaster 47
Cotoneaster franchetii 47, 189
Cotoneaster glaucophyllus 47, 189
Cotoneaster pannosus 47, 189
Cotoneaster salicifolius 47, 189
Cotoneaster simonsii 47, 189
Cotoneaster spp. 163
cottonwood 156
cotton wool grass 85
cotton-leaf physic nut 50
cotula 214, 317
Cotula australis 308, 317, 318
Cotula spp. 317
Cotula tenella 317
Cotula turbinata 317, 318
couch grass 48, 76
couch paspalum 93
cowhop clover 333
crab finger-grass 79
crack willow 56, 157
CRASSULACEAE 381
Crataegus monogyna 41
Crataegus pubescens 47
creeping burhead 48
creeping crofton weed 211
creeping goosefoot 324
creeping inch plant 46, 325
creeping mallow 336
creeping prickly pear 378
creeping sorrel 338
creeping thistle 223
crested wattle 131
crimson bottlebrush 139
crimson cestrum 47, 158
crimson oak 151
crofton weed 44, 210, 211
Crotalaria 47, 258, 269
Crotalaria agatiflora 47, 269
Crotalaria sphaerocarpa subsp. *sphaerocarpa* 197, 269
crowfoot 78
cruel plant 268, 353
Cryptostegia grandiflora 47
Cryptostemma calendulaceum 214
Cucumis myriocarpus 308, 329, 330
CUCURBITACEAE 329, 330
cudweed 235
curled sorrel 296
curly dock 56, 296
Cuscuta campestris 47, 350, 360
Cuscuta suaveolens 47, 360
cut leaf evening primrose 283
cutleaf groundcherry 300
Cyanella lutea 38

Cyathea cooperi 41
Cynodon aethiopicus 77
Cynodon dactylon 9, 48, 76
Cynodon nlemfuensis 77
Cynodon plectostachyus 77
CYPERACEAE 60, 62
Cyperus distans 38
Cyperus esculentus 60, 61, 62
Cyperus natalensis 62
Cyperus rotundus subsp. *rotundus* 60, 61, 62
Cyperus rotundus subsp. *tuberosa* 62
Cytisus monspessulanus 48, 183
Cytisus scoparius 48, 163, 183

D

Dactylis glomerata 77, 78
Dactyloctenium aegyptium 78, 79
Dactyloctenium giganteum 78, 79
dagga 257, 276
dagger cocklebur 243
dallis grass 92
dalmatian toadflax 51
damba 326
damslyk 388
darnel 87
Datura ferox 48, 299
Datura innoxia 48, 299
Datura metel 299
Datura spp. 197, 258
Datura stramonium 48, 299
Datura tatula 299
Deccan hemp 276
dekgras 85
dekriet 98
Demoina-bossie 232, 233
denne 15
DENNSTAEDTIACEAE 265
dense water weed 48
dense-thorned bitter apple 57, 193
desert tamarisk 162
deurmekaarbos 18
devil's fig 286
devil's pumpkin 366
devil's thorn 340, 349
devil's tongue 378
devil's trumpet 299
dewdrop lawn 328
Dichondra micrantha 309, 313, 328
Dichondra repens 328
Dichrostachys cinerea 18, 22
digdoringbitterappel 193
Digitaria eriantha 79
Digitaria sanguinalis 79, 80, 82,

91, 110
Diheteropogon filifolius 20
dikboom 146
dinjinsa 193
Diospyros lycioides 18
Diplocyclos palmatus 48
disseldoring 216, 223
ditch weed 299
Dittrichia graveolens 198, 227, 236
Dizygotheca elegantissima 117
dock 56, 295, 296
dodder 360
Dodonaea angustifolia 18
dog-bane 165
dog-daisy 210
dog-fennel 214
dog-rose 55
Dombeya rotundifolia 18
donkieklits 316
donsgras 85
doobgras 76
dooryard knotweed 341
doringbos 19
doringmisbredie 206
doringpapawer 288
doringsydissel 239
doringtamatie 57, 193
double oleander 165
doublegee 340
downy thorn apple 48
drabok 87
driedoring 20
driedoringboom 130
drooping prickly pear 378
drug fumitory 272
dubbeltjie 311, 349
dubbeltjiegras 72
Duchesnea indica 48, 308, 344
duck grass 78
duck's meat 388
duckweed 388
duingras 38
duisendblaar-achillea 210
duiwekerwel 272
duiwelsdis 340
duiwelskerwel 272
dune grass 38
Dunn's white gum 142
Duranta erecta 48, 163, 195
Duranta plumieri 195
Duranta repens 195
Durban guava 55, 144
Durbanse koejawel 55, 144
Dutchman's pipe 45, 352
dwaalbos 257
dwarf buffalo thorn 19
dwarf cactus 377

dwarf marigold 235
dwarf meadow grass 100
dwarf umbrella tree 56, 117
dwarf yellow-striped bamboo 56
dwerg-blinkblaar-wag-'n-bietjie 19
dwerggeelstreepbamboes 56
dwergmispel 47
dyer's rocket 296
dyer's weed 296

E
eagle fern 265
earth almond 60
East African couch 77
East African dock 56
Echinochloa colona 80, 81
Echinochloa crus-galli 80, 81
Echinodorus cordifolius 48
Echinodorus tenellus 48
Echinopsis spachiana 48, 371, 374
Echium lycopsis 246
Echium plantagineum 48, 200, 234, 246, 247
Echium vulgare 48, 201, 246, 247
eclipta 318
Eclipta alba 318
Eclipta prostrata 309, 318, 319
edible fig 43
edible galingale 60
eendekroos 388
eendjiesgras 62
eenjarige blougras 100
eenjarige geelstinkklawer 271
Egeria densa 48, 387
eglantine 56, 191
Egyptian millet 106
Ehretia rigida 18
Ehrharta brevifolia 82
Ehrharta calycina 82
Ehrharta longiflora 82
Ehrharta villosa 82
Eichhornia azurea 390
Eichhornia crassipes 26, 48, 382, 390, 391
eierbossie 260
eight-seeded starbur 316
elataboom 125
elbow buffalo grass 91
elder bush 120
elderberry 120
elephant grass 95
elephant's ear 40
Eleusine africana 82
Eleusine coracana 79, 91, 110
Eleusine coracana subsp. *africana* 78, 82, 83

Eleusine indica subsp. *africana* 82
Elionurus muticus 20
elmboogbuffelsgras 91
Elodea canadensis 48, 387
Elytropappus rhinocerotis 19
Emex australis 308, 340
emexdubbeltjie 340
empress tree 53
Engelse hedera 49, 352
Engelsmannetjies 245
English hawthorn 41
English ivy 49, 352
English oak 43
English plantain 291
enkelblaar-cleome 258
enkeldoring 17
Enneapogon cenchroides 20
Equisetum hyemale 48
Eragrostis ciliaris 83
Eragrostis curvula 83, 107
Eragrostis plana 83, 107
Eragrostis spp. 20, 91
erect sword fern 52
Erica-leaved yellow bush 20
Erigeron bonariensis 224
Eerigeron canadensis 224
Eriobotrya japonica 48, 113, 154
Eriocephalus africanus 20
Erodium moschatum 199, 272, 273
Erucastrum austroafricanum 254, 255
Erucastrum strigosum 253
Erytenna consputa 27
essenblaarhorn 44, 114
Eucalyptus camaldulensis 48, 140
Eucalyptus cinerea 140, 141
Eucalyptus cladocalyx 48, 142
Eucalyptus diversicolor 48, 142
Eucalyptus dunnii 142
Eucalyptus globulus 142
Eucalyptus gomphocephala 142
Eucalyptus grandis 48, 141, 142
Eucalyptus lehmannii 48, 140, 141
Eucalyptus macarthurii 142
Eucalyptus maculata 142
Eucalyptus paniculata 48, 142
Eucalyptus sideroxylon 48, 142
Eucalyptus smithii 142
Eucalyptus spp. 112
Eucalyptus tereticornis 48, 142
Euclea crispa 18
Euclea divinorum 18
Euclea undulata 18
Eugenia uniflora 41, 48
Eupatorium 168
Eupatorium adenophorum 210
Eupatorium macrocephalum 219
Eupatorium odoratum 168

Euphorbia chamaesyce 331
Euphorbia geniculata 268
Euphorbia helioscopia 266, 267
Euphorbia heterophylla 268
Euphorbia hirta 330, 331
Euphorbia inaequilatera 331, 332
Euphorbia inaequilatera var. *inaequilatera* 331
Euphorbia leucocephala 48
Euphorbia peplus 266, 267, 323
Euphorbia prostrata 331, 332
Euphorbia pulcherrima 268
Euphorbia spp. 179, 308
EUPHORBIACEAE 174, 266, 268, 330, 331
eurabbie 142
European blackberry 56, 192
European elder 56, 120
European gorse 58, 184
European heliotrope 248
European turnsole 248
European verbena 306
Europese vlier 56, 120
evening primrose 52, 283
evergreen millet 106

F

FABACEAE 15, 121, 122, 123, 125, 126, 127, 129, 131, 132, 133, 134, 175, 177, 178, 179, 180, 181, 182, 183, 184, 269, 271, 332, 333, 364, 365
fairy bells 69
fairy grass 88
Fallopia convolvulus 350, 368
Fallopia sachalinensis 49
false acacia 133
false aralia 56, 117
false barley 84
false capsicum 304
false daisy 318
false dandelion 231, 320
false jalap 281
false lebbeck 45, 127
false signal grass 68
false strawberry 344
false umbrella thorn 17
fat hen 261
feathertop 54, 97
feathertop chloris 74
Felicia muricata 38
fern tree 118
ferweelgras 88
fever tea 18
Ficinia filiformis 38
Ficinia indica 38

Ficus carica 43
fiddleneck 245
field bindweed 47, 358
field poppy 288
field sorrel 295
field speedwell 346
field wall-flower 252
Fimbristylis dichotoma 38
Fimbristylis hispidula 38
fine-leaved verbena 348
finger sorrel 338
fire-and-ice plant 193
firethorn 55, 190
fir-leafed celery 208
five-leaved black jack 217
flaky thorn 17
flame thorn 17
Flanders poppy 288
flannel weed 161, 279
Flaveria bidentis 49, 200, 228
flax-leaf fleabane 224
florist's gum 140
floss flower 212
flower-of-an-hour 276
fluff bush 18
fluitjiesbossie 245
fluitjiesriet 64, 98
fluweelboontjie 132
fluweelprosopis 55
Foeniculum vulgare 49, 199, 209
fog grass 63
fonteinbultvaring 265
forest buffalo grass 102
forest inkberry 54, 290
forest red gum 48, 142
forget-me-not-tree 195
formosa firethorn 55
Formosa lily 274
fountain grass 54, 97
four o'clock 51
four-leaved allseed 259
fragrant false-garlic 202
fragrant weed 257
Franklin weed 246
Franse silene 260
Franse tamarisk 57, 162
Fraxinus americana 49
Fraxinus angustifolia 49
Free State daisy 215
French catchfly 260
French silene 260
French tamarisk 57, 162
fringed dodder 360
fringed water-lily 389
frog-buttons 388
Fumaria muralis 200, 272
Fumaria officinalis 272
FUMARIACEAE 272

fumitory 272
furze 184
fynbiesie 38
fynblaarverbena 348
fyndoring 17
fyngras 83
fynkatbos 167

G

galbessie 303
Galenia secunda 38
Galinsoga ciliata 229
Galinsoga parviflora 199, 229
galinsoga weed 229
Galium spurium 201, 298
Galium spurium subsp. *africanum* 298
Galium tricornutum 49
gallant soldier 229
gallow grass 257
galsiektebos 263
gammock 185
Gamochaeta pensylvanica 201, 230
gansevoet 263
ganskos 317
gansvoete 57, 351
garden ageratum 212
garden bristle grass 20
garden cactus 377
garden canna 255
garden setaria 103
garden sorrel 338
garden spurge 330
garden urochloa 110
garingboom 44, 372
gaspeldoring 58, 184
Gaura coccinea 49
gebrokenhartjieboom 49
geel branddoringbos 190
geel iris 50
geel lupine 269
geelbessie 195
geelblombloudissel 45, 286
geelbokbaard 242
geelbranddoring 55, 190
geeldissel 220
geelgemmerlelie 49, 307
geelhaak 17
geelklawer 333
geelklits 44, 297
geelklokkies 57, 172
geeloleander 57
geelopslag 236, 238
geelsuring 338
geeluintjie 31, 60

geelwaterlelie 52, 389
gekleurde euphorbia 268
geldbeursie 250
Genista monspessulana 49, 183
GERANIACEAE 272
German psyllium 291
gevlekte bloekom 142
gewone bamboes 66
gewone brandnetel 304
gewone buffelsgras 89, 91
gewone dodder 47
gewone ehrharta 82
gewone fluitjiesriet 64
gewone ghwarrie 18
gewone kanariegras 97
gewone knapsekêrel 217
gewone lantana 28, 196
gewone liguster 51, 188
gewone opklim 58
gewone papkuil 98
gewone paspalum 92
gewone stinkblaar 48
gewone wildehawer 65
gewone wildesorghum 104
gewone wortelsaadgras 21
giant crowfoot 78
giant devil's fig 57, 193
giant knotweed 49
giant milkweed 46
giant paspalum 92
giant reed 64
giant sensitive plant 51, 178
gifappel 193
gifbessie 304
gifgras 77
ginger-thomas 172
gisekia 334
Gisekia pharnacioides 308, 334
GISEKIACEAE 334
gladde kruipmelkkruid 331
gladdee turksvy 25
gladdesumak 55, 115
gladde-rooi-opslag 331
glasswort 264
Glaucous cotoneaster 189
Gleditsia triacanthos 49, 112, 129, 130
glossy privet 188
Glyceria maxima 49
Gnaphalium luteo-album 235
Gnaphalium pensylvanicum 230
goat's head 316
goatsbeard 185
goedkaroo 38
golden ball wattle 121
golden crown grass 92
golden dewdrop 195
golden wattle 44, 125

golden willow 126
gomdagga 196
Gomphrena celosioides 309, 312
Gomphrena globosa 312
goose daisy 317
gooseberry cucumber 330
goosefoot plants 351
goosegrass 298
goudsblom 221
goudwilger 126
goue wattel 27, 33, 44, 125
gousblom 215
Gozard's curse 192
granadina 53, 366
great brome 71
great duckweed 388
greater periwinkle 58, 315
greater plantain 291
green cestrum 158
green goosefoot 324
green mother of millions 46, 381
green wattle 44, 123
grenadella 53, 366
Grevillea banksii 49, 151, 152
Grevillea juniperina 151
Grevillea lanigera 151
Grevillea robusta 49, 151, 152
Grevillea rosmarinifolia 151
Grevillea spp. 112
Grewia bicolor 18
Grewia flava 18
Grewia flavescens 18
Grewia monticola 18
grey bitter apple 193
grey iron bark 48, 142
grey poplar 54, 155
grey raisin 18
gringed water lily 52
groenbasboom 123
groenbossie 235
groenhondebossie 324
groenwattel 44, 123
grondboontjiebotterkassia 56, 178
groot bitterappel 161
groot wildehawer 65
grootrondeblaarturksvy 53, 377
grootdoringturksvy 377
grootklits 44, 203
grootsoutbos 173
grootstinkblaar 48, 299
grootweëblaar 291
grootwortelsaadgras 21, 109
growwe medicago 333
gryspopulier 155
grysysterbasbloekom 48, 142
guava 55, 144
Guilleminea densa 309, 313
guinea grass 90

guinea-fowl grass 101
gunpowder weed 260
Gymnosporia polyacantha 19
Gymnosporia senegalensis 19

H
haakdoring 17
haak-en-steek 18
haak-en-steek-bossie 203
haarwurmgras 97
haasgras 38, 108
haaspootvaring 54, 280
haasstert (gras) 86
hairy creeping milkweed 331
hairy spurge 330
hairy tare 365
hairy thorn-apple 299
hairy wild lettuce 231
Hakea drupacea 49, 152
Hakea gibbosa 49, 152
Hakea salicifolia 49, 152, 153
Hakea sericea 33, 49, 152, 153
Hakea spp. 15, 112
Hakea suaveolens 49, 152
HALORAGACEAE 386
hanekam 204, 205, 294
hanepootmanna 80
hardebossie 236
hare's foot 86
hare's tail 86
hare's-foot clover 333
haregrass 38
harige hakea 49, 152
harige kruipmelkkruid 331
harige skaapslaai 231
harige speldebos 152
harige stinkblaar 48
Harrisia martinii 49, 371, 375
hartblaartaaiman 279
Hawaiiese dwerg 56, 117
hawersaadgras 82
hawkweed oxtongue 234
hay fever weed 213
hay grass 74
heart pea 369
heart seed 369
heart-leaf sida 279
Hedera helix 49, 350, 352
Hedera helix subsp. *canariensis* 49, 352
Hedera helix subsp. *helix* 49, 352
hedge bamboo 66
hedge hakea 152
hedge mustard 254
hedge spike-thorn 19
hedge water berry 145

Hedychium coccineum 49, 307
Hedychium coronarium 49, 307
Hedychium flavescens 49, 307
Hedychium gardnerianum 49, 307
Hedychium spp. 198
heideblaargeelbos 20
heiningwildemosterd 254
Helianthus annuus 14, 242
Helichrysum auriculatum 20
Heliotropium europaeum 200, 248
hemelboom 44, 158
hemp plant 372
hen's eyes 187
henbit deadnettle 274
hennep 257, 276
he-oka 151
Heptapleurum arboricolum 117
herb-john 185
herderstassie 250
herringbone grass 110
heuningprosopis 55, 132
hexham scent 271
Hibiscus cannabinus 276
Hibiscus spp. 198
Hibiscus trionum 276
hill raspberry 192
Himalayan cotoneaster 47, 189
Himalayan firethorn 55, 190
hlaba 223
hlabo 331
hlaba-hlabane 243, 244
hoary cardaria 251
hoenderboom 122
hoenderspoor 78
hoenderuintjie 60
hogweed 341
Hollandsche gras 322
Homalanthus populifolius 49
hondebosch 204
hondekruid 266
hondepisbossie 261
hondsroos 55
honey locust 49, 129
honey mesquite 55, 132
honcyshuck 130
hongerbos-senecio 236
hooked bristlegrass 104
hop trefoil 333
hop wattle 44
Hordeum murinum subsp. *murinum* 84
horingdoring 17
horned thorn 17
horse grass 103
horsetail tree 46
horseweed fleabane 224
Houttuynia cordata 49
huilboom 19

huilwigeboom 157
hunyani grass 73
hydrilla 49, 386, 387
Hydrilla verticillata 49, 382, 386, 387
HYDROCHARITACEAE 386
Hydrocleys nymphoides 50, 382, 389
Hydrocotyle americana 309, 313, 314
Hylocereus undatus 50
Hyparrhenia dregeana 85
Hyparrhenia hirta 85
Hyparrhenia tamba 85
HYPERICACEAE 184
Hypericum androsaemum 50, 185
Hypericum forrestii 185
Hypericum patulum 185
Hypericum perforatum 26, 50, 163, 184, 185
Hypochaeris glabra 231
Hypochaeris radicata 198, 231
Hypochaeris spp. 320
Hypoestes phyllostachya 309, 310
hyssop loosestrife 51

I

ibhoqo 363
ibhuma 98
ibutha 167
idambizo 326
idlebendlele 326
idololenkonyane 296
igwanitsha 341
ihahabe 239
ihlaba 239
ihlambe 363
ijalamu 363, 370
ijikijolo 192
ijingijoye 192
ikhabe 329
ikhambi lakwangcolosi 318
ikhungele 311
iligcume 243
iloqi 299
imbathi 304
imbozisa 209
imbricate cactus 53
imbricate prickly pear 376
imbuyu 204
imotyikatsana 363
Imperata cylindrica 85, 86
inabulele 359
indawo 60
Indian almond 150
Indian cress 346

Indian fig 377
Indian hedge mustard 254
Indian hemp 257
Indian jute 275
Indian laurel 51, 136
Indian shot 46, 255
Indian strawberry 344
Indian three-leaf vitax 58
Indian weed 238
Indiese kanna 46
Indiese lourier 51, 136
indigo berry 53, 366
indlebe-ka-tekwane 291
indlebe-kathekwane encane 291
indringer-ageratum 44, 212
ingabe 239
inkberry 47, 158, 159, 289, 303
inkbessie 47, 54, 158, 159, 289
inkbos 54
inkunzane 349
insangwana 241
insimbephuzi 172
insukumbili 263
intandangulube 312
intandela 354, 369
intfuma 193
intunga 85
Inula graveolens 227, 311
invading ageratum 44, 212
iphamphuce 318
Ipomoea alba 50, 350, 361
Ipomoea cairica 363
Ipomoea carnea subsp. *fistulosa* 363
Ipomoea coscinosperma 363
Ipomoea fistulosa 363
Ipomoea hederifolia 350, 362
Ipomoea indica 50, 363
Ipomoea purpurea 50, 350, 358, 363, 364, 368
Ipomoea quamoclit 362
iquangoboto 106
Iris pseudacorus 50
Irish shamrock 333
irwabe lenyoka 320
isangu 257
isheke 205
isibathi 304
isihlungu 216
isinama 203
isishushlungu 315
isisinini 319
isithathe 338
isiwisa esiluhlaza 258
Isolepis antarctica 38
isona 285
isona weed 285
Italiaanse populier 54, 156

Italiaanse raaigras 51, 87
Italian ryegrass 51, 87
itch grass 101
itsy bitsy inch vine 325
ivivane 279
ixabaxaba 303
iyeza 236
iyeza-lentshulube 276

J

jacaranda 50, 118
Jacaranda mimosifolia 50, 112, 118, 119
jakaranda 50, 118
jamboes 57, 145
jambolan 57, 145
Jan-tak 160
Japanese barberry 45
Japanese berberis 45
Japanese honeysuckle 51, 358
Japanese liguster 50, 187
Japanese medlar 154
Japanese mispel 154
Japanese plum 154
Japanese poinsettia 268
Japanese thistle 47
Japanese wax-leaved privet 50, 187
Japanse kanferfoelie 51, 358
Jatropha curcas 50
Jatropha gossypiifolia 50
java bramble 192
jeremane 241
jersey cudweed 235
Jerusalem cherry 57, 303
Jerusalem thorn 53
Jerusalemkersie 57, 304
jimson weed 299, 338
joang-ba-lintja 100
joang-ba-masimo 80
jobskrale 260
Jodeluis 316
Johanneskruid 50, 184
Johnson grass 57, 106
Johnsongras 57, 106
jointed cactus 53, 376
jointed prickly pear 376
jongosgras 82
jongsnoekkruid 54, 392
jungle rice 80

Juniperus virginiana 50

K

Kaapse dubbeltjie 340
Kaapse kakiebos 227
Kaapse madeliefie 214
Kaapse wilger 157
kaate 329
kaatjie 376
kabelkaktus 375
kabelturksvy 53, 376
kahili ginger lily 49, 307
kahiligemmerlelie 49, 307
kakiebos 227, 236, 241, 264, 311
kakiedubbeltjie 236, 311
kaktus 25, 356, 371, 373, 374, 375, 376, 377, 380
Kalanchoe delagoensis 381
Kalanchoe pinnata 381
Kalanchoe proliferum 381
Kaliforniese esdoring 114
Kaliforniese liguster 51, 187
kalkoengif 300
kalkoenslurp 204, 205
kameeldoring 17
kameeldoringbos 45, 175
kameelspoor 129
Kanadese skraalhans 224
Kanadese vlier 56, 120
Kanadese waterpes 48
kanariesaadgras 251
Kanariese den 54, 148
kandelaarplant 45, 381
kanferboom 47, 135
kanferbos 19
kanferfoelie 341
kanfer-inula 227
kangaroo wattle 44
kankerroos 58, 244
kanniedood 18, 341
kanola 14
kantblom 207
kapokbossie 20
kappertjie 346, 347
Kariba weed 56, 393
karie 48, 142
karkoer 329
karlieboom 268
karmedik 223
karmosynbos 290
karmosynsestrum 47, 158
karo 54, 150
Karoo daisy 215
karoo-aster 38
karooklits 332
karri 48, 142
kasies 335
kasterolieboom 55, 174
kasuarisboom 46
katbos 18, 167
katoor 231
katstert 19, 296
katstertgras 20

katstert-fynsaadgras 20
kattteklouranker 51, 355
Kaukasiese vleuelneut 42
keeled goosefoot 324
keiserinboom 53
kenaf 276
kerese 385
kersiepruim 55
kerwel 272
khaki burweed 311
khakiweed 227, 311
khlane 68
Khoi-kooigoed 170
khola-bosiu 263
khotolia 238
khotswana 326
kidney grass 328
kidney weed 328
kiesieblaar 50, 51, 277, 335
kikoejoegras 54, 94
kikuyu 54, 94
Killarney ash 142
Kirkia acuminata 137
Kirkia wilmsii 137
Klaaslouw bush 19
Klaaslouwbos 19
klamath weed 185
klapbessie 300
klawerbesemraap 53
klawergras 104, 332
klawersuring 338
kleefdoring 17
kleefgras 238
kleefkruid 298
klein rondeblaarturksvy 53
kleinafrikaner 241
kleinbewertjies 69
kleinbrakbossie 260
kleinbrandnetel 304
kleingaringboom 372
kleinkakiebos 235
kleinkankerroos 316
kleinrondeblaarturksvy 380
kleinsaadkanariegras 97
kleinskraalhans 224
kleinwatergras 80
kleinwortelsaadgras 108
klewerige appelliefie 301
klimop 358, 359
klitsborselgras 104
klitse 38, 72, 243, 244, 308, 309, 311, 316, 333
klitsklawer 332
klossiesgras 86
knawel 259
knob thorn 17
knoopkruid 54
knopherik 252

knopklitsgras 72
knopkruid 229
knoppiesdoring 17
knoppiesgras 62
knotweed 294, 341, 368
koejawel 40, 55, 144
koekbossie 279
koerajong 45
koffieboom 47
kokoma grass 101
koksvoetgras 77
koorsbossie 18
koperdraad 20, 341
kopersaadgras 79
koraalbessieboom 45, 187
koraalklimop 45
koringblom 220
koringdissel 220
koringkrisant 221
koringpapawer 288
koringroos 288
kosmos 226
kousgras 108
kousklits 108
kraaldoring 19, 46, 175, 176, 195
kraanvoëlbossie 204
krakerbossie 236
kropaar 77
kruidjie-roer-my-nie 178
kruiehondebossie 263
kruipgras 84
kruip-knopamarant 312
kruipsoutbos 324
kruipsterklits 316
kruisbessie 18
kruisgras 79
kruisvingergras 79
kruitbossie 260
krultongblaar 296
kudzu vine 55, 364
kudzuranker 55, 364
Kunzea ericoides 50
kuri-millet 110
kusebere 301
kwarrelgras 98
kwasgras 74
kweek 48, 76
kweekpaspalum 93

L

lace flower 207
Lactuca indica 232
Lactuca serriola 197, 232
lady's hand 38
lady's heart 69
laeveld-fluitjiesriet 64, 98

Lagurus ovatus 86
lamb's quarters 261
lamb's tails 354
LAMIACEAE 186, 274
Lamium amplexicaule 200, 274
langbeenkatdoring 167
langbaardswenkgras 111
langbeenpaspalum 92, 94
langbeenwatergras 94
langblaarwattel 26, 44, 125
langdoringkaktus 53
langkakiebos 241
langnaaldbromus 71
lantana 15, 28, 39, 40, 50, 196
Lantana camara 28, 39, 40, 163, 196
Lantana montevidensis 40, 196
Lantana rugosa 196
Lantana spp. 50
large carrot-seed grass 21
large cocklebur 58, 244
large quaking grass 69
large round-leaved cactus 377
large round-leaved prickly pear 53
large thorn-apple 48, 299
large water grass 92
large-flowered prickly pear 53, 378
Lasiochloa longifolia 38
late cotoneaster 47, 189
late weed 249
LAURACEAE 135, 136
Lavatera arborea 50, 198, 277
Lavatera assurgentiflora 277
Lavatera cretica 277
Lavatera trimestris 277
lawn celery 208
lawn paspalum 93
lead tree 177
leafy cactus 356
lebbeck tree 45, 127
lebbeckboom 45, 127
lechoe 299
lehaha-la-tsela 215
Lehman's gum 140
lehola 90
lehola-la-lipere 103
lekhala 372
lekkerbreek 19
lekkerruikpeul 17
Lemna gibba 382, 388
Lemna spp. 388
LEMNACEAE 388
lemoenperdestert 46
lemon bottlebrush 46, 139
lemon vine 356
lenamo 203
lenjana 341
leoka 130

lepero 243
Lepidium africanum 251
Lepidium bonariense 251
Lepidium draba 50, 251
Lepidium spp. 198
Leptospermum laevigatum 27, 50, 113, 143
Leptospermum scoparium 50, 143
lerotho 258
lesabe 239
lesese 239
lesser balloon vine 46
lesser broomrape 53
lesser canary grass 98
lesser periwinkle 58
lesser trefoil 333
letapiso 236
letjotjo 38
Leucadendron ericifolium 20
Leucadendron rubrum 20
leucaena 50, 177
Leucaena glauca 177
Leucaena leucocephala 50, 177
Leucaena leucocephala subs. *leucocephala* 177
Leucaena leucophylla 163
Leucas martinicensis 38
Leucosidea sericea 18
life plant 381
liguster 50, 51, 187, 188
Ligustrum japonicum 50, 187, 188
Ligustrum lucidum 51, 188
Ligustrum ovalifolium 51, 187, 188
Ligustrum sinense 51, 188
Ligustrum spp. 163
Ligustrum vulgare 51, 188
LILIACEAE 274
Lilium formosanum 51, 198, 274, 275
Limonium sinuatum 51, 201, 293
Linaria dalmatica 51
Linaria vulgaris 51
lindenleaf sage 56
Lindley's saltbush 260
linyooko 300
lion's tooth 320
lipii 245
Lippia javanica 18
lira-ha-li-bone 341
litjiesgras 341
litjiesinjaalgras 68
litjieskaktus 25, 53, 376
litjiesturksvy 376
litjoti 329
Litsea glutinosa 51, 113, 136
Litsea sebifera 136
little burweed 332

little-seeded canary grass 97
loblolly pine 54, 148
loblollyden 54, 148
locust tree 133
Lolium multiflorum 51, 87
Lolium perenne 51, 87
Lolium sp. 36
Lolium temulentum 87
lollipop-climber 48
Lombardy poplar 54, 156
long pricklyhead poppy 288
long spine cactus 53
longflower tobacco 160
longifolia pine 148
long-leaved wattle 44, 26, 125
long-spined thorn-apple 299
Lonicera japonica 51, 350, 358
Lopholaena coriifolia 18
loquat 48, 154
Lotus subbiflorus 308, 332
love grasses 20, 104
low spear grass 100
lowveld reed 64, 98
lube 258
lucerne 14, 71, 84, 111, 220, 247, 259, 272, 292, 328, 333, 359, 360
lucerne dodder 47, 360
lucky nut 166
Ludwigia peruviana 51
luisboom 27, 57, 161
luisgras 108
luisiesturksvy 378
lukwart 48, 154
lupien 14
Lupinus angustifolius 198, 269, 270
Lupinus luteus 14, 198, 269, 270
Lupinus spp. 14, 270
lusern 14, 71, 84, 111, 220, 247, 259, 272, 292, 328, 359, 360
luserndodder 47
Lycium ferocissimum 20
Lythrum hyssopifolia 51
Lythrum salicaria 51

M

maagbossie 38
maagdepalm 315
maanblom 50, 361
maandrosie 315
Macfadyena unguis-cati 27, 51, 350, 355, 356
macopo grass 102
Madagascar periwinkle 46, 315
Madagaskar saliehout 46
Madascar sagewood 46, 193
Madeira vine 45, 354

Madeira winter cherry 304
Madeiraklimop 49
Madeira-ranker 45, 354
madeliefie 319
madumbe 40
magic guarri 18
maidenhair 280
maidenhair fern 280
mak-kiesieblaar 50, 277
makha 82
mak-kiesieblaar 277
maksering 51, 137
makspeldebos 152
makvy 43
mallow 51, 277
malpitte 299
Malta centaurea 220
Maltadissel 220
Malva aegyptia 335
Malva parviflora 277, 309, 335, 336
Malva pusilla 335
Malva verticillata 51, 335
MALVACEAE 275, 276, 277, 278, 279, 335, 336
Malvastrum coromandelianum 51, 200, 278, 279
manakalali 60
Mangifera indica 43
mango 43
manitoka 52, 138
mankamirt 50, 143
manku 235
manna 79
mannetjiesden 149
mantisalca 220
Mantisalca salmantica 200, 220
manuka myrtle 50, 143
marijuana 257
maritime pine 149
marotlo-a-mafubelu 108
marram grass 45
marsh grass 80
marsh parsley 208
marsh pennywort 313
marsh rosemary 293
marvel-of-Peru 281
maseka 82
Master John-Henry 241
mat grass 341
mat sandbur 72
Matabele flower 285
match poplar 54, 156
matjiesgoed 98
matlepelo 258
matokwane 257
matolo 83
matricaria 38

Mauritius thorn 46, 175
mayweed 214
mbanje 241
mealy stringybark 140
Medicago aschersoniana 332
Medicago falcata 333
Medicago hispida 332
Medicago laciniata 332
Medicago lupulina 333
Medicago polymorpha 332, 333
Medicago sativa 14, 333, 359
Medicago spp. 308
meerjarige misbredie 204
meerjarige raaigras 51, 87
meidoring 41, 47
Meksikaanse ageratum 44
Meksikaanse groenhaarboom 53
Meksikaanse klawer 345
Meksikaanse meidoring 47
Meksikaanse papawer 286
Meksikaanse richardia 345
Meksikaanse sonneblom 58, 171
Melaleuca hypericifolia 51
Melaleuca spp. 139
Melia azedarach 51, 112, 137
MELIACEAE 137
melilot 271
Melilotus alba 271
Melilotus albus 199, 271
Melilotus indicus 199, 271
Melilotus officinalis 271
Melinis repens subsp. *repens* 88
melkbossiegras 266
melkdissel 232
mesquite 132
Messina creeper 363
Metalasia muricata 20
Metrosideros excelsa 51
Mexican ageratum 44, 212
Mexican aster 226
Mexican clover 345
Mexican devil 210, 211
Mexican fireplant 268
Mexican hawthorn 47
Mexican marigold 241
Mexican oleander 166
Mexican poppy 45, 286
Mexican richardia 345
Mexican sunflower 58, 171
Mexican tea 263
Mexican tobacco 160
mgilane 235
mielie-crotalaria 269
mieliegif 285
mieliepes 226
mier 323
mierbossie 312
Mikania cordata 168

milfoil 210
milkthistle 239
milkweed 268, 353
Mimosa asperata 178
Mimosa pellita 178
Mimosa pigra 51, 163, 178
Mirabilis jalapa 51, 199, 281, 282
miracle leaf 381
misblom 44
misbredie 205, 206, 261, 341
mission grass 108
mission mallow 277
mission prickly pear 377
mistflower 44, 211
mock orange 150
Modiola caroliniana 308, 336, 337
moerasgras 80
mofantsoe-o-moholo 91
mofula-tsephe 85
mohlafotha 160, 174
mohloa 76
mohlomo 85
mohlorumo 85
mohwa 204
mohwa-guru 206
mokhura 174
mokolonyane 217
mole plant 268
Molteno disease senecio 238
moluoane 157
monk's cress 346
monkey grape 146
monkey thorn 17
monnikbaard 358
monokotswai-wa-banna 192
montanoa 52, 169
Montanoa hibiscifolia 52, 164, 169
Monterey pine 148
Montpellier broom 48, 49, 183
Montpellierbrem 48, 49, 183
monyaku 330
monyane 217
mooinooientjie 258
moon cactus 49, 375
moonflower 50, 299, 361
moopeli-o-mosoeu 284
mo-ora-tsatsi 335
moqhobo-o-monyenyane 346
moqopshoe 79
MORACEAE 138
morarana 298, 368
morning glory 50
morning glory bush 50, 363
morola 193
Morus alba var. *alba* 52, 138
Morus nigra 52, 138
Morus spp. 112
moseli 82

mosquito fern 45
moss verbena 348
moth catcher 45, 353
mothepetelle 235
mother of millions 381
mother of thousands 381
mothowa 76
moth-vine 353
motsitla 98
motvanger 45, 353
mountain cedar wattle 125
mountain ebony 129
mountain thorn 17
mouse barley 84
mousehole tree 138
moxato 203
mphaga 90
mtshiki 83, 107
muchize 217
mud plantain 45
mufhafha 102
mufufu 95
mufuta 174
muggiesgras 323
muisvygie 310
muiswildegars 84
multiflora rose 191
munwahuku 179
munyenyae 258
mupunganini 92
Murray red gum 140
Murraya paniculata 52
musaka 303
mushangishangi 258
mushiji 217
musk clover 272
musk heron's bill 272
muskietboom 132
muskuskruid 272
mustard tree 160
musuwane 235
muurhondebossie 263
MYOPORACEAE 138
Myoporum insulare 52, 138
Myoporum laetum 52, 138
Myoporum montanum 138
Myoporum spp. 113
Myoporum tenuifolium subsp. *montanum* 52, 138, 139
Myriophyllum aquaticum 52, 382, 386
Myriophyllum spicatum 52, 386
MYRSINACEAE 187
Myrsine africana 20
MYRTACEAE 139, 140, 141, 143, 144, 145
myrtle 315

mysore thorn 176

N

naaldbos 18, 152
naaldvrug 259
nagblom 47, 52, 282, 283, 373
nama 279
napa thistle 220
napier fodder 95
napier grass 54, 95
narrow-leaved clover 333
narrow-leaved purple vetch 365
Nassella-polgras 52, 89
Nassella tenuissima 52, 89
Nassella trichotoma 15, 52, 89
Nassella tussock 52, 89
nastergal 303
nasturtium 346, 347
Nasturtium officinale 52, 347, 382, 385
Natal cherry 304
Natal red-top 88
Natalkweek 78
Natalse rooipluim 88
natterkop 246
navelwort 313
needle burr 206
needle bush 18, 152
negenaaldgras 20
Nepal privet 188
NEPHROLEPIDACEAE 280
Nephrolepis cordifolia 52, 280
Nephrolepis exaltata 52, 201, 280
Nerium oleander 52, 163, 165
netelboom 46
nettle tree 46
nettleleaf velvetberry 305
nettle-leaved goosefoot 263
New Zealand Christmas tree 51
New Zealand manitoka 52
ngaio 138
ngongoni 20
ngwengwe 76
Nicandra physalodes 52, 197, 300
Nicotiana glauca 52, 112, 159, 160
Nicotiana longiflora 160
Nicotiana tabacum 160
Nieu-Seelandse manitoka 52
Nieu-Seelandse perdestert 51
night-blooming cereus 50
nightshade 303
Nile cabbage 383
Nile lettuce 383
Nile lily 390
nine-awned grass 20
Nothoscordum borbonicum 202
Nothoscordum entrerianum 202

Nothoscordum gracile 200, 202
nsangu 257
ntsoa-ntsane 223
nyankomo 73
NYCTAGINACEAE 281, 337
Nyllelie 390
Nymphaea lotus 389
Nymphaea mexicana 52, 382, 389
NYMPHAEACEAE 389
Nymphoides indica 389
Nymphoides peltata 52, 389

O

oatmeal cassia 179
oat-seed grass 82
Ochna pulchra 19
octopus tree 117
Oenothera biennis 282, 283
Oenothera glazioviana 282
Oenothera indecora 52, 283, 284
Oenothera jamesii 282
Oenothera laciniata 283
Oenothera rosea 52, 198, 284, 285
Oenothera spp. 198, 284
Oenothera stricta subsp. *stricta* 52, 283, 284
Oenothera tetraptera 53, 198, 284, 285
old man saltbush 45, 173
oldlandgrass 74
oldmaid 315
oldwood 18
OLEACEAE 187
oleander 52, 165, 166
olieboom 299
oliebossie 289
olifantsgras 54, 95
olifantsoor 40
olumbungu 98
ombu 146
ONAGRACEAE 282, 283, 284
Oncosiphon suffruticosum 38
onion weed 202
oorlosie 272
oorpynhoutjie 291
oorpynpeultjie 258
Oos-Afrikaanse tongblaar 56
opblaasboontjie 369
Opuntia aurantiaca 25, 53, 371, 376
Opuntia exaltata 53
Opuntia ficus-indica 25, 53, 371, 377, 378
Opuntia fulgida 53
Opuntia humifusa 53, 371, 378, 379

Opuntia imbricata 53, 371, 376
Opuntia lindheimeri 53, 371, 380
Opuntia microdasys 53, 377
Opuntia monocantha 53, 371, 378, 379
Opuntia robusta 53, 371, 380
Opuntia spinulifera 53, 377
Opuntia spp. 15, 371, 375, 377
Opuntia stricta 53, 377
Opuntia vulgaris 25
Orange cestrum 47
orange cotoneaster 189
orange Jessamine 52
oranjejasmyn 52
oranjesestrum 47, 158
orchard grass 70, 73, 77
orchid tree 45, 129
orgideeboom 45, 129
oriental hedge 66
OROBANCHACEAE 285
Orobanche minor 53
Orobanche ramosa 53
orrelkaktus 48, 374
osgras 82
osipundula 269
ostong 234
ouhout 18
oulandsgras 83
oumansoutbos 45, 173
oupa-se-hoed 352
OXALIDACEAE 338
Oxalis corniculata 308, 338, 339
Oxalis latifolia 308, 338, 339
Oxalis luteola 338
Oxalis pes-caprae 308, 338, 339
Oxalis polyphylla 338
Oxalis repens 338

P

paardgras 74
paint leaf 268
painted euphorbia 268
pale persicaria 294
pale smartweed 294
palmiet 98
pampasgras 47, 75
pampas grass 47, 75
Panicum deustum 90
Panicum laevifolium 91
Panicum maximum 89, 90, 91
Panicum schinzii 79, 82, 90, 91, 110
Panicum subalbidum 91
Papaver aculeatum 288
Papaver argemone 288
Papaver hybridum 288

Papaver rhoeas 288
Papaver spp. 197
PAPAVERACEAE 286, 288
paperbark thorn 17
paperbarks 139
paperthorn 311
papierbasdoring 17
papierblom 51
papierblommetjies 293
papies 98
paradysvoëlblom 46, 176
paraffin weed 168
paraffienbos 15, 24, 28, 168
Paraguay bur 316
Paraserianthes lophantha subsp. *lophanta* 27, 53, 112, 131
Parkinsonia aculeata 53
parrot's feather 52, 386
parthenium 53, 213, 232, 233
Parthenium hysterophorus 53, 200, 213, 232, 233
Paspalum dilatatum 92, 94
Paspalum distichum 93
Paspalum hysterophorus 53
Paspalum notatum 93
Paspalum paspalodes 93
Paspalum urvillei 92, 94
Passiflora caerulea 53, 366, 367
Passiflora edulis 53, 366
Passiflora foetida 366
Passiflora mollissima 53, 366, 367
Passiflora montana var. *lobata* 364
Passiflora spp. 350, 366
Passiflora suberosa 53, 366, 367
Passiflora subpeltata 53, 366
PASSIFLORACEAE 366
passion fruit 366
pasture brake 265
patrysuintjie 60
Patterson's curse 48, 246
patula pine 54, 148
Paulownia tomentosa 53
Pauluskruid 238, 239
peach tree 155
peanut butter cassia 56, 178
pearl acacia 44, 121
PEDALIACEAE 289
peeling plane 19
Peking willow 157
Peltophorum africanum 19
pendula 52
Pennisetum clandestinum 54, 94, 95
Pennisetum purpureum 54, 95, 96
Pennisetum setaceum 54, 97
Pennisetum villosum 54, 97
penny gum 140
penny-john 185

Pentaschistis thunbergii 38
Pentzia globosa 20, 38
Pentzia grandiflora 38
Pentzia suffruticosum 38
peperboom 56, 116, 125
peperboomwattel 44, 125
peperbossie 50, 251, 321
peperkruid 321
pepper cress 50, 251
pepper hedge 116
pepper plant 250
pepper tree 56, 116
pepper tree wattle 44, 125
peppergrass 251
pepperweed 251
pepperwort 251
perdeblom 242, 320
perdegras 322
perdekloutjies 313
perdesoetgras 103
perdestert 46, 139
perdestertboom 46
pêrel-akasia 121
perennial pigweed 204
perennial ryegrass 51
Pereskia aculeata 15, 54, 350, 356, 357
Perotis patens 20
persbokbaard 242
pers-echium 48, 234, 246
Persian lilac 137
Persicaria capitata 54, 294
Persicaria hydropiper 294
Persicaria lapathifolia 294
Persicaria limbata 294
Persicaria spp. 199
perskeboom 155
perstamarisk 57, 162
Peruvian apple cactus 373
pest pear of Australia 53
petty spurge 266
peulbos 180
Phalaris 36
Phalaris aquatica 98
Phalaris canariensis 97, 98
Phalaris minor 97, 98
Philenoptera violacea 18
Phlebodium aureum 280
Phragmites australis 64, 98, 99
Phragmites mauritianus 64, 98
Physalis angulata 300, 301
Physalis peruviana 301
Physalis spp. 197
Physalis viscosa 301
physic nut 50
Phytolacca americana 54, 290
Phytolacca dioica 54, 112, 146
Phytolacca octandra 54, 159, 199,

289, 290
PHYTOLACCACEAE 146, 289, 291
pickerel weed 54, 392
Picris echioides 197, 234
Picris hieracioides 234
pienkaandblom 52
pienkporseleinlelie 307
pienksuring 338
piesangdilla 53, 366
pigeon berry 48, 195
pigsty daisy 214
pigweed 204, 205, 206, 261, 341
pill-pod 330
pimpernel 267, 323, 343
PINACEAE 147, 148, 149
pine cone cactus 57
pink evening primrose 52
pink knotweed 294
pink periwinkle 315
pink porcelain lily 307
pink sorrel 338
pink tamarisk 57, 162
pink weed 272
pinotiebossie 243
Pinus canariensis 54, 148
Pinus elliottii 54, 148
Pinus halepensis 54, 147, 148
Pinus longifolia 148
Pinus patula 54, 112, 148
Pinus pinaster 54, 112, 147, 149
Pinus pinea 54, 148
Pinus radiata 54, 148
Pinus roxburghii 54, 148
Pinus spp. 15
Pinus taeda 54, 148
pipevine 352
Pistia stratiotes 54, 382, 383
pitanga 41, 48
pitchfork 217
PITTOSPORACEAE 150
Pittosporum crassifolium 54, 150
Pittosporum undulatum 54, 113, 150
PLANTAGINACEAE 291
Plantago lanceolata 36, 291, 292
Plantago major 291, 292
Plantago spp. 201
platdissel 320
platdubbeltjie 349
platturksvy 376
platvoet 291
Plectranthus barbatus 186
Plectranthus comosus 54, 163, 186
pluisbossie 18
pluisiesgras 86
PLUMBAGINACEAE 293
plume thistle 223

Poa annua 100
POACEAE 63, 64, 65, 66, 68, 69, 70, 71, 72, 74, 75, 76, 77, 78, 79, 80, 82, 83, 84, 85, 86, 87, 88, 89, 91, 92, 93, 94, 95, 97, 98, 100, 101, 102, 103, 104, 106, 107, 108, 110, 111
poea 205
poinsettia 268
poison berry 158
poison daisy 214
poke berry 146
poke weed 54, 110, 290
poker plant 98
pokwana 82
polka-dot plant 310
Polycarpon tetraphyllum 198, 259
POLYGONACEAE 150, 294, 295, 296, 340, 341, 368
Polygonum aviculare 309, 341
Polygonum capitatum 294
Polygonum convolvulus 368
Polygonum lapathifolium 294
Polygonum limbatum 294
Polypodium aureum 54, 201, 280
Polypodium exaltatum 280
Polypogon monspeliensis 100, 101
pompom weed 46
pompom-bossie 46, 219
Pontederia cordata 54, 382, 392
PONTEDERIACEAE 390, 392
poor man's spinach 204
poor man's weather glass 343
popcorn senna 178
poplar 156
popoliri 156
populier 156
Populus alba var. *alba* 54, 155
Populus deltoides 54, 156
Populus nigra var. *italica* 54, 156
Populus simonii 54
Populus tremula 156
Populus x canescens 54, 112, 155, 156
porslein 345
porselein 342
Port Jackson 28, 44, 126
Port Jackson acasia 125
Port Jackson willow 44, 126
portjeksonwilg 126
Portulaca oleracea 308, 341, 342
Portulaca quadrifida 342
potato creeper 57, 370
Potentilla indica 344
prairie grass 70
predikantsluis 71
Pretoria sida 279
pretty lady 258

414

prickly bush 19
prickly careless weed 206
prickly jack 340
prickly lettuce 232
prickly malvastrum 51, 278
prickly pear 53, 376, 377, 378, 380
pride of Bolivia 134
primitive cactus 356
primrose 282, 283, 284
primroseleaf fleabane 224
PRIMULACEAE 343
pronkbessiebossie 189
pronkgras 54, 97
Prosopis pubescens 132
Prosopis chilensis 132
Prosopis glandulosa 40, 55, 112
Prosopis glandulosa var. *torreyana* 132
Prosopis pubescens 132
Prosopis velutina 55, 132
prostrate globe amaranth 312
prostrate knotweed 341
prostrate spurge 331
prostrate starbur 316
Protasparagus laricinus 167
PROTEACEAE 151, 152
Prunus cerasifera 55
Prunus persica 112, 155
Prunus serotina 55
Pseudobrachiaria deflexa 68
Pseudognaphalium luteo-album 199, 235
Pseudognaphalium undulatum 235
Psidium cattleianum 55, 144
Psidium guajava 40, 55, 113, 144
Psidium guineense 55, 144
Psidium x durbanensis 55, 144
Pteridium aquilinum 265, 266
Pterocarya fraxinifolia 42
Pueraria lobata 350, 364
Pueraria lobata var. *lobata* 364
Pueraria montana var. *lobata* 55, 364
puniyi 263
purgeerboontjie 50
purperwind 50, 363
purple clover 334
purple cluster-leaf 19
purple echium 246
purple goat's beard 242
purple heart 327
purple loosestrife 51
purple nutsedge 62
purple pampas grass 75
purple spurge 266
purple star thistle 220
purple top 306
purslane 341, 342

puzzle bush 18
pylgras 89
pypblom 352
pypgras 82
Pyracantha angustifolia 55, 190
Pyracantha coccinea 55, 190
Pyracantha crenatoserrata 55, 190
Pyracantha crenulata 55, 190
Pyracantha koidzumii 55, 190
Pyracantha rogersiana 55
Pyracantha spp. 163

Q
qoqobala 323
Queen Anne's lace 207
queen of the night 47, 373
Queensland silver wattle 121
Queensland umbrella tree 117
Quercus robur 43
quickgrass 76

R
raaigras 87
raak-my-nie 51, 178
raap 14
rabbit's foot 100
rabbit's-foot fern 54, 280
racehorse tree 134
radiata pine 54, 148
radiataden 54, 148
radiate finger grass 73
radiatorbossie 224, 236
ragwort 236
rambling cassia 56, 180
ramenas 252, 254
ramsammy grass 107, 108
rankdissel 223
Raoul grass 101
Raphanus raphanistrum 36, 198, 252, 253, 254
Rapistrum rugosum 199, 253, 254
rapoko grass 82
rat's tail dropseed 20, 107
rat's tail fescue 111
rattlepod 182
red bristle grass 103
red bush willow 18
red caustic creeper 331
red cedar 50
red chickweed 343
red clover 334
red devil 204
red firethorn 55, 190
red garden sorrel 338
red ginger lily 49, 307

red gum 140, 142
red iron bark 142
red milkweed 330
red nut-grass 62
red oats 65
red pigweed 204
red porter weed 305
red river gum 48, 140
red sesbania 26, 57, 182
red shank 294
red spike-thorn 18
red sunflower 58, 171
red thorn 17
red umbrella thorn 17
reddingsgras 70
redeye 44, 122
red-flowered mallow 336
red-flowering tea tree 51
redshank 205
redstar zinnia 245
redwreath acacia 122
reed meadow grass 49
reedmace 98
regopboerhavia 337
regopsterklits 316
reiersbek 272
renosterbos 19
renostergras 83
rescue brome 70
rescue grass 70
Reseda lutea 200, 296, 297
RESEDACEAE 296
reusehoenderspoor 78
reusekweekgras 77
reusewattel 50, 177
Rhigozum trichotomum 19
rhinoceros bush 19
Rhipsalis baccifera 376
Rhodes grass 72
Rhodesgras 72
Rhodesian blue grass 73
Rhus glabra 55, 115
Rhus succedanea 55, 112, 115
Rhynchelytrum repens 88
ribbon bristle grass 102
ribwort 291
Richardia brasiliensis 309, 345
Ricinus communis 55, 163, 174
Ricinus communis var. *communis* 174
ripgut brome 71
ripplegrass 291
rippleseed plantain 291
rivabe 300
rivina 291
Rivina humilis 55, 199, 291
Robertsonbrak 324
Robinia pseudoacacia 55, 112,

415

130, 133
robust stargrass 77
robusta-turksvy 53, 380
rock hakea 49, 152
roerkruid 230, 235
roggras 87
rolbossie 56, 264
rolypoly 264
rooibloekom 48, 140
rooiblom 285
rooiborselgras 103
rooibos (wilg) 18
rooibossie 205
rooibranddoring 55
rooidoring 17
rooigemmerlelie 49, 307
rooihaak-en-steek 17
rooihawer 65
rooiklosgras 73
rooikrans 27, 44, 122
rooikrans-saadkalander 27
rooimelkbossie 330
rooimelkkruid 330
rooimisbredie 204
rooimuur 343
rooipendoring 19
rooipit 122
rooiseder 50
rooisesbania 26, 57, 182
rooisonneblom 58, 171
rooisuring 295, 338
rooituinsuring 338
rooi-uintjie 31, 62
rooivuurdoring 55, 190
rooiwatervaring 45, 384
Rorippa nasturtium-aquaticum 347, 385
Rosa canina 55
Rosa eglanteria 163, 191
Rosa multiflora 191
Rosa polyantha 191
Rosa rubiginosa 56, 163, 191
ROSACEAE 154, 155, 189, 190, 191, 192, 297, 344
rose apple 57, 145
rose bay 165
rose evening primrose 284
rose gum 142
rose laurel 165
rose mallow 277
rosea cactus 53
roseakaktus 53
rosemary 20
rosewood 134
rosin rose 185
rostrata gum 140
rotstertfynsaadgras 107
Rottboellia cochinchinensis 101, 102
Rottboellia exaltata 101
rough horsetail 48
rough medic 333
rough-leaved raisin 18
round pricklyhead 288
rubberklimop 47
rubber vine 47
RUBIACEAE 298, 345
Rubus albescens 192
Rubus baileyanus 192
Rubus canadensis 192
Rubus cuneifolius 56, 191, 192,
Rubus eglanteria 191
Rubus flagellaris 56, 192
Rubus fruticosus 56, 192
Rubus lasiocarpus 192
Rubus niveus 56, 192
Rubus procumbens 192
Rubus proteus 192
Rubus rigidus 192
Rubus spp. 163
Rubus x proteus 56, 192
Rumex acetosella subsp. *angiocarpus* 295
Rumex angiocarpus 295
Rumex crispus 56, 296
Rumex spp. 201
Rumex usambarensis 56
running myrtle 315
rush grass 107
rusperbossie 258
Russian tumbleweed 56, 264
Russian vetch 365
Russiese rolbossie 56, 264
ryegrass 87

S

safsaf willow 157
safsafwilger 157
SALICACEAE 155, 157
saligna 142
saligna gum 48, 141
saligna wattle 126
salignabloekom 48, 141
Salix babylonica 56, 157
Salix fragilis 56, 157
Salix mucronata 157
Salix spp. 112
Salix subserrata 157
sallow wattle 125
salsify 242
Salsola australis 264
Salsola kali 56, 197, 264
Salsola tragus 56
saltcedar 162
saltwort 264
salvation Jane 246
Salvia tiliifolia 56
salvinia 393
Salvinia hastate 393
Salvinia molesta 26, 56, 382, 384, 393
SALVINIACEAE 393
Salviniaceae 56
sambreelden 54, 148
sambreelmelkkruid 266
Sambucus canadensis 56, 120
Sambucus nigra 56, 113, 120
sand bramble 192
sand bur grass 72
sand olive 18
sandbraam 192
sandgeelhout 19
sandklitsgras 72
sand-mat 330
sandolien 18
sandworm plant 263
SAPINDACEAE 369
Sasa ramosa 56
satansbos 27, 57, 302
satin flower 323
saucepan cactus 377
saucy jack 220
scarlet firethorn 190
scarlet gaura 49
scarlet lobelia 285
scarlet morning glory 362
scarlet pimpernel 343
scarlet silky oak 151
scarlet sumac 55, 115
scented agrimony 44
scented thorn 17
Schefflera actinophylla 56, 112, 117
Schefflera arboricola 56, 117
Schefflera elegantissima 56, 117
Schinus molle 56, 116
Schinus spp. 112
Schinus terebinthifolius 56, 116
Schkuhria pinnata 199, 235, 236
Scleranthus annuus 201, 259
Scotch broom 48, 183
Scotch thistle 47, 223
scrambled eggs 180
screw bean 132
screw-pod wattle 44
SCROPHULARIACEAE 193, 346
sea lavender 293
sebitsa 250
sedge 38
seepbossie 261
seeroogblaar 289
sehabane 74

sehlolo 288
sekelbos 18
selatsi 341
selonsroos 52, 165
sendelenja 330
Senecio burchellii 238
Senecio consanguineus 236, 237
Senecio ilicifolius 238
Senecio inaequidens 238
Senecio latifolius 236, 237
Senecio madagascariensis 236, 237
Senecio spp. 199, 238
Senecio vulgaris 36
Senna bicapsularis 56, 180
Senna corymbosa 180
Senna didymobotrya 56, 178, 179
Senna hirsuta 56, 179
Senna multiglandulosa 56, 180
Senna occidentalis 56, 180
Senna pendula var. *glabrata* 56, 180
Senna septemtrionalis 57, 180
Senna spp. 163
Senna tomentosa 180
seona 285
seona-se-seholo 306
sere pelêlê 206
sering 137
Seriphium plumosum 20, 164, 170
seritsoana 83
scrue 261
sesame 289
Sesamum indicum 289
Sesamum triphyllum 198, 289
Sesbania bispinosa 163, 181
Sesbania punicea 26, 57, 163, 182
Sesbania spp. 15
seshoa-bohloko 303
Setaria chevalieri 102
Setaria megaphylla 102, 103
Setaria pallide-fusca 20, 103
Setaria verticillata 104
setla-bocha 316
shama millet 80
shamrock 333
shamva grass 101
shashe 239
sheep sorrel 295
shell ginger 45, 307
shellflower 383
shepherd's calendar 343
shepherd's purse 250
shoebutton ardisia 45
shoofly 176
shoo-fly plant 300
Siam weed 168
sickle bush 18

sickle medic 333
Sida cordifolia 278, 279
Sida cordifolia subsp. *cordifolia* 279
Sida rhombifolia subsp. *rhombifolia* 279
Sida spp. 198, 275, 278
Sida spinosa var. *spinosa* 279
siergras 100, 104
siergrenadella 53, 366
Sigesbeckia orientalis 200, 238, 239
sigorei 47, 222
silele 341
Silene gallica 198, 260
silk tree 44, 131
silky hakea 49, 152
silver cluster-leaf 19
silver oak 151
silver spike 85
silver wattle 44, 123
silver-leaf bitter apple 57, 302
silver-leaf cotoneaster 47, 189
silweraargras 85
silwerblaarbitterappel 302
silwerdwergmispel 47
silwergras 75
silwerwattel 44, 121, 123
SIMAROUBACEAE 158
Simon poplar 54
Simon's cotoneaster 189
Simon-populier 54
Singapoer-madeliefie 57, 319
Singapore daisy 57, 319
single-leaved cleome 258
Sintjosefslelie 51, 274
sirus 131
sisal hemp 44, 372
sisblom 45, 352
Sisymbrium capense 254
Sisymbrium officinale 254
Sisymbrium orientale 254, 255
Sisymbrium spp. 198, 253, 254
Sisymbrium thellungii 254
six weeks grass 100
skaapbossie 38
skaapdissel 220, 223
skeleton weed 47
skilferdoring 17
skoenlapperorgideeboom 45, 129
Skotse brem 48, 183
Skotse dissel 47, 223
skraal Cyperus 38
skraal misbredie 206
skulpgemmer 45, 307
sky flower 195
slangbos 20, 170
slanghoutjie 170

slapdoring 17
slash pine 54, 148
slender amaranth 206
slender celery 208
slender Cyperus 38
slender sedge 38
slender thorn 17
slender wild oats 65
slingerduisendknoop 368
smalblaarblougras 20
smalblaarperswieke 365
smalblaarplantago 291
smalblaartaaiman 279
small canary grass 98
small carrot-seed grass 108
small mallow 335
small purple vetch 365
small quaking grass 69
small round-leaved prickly pear 53, 380
small stinging nettle 304
smaller tree mallow 277
small-flowered quickweed 229
small-leaf fluff bush 18
small-leaved tea tree 143
smalweëblaar 291
smelter's bush 49, 228
smelterbossie 49, 228
smooth cat's ear 231
smooth creeping milkweed 331
smooth prostrate euphorbia 331
smooth senna 180
smooth sumac 115
snake root 210, 211
snakeweed 57, 305, 330
sneezeweed 210
snotterbel 341, 390
snotterbelletjie 258
sobosobo berry 303
soet pittosporum 150
soetdoring 17
soetgousblom 214
soetgras 91
soethakea 49, 152
soetnagblom 52
soetpeulboom 29, 130
soetspeldebos 152
SOLANACEAE 158, 160, 161, 193, 299, 300, 301, 302, 303, 370
Solanum betaceum 57, 161, 162
Solanum chrysotrichum 57, 193, 194
Solanum elaeagnifolium 27, 57, 302
Solanum incanum 194
Solanum incanum subsp. *incanum* 193
Solanum mauritianum 27, 57,

113, 161
Solanum nigrum 303
Solanum panduriforme 302
Solanum pseudocapsicum 57, 303, 304
Solanum retroflexum 303
Solanum seaforthianum 57, 350, 370
Solanum sisymbriifolium 57, 163, 193, 194
Solanum spp. 199, 303
soldatenblom 315
soldier weed 206
solitz grass 102
solwane 276
somerlila 193
Sonchus asper var. *asper* 239, 240
Sonchus oleraceus 239, 240
Sonchus spp. 197
son-euphorbia 266
sonneblom 14, 58, 171
sono 285
sonquasriet 98
sorghum 14, 104, 106, 286
Sorghum almum 104
Sorghum bicolor subsp. *arundinaceum* 104, 105, 106
Sorghum bicolor subsp. *drummondii* 104, 105
Sorghum halepense 57, 106
Sorghum versicolor 106
Sorghum verticilliflorum 104
Sorghum vulgare 14
sosoori 276
sour sobs 338
South American pepper 116
South Sea rose 165
southern blue gum 142
sowthistle 239
Spaanse besem 57, 183, 184
Spaanse knapsekêrel 217
Spaanse riet 45, 64
Spanish blackjack 217
Spanish broom 57, 183
Spanish gooseberry 356
Spanish reed 45, 64
sparden 149
Spartium junceum 57, 163, 183
Spathodea campanulata 57
spear grass 89
spear thistle 223
speerdissel 223
speldedoring 18
Spergula arvensis 308, 322, 323
Sphagneticola trilobata 319, 320
spiceberry 187
spider flower 151, 258
spider gum 48, 140

spiderling 337
spiderweb chloris 72
spider-wisp 258
spiked water-milfoil 52, 386
spilanthes 319
Spilanthes decumbens 309, 319, 320
Spilanthes mauritiana 319
spindlepod 258
spineless monkey-orange 19
spinnekopbloekom 48, 140
spinnerak-chloris 72
spiny bur grass 72
spiny cocklebur 58, 243
spiny emex 340
spiny sesbania 181
spiny sowthistle 239
Spirodela spp. 388
splendid acacia 17
sponge-fruit salt bush 45, 260
Sporobolus africanus 20, 107
Sporobolus fimbriatus 107
Sporobolus pyramidalis 20, 107
sporrie 322
spotted cat's ear 231
spotted gum 142
spotted knotweed 294
spreading century plant 44
sprinkaanboom 130
sprinkaanbossie 204, 205, 238
sprinkaan-senecio 238
spurge 266
St Augustine grass 107
St Joseph's lily 51
St Joseph's trumpet 274
St Paul's wort 238
St. John's wort 50, 184
Stachytarpheta mutabilis 305
Stachytarpheta spp. 57
Stachytarpheta urticifolia 201, 305
star grass 38, 77
starfish grass 78
starhair groundcherry 301
starvation senecio 236
starwort 323
statice 51, 293
steekgras 20, 104
steeleik 43
steenboksuring 295, 338
stekelklawer 333
stekel-picris 234
stekelrige beestong 234
stekelrige malvastrum 278
stekelsesbania 181
Stellaria media 36, 267, 308, 323, 343
Stenotaphrum secundatum 107, 108

stercors 385
sterdissel 220
stergras 38, 77
sterkbos 19
sterkgras 77, 251
sterkkos 251, 385
sterremuur 323
stick grass 71
sticky bristle grass 104
sticky gooseberry 301
sticky nightshade 193
sticky physalis 301
sticky thorn 17
stiff-leaved bottlebrush 46
stiff-leaved cheesewood 150
stinging milkweed 266, 323
stink bean 27, 53
stinkboon 27, 53, 131
stinkboontjie 122
stinking cedar 158
stinking mayweed 214
stinking Roger 241
stinking sumac 158
stinking weed 56, 180
stinkkamille 214
stinkkruid 38
stinkpeul 122
stinkweed 38, 299
stinkwort 299
Stipa neesiana 89
Stipa tenuissima 89
Stipa trichotoma 89
stitchwort 323
Stoebe plumosa 170
Stoebe spiralis 20
Stoebe vulgaris 170
stone pine 54, 148
straatgras 100
strandbuffelsgras 107, 108
strandwildemosterd 254
stranglehold plant 353
strawberry guava 55, 144
streepwildekommer 329, 330
Striga asiatica 198, 285, 286
Striga forbesii 285
Striga gesnerioides 285
Striga hermonthica 285
striped wild cucumber 330
Strychnos madagascariensis 19
stuipboom 177
styweblaarkasuur 54, 150
succory 222
sugar gum 48, 142
Suidwesdoring 132
suikerbloekom 48, 142
sumba 196
summer lilac 193
sun euphorbia 266

surinam cherry 41
suurturksvy 53, 377, 378
swaardvaring 52, 280
swartapiesdoring 17
swartels 45
swarthaak 17
swarthout 27, 122
swartkersie 55
swartklapper 19
swartkop-biesie 38
swartmoerbei 52, 138
swartsaadwildesorghum 106
swartstampepperment 142
swartwattel 26, 27, 33, 44, 123
swartysterbasbloekom 48, 142
Swedish hemp 304
sweet buffalo grass 91
sweet clover 271
sweet hakea 49, 152
sweet locust 130
sweet pittosporum 150
sweet prickly pear 53, 377
sweet signal grass 68
sweet sundrop 52
sweet thorn 17
sweetbriar 191
sweetgrass 74
swinecress 321
sword fern 52, 280
sword grass 85
syboom 44
sydissel 239
Sydney golden wattle 125
syerige hakea 33, 49, 152
Syngonium podophyllum 350, 351
Syngonium spp. 57
syringa 51, 137
Syzygium cuminii 57, 145
Syzygium jambos 57, 145
Syzygium paniculatum 57, 145
Syzygium spp. 113

T

taaibos 264
taaiman 279
taaipol 20, 107
taaipol-eragrostis 83
tabakboom 160
tabaka bume 160
Tagetes minuta 199, 227, 236, 241, 311
tajoe 95
tall fleabane 224
tall khaki weed 241
tall nettle 304
tall paspalum 94

tall verbena 306
tall wild oats 65
TAMARICACEAE 162
Tamarix aphylla 57, 162
Tamarix chinensis 57, 162
Tamarix gallica 57, 162
Tamarix ramosissima 57, 162
Tamarix usneoides 162
tamatiedissel 193
tamboekiegras 85
tamarillo 161
tango 182
Taraxacum officinale 231, 308, 320, 321
Tarchonanthus camphoratus 19
tarentaalgras 101
Tecoma stans 57, 163, 172
Tephrocactus articulatus 57
Terblansbossie 276
Terminalia prunioides 19
Terminalia sericea 19
Texas poppy 286
theepe 204
Thelechitonia trilobata 57, 319
Thevetia neriifolia 166
Thevetia peruviana 57, 163, 166
Thevetia yccotlii 166
thibapitsa 335
thlare-sa-mpja 330
thorn apple 193
thorny pigweed 206
thorny poppy 288
thousand-leaf 210
thread-leaved bluestem 20
three thorn rhigozum 19
three-awn grasses 20
three-hook thorn 17
thunderbolt flower 289
tickberry 196
tickseed 47, 225
tipoeboom 57, 134
Tipton's weed 185
tipu tree 57, 134
Tipuana tipu 112, 57, 134
Tithonia diversifolia 58, 164, 171
Tithonia rotundifolia 58, 163, 171
tjiekoriebos 222
tjirden 54, 148
Todd's curse 212
tolbos 264
tolbossie 38
tolletjiesbos 20
tolo-la-khongoana 294
tonguegrass 251, 323
tonteldoek 214
toon tree 58
Toona ciliata 58
toonboom 58

toothed medic 333
torch cactus 48, 374, 375
touch-and-heal 185
tough dropseed 107
toukaktus 49, 375
towerghwarrie 18
Toxicodendron succedaneum 115
Tradescantia fluminensis 58, 327
Tradescantia pallida 327
Tradescantia spp. 308
Tradescantia zebrina 58, 327
Tragopogon dubius 242
Tragopogon porrifolius 242
Tragopogon spp. 197
Tragus berteronianus 21, 108, 109
Tragus racemosus 21, 109
trassiedoring 17
tree daisy 52, 169
tree mallow 50, 277
tree marigold 171
tree tabacco 160
tree tomato 57, 161
tree-horned bedstraw 49
tree of heaven 44
treurboom 157
treurden 54, 148
treurwilger 56, 157
Tribulus terrestris 308, 349
Trichodesma zeylanicum 201, 249
Tricholaena monachne 21
tridax daisy 319
Tridax procumbens 308, 319, 320
trident maple 114
triffid weed 47, 168
Trifolium africanum 333
Trifolium angustifolium 333, 334
Trifolium arvense 333
Trifolium campestre 333
Trifolium dubium 333
Trifolium pratense 334
Trifolium repens 333, 334
Trifolium spp. 309
Trifolium tomentosum 334
trilgras 69
triplaris 58, 150, 151
Triplaris americana 58, 112, 150, 151
Triumfetta pilosa 38
trompetdoring 18
trompetlelie 274
TROPAEOLACEAE 346
Tropaeolum majus 308, 346, 347
Tropaeolum speciosum 58
tropical richardia 345
tropical whiteweed 212
trosden 54, 149
trumpet evening primrose 282
trumpet thorn 18

tshamma 329
Tsumeb onkruid 316
tsuna 258
tuart 142
tuber vervain 348
tuberous sword fern 280
tuinageratum 212
tuindissel 239
tuinkanna 255
tuinmanna 80
tuinmannagras 20
tuinranksuring 338
tuinsetaria 103
tuinurochloa 110
tumbleweed 38, 56
turf thorn 17
turknael 272
turksenael 274
turksvy 25
turtle vine 325
tussock 89
tussock paspalum 53
tutsan 50, 185
twitch grass 76

Typha capensis 98, 99

U

ubabe 90
ubatata wentaba 363
ubhongobhongo 169
uBhoqo 161
ubima 316
ubobo 177
ubobo-encane 176
ubukhwebezane 196
ubuklunga 296
ubutywala bentaka 196
ucucuza 217
udekane 276
udombo 203
udumbedumbe 255
ufenisi 176
ugamfe 217
Uganda grass 95
ugudluthukela 286
uhlotshane 326
ukwatyapheya 136
Ulex europaeus 58, 163, 184
Ulmus parviflora 43, 58
ulusina 177
umadolwana 74
umakhuthula 297
umancibikela 294
umankunkunku 360
umashwababa 320
umavayi 171

umazifisa 178
umbabane 171
umbabazane 304
umbanga banga 161
umbhido 205
umbikicane 261
umbra tree 146
umbrella milkweed 266
umbrella pine 148
umbrella thorn 18
umbrella tree 117
umbuya 205
umdulukwa 193
umesisi 217
umfino 205
umfisane 68
umfude 174
umgabaganga omncane 300
umgangampunza 276
umhatji 90
umhlabangubo 217
umhlakanye 276
umhlakuva 174
umhlavuthwa 299
umkhanzi 98
umkhothane 231, 320
umkoka 359
umnaka 285
umnanja 290
umnyandla 290
umpungempu 300
umsilinga 137
umsobo 303
umsobosobo 303
umthelekisi 377
umthente 85
umtombo 108
umtyutu 205
umya 257
uMzimuka 146
umzonde 258
ungcolozi 318
ungwengwe 78
unomolwana 335
unyankomo 82
unyawo-lwengulube 312
unyendenyende 360
uphaphe 386
upright paspalum 94
upright starbur 316
uqadolo 217
uqaqaqa 76
uqedizwe 236
uqethu 76
uqupose 204
uqwaningi 356
Urochloa mosambicensis 21, 110
Urochloa panicoides 79, 82, 91,

110
Urtica dioica 304, 305
Urtica spp. 201
Urtica urens 304, 305
URTICACEAE 304
usandanezwe 168
usibambangubo 203
uvemvane olukhulu 276
uwatela 123
uyinki 158
uzipho 369

V

vaalbos 170
vaaljacob 220
vaalmimosa 44, 121
vaalpopulier 54, 156
vaalrosyntjie 18
vaalskraalhans 224
vaalwattel 123
vaderlandsriet 98
valsakasia 133
valsaralia 56, 117
valswitstinkhout 46
Vandermerwegras 70
Vanwyksbossie 38
varkbossie 261
varkkos 341
varksuring 338
varnish tree 158
vasey grass 94
veergras 54
veld shamrock 332
veldgras 82
velvet mesquite 55, 132
verbain 306
Verbena aristigera 309, 348
Verbena bonariensis 58, 200, 219, 306
Verbena brasiliensis 306
Verbena officinalis 306
Verbena tenuisecta 348
VERBENACEAE 195, 196, 305, 306, 348
Verbesina encelioides 200, 242
verdompsterk 279
vergeet-my-nie-boom 48, 195
Veronica agrestis 346
Veronica anagallis-aquatica 346
Veronica persica 308, 346
veselperske 43
Vicia benghalensis 365
Vicia hirsuta 365
Vicia sativa subsp. *sativa* 365
Vicia spp. 350
Vicia villosa 365

Victorian box 150
vieruurblom 281
vieruurtjie 51
viltige duisendknoop 294
vinca 315
Vinca major 58, 315, 316
Vinca minor 58, 315
Vinca rosea 315
Vinca spp. 308, 315
vingerblaartee 258
vingergras 79
vingerhoedblom 289
vingersuring 338
vinkel 49, 209
vinkriet 98
vioolnek 245
viper's bugloss 246
Vitex trifolia 58
vlakkiesriet 98
vlamdoring 17
vlam-van-die-vlakte 129
vleibuffelsgras 91
vlei-panicum 91
vlooibossie 263
voëlduisendknop 341
voëltjiebos 47, 269
volstruisdoring 175, 340, 349
volstruisgifboom 160
Vulpia bromoides 111
Vulpia myuros 111
vuurbossie 285
vuurhoutjiepopulier 54, 156

vyfvingerblaar 38

W

waaibossie 38, 236
wagter-se-sakkie 250
wall fumitory 272
wandelende Jood 58, 326, 327
wandering Jew 58, 326, 327
warkruid 358
wartwort 266
wasboom 55, 115
water cabbage 383
water feather 386
water fern 383, 384, 393
water flaxseed 388
water hyacinth 26, 48, 390
water lettuce 54, 383
water milfoil 386
water pepper 294
water poppy 50, 389
water primrose 51
water speedwell 346
water thorn 17
wateralisma 45

watercress 52, 385
waterdoring 17
waterduisendblaar 52, 386
water-ereprys 346
watergrass 60
waterhiasint 26, 48, 390
waterpes 48
waterslaai 54, 383
watervaring 26, 56, 384, 393
wattle 15, 19, 33
wax tree 55, 115
Wedelia trilobata 319
wedevrouens 217
weeblaar 56, 296
weeping bottlebrush 139
weeping grass 83
weeping love grass 83
weeping myall 44
weeping willow 56, 157
weld 296
wheat bush 263
whin 184
white bird's eye 323
white burr 279
white cedar 137
white clover 333
white Dutch clover 333
white evening primrose 53, 284
white eye 345
white ginger lily 49, 307
white goosefoot 261
white grass 74
white mulberry 52, 138
white poinsettia 48
white popinac 177
white poplar 54, 155
white shroud 354
white stinkweed 299
white sweet clover 271
white tea tree 50
white thorn apple 299
white tussock 52, 89
white water lily 389
white wax tree 188
white-flowered 286
white-flowered Mexican poppy 45, 286
white-stemmed filaree 272
white thorn-apple 299
wigandia 58
Wigandia urens 58
wild asparagus 18, 167
wild aster 38
wild barley 84
wild bindweed 359
wild buckwheat 368
wild camomile 214
wild canna 255

wild carrot 321
wild celery 208
wild coffee 180
wild fennel 49, 209
wild foxglove 289
wild gooseberry 300, 301
wild grain sorghum 104
wild grenadilla 53, 366
wild hollyhock 276
wild jute 275
wild lettuce 231, 232, 320
wild lucerne 269, 324
wild mignonette 296
wild millet 79
wild morning glory 358
wild mustard 252, 253, 254
wild myrtle 20
wild pear 18
wild physalis 300
wild poinsettia 268
wild poppy 288
wild purslane 342
wild radish 252
wild raisin 18
wild rocket 296
wild sago 291
wild senna 178, 180
wild seringa 18
wild sesame 289
wild spinach 261
wild stockrose 276
wild strawberry 48
wild sunflower 242
wild tamarisk 162
wild tobacco 52, 160
wild tomato 193
wild verbena 58, 306, 348
wild water lemon 366
wild watermelon 329
wilde anyswortel 209
wilde radys 252
wilde waatlemoen 329, 330
wilde-aarbei 48, 344
wilde-akkerwinde 359
wilde-appelliefie 300
wildebaardhawer 65
wildebitter 300
wildebokwiet 368
wildegars 84, 111
wildegraansorghum 104
wildegranaat 19
wildegrenadella 366
wildehawer 36, 66, 87
wildejakobregop 15, 245
wildejute 275
wildekamille 214
wildelusern 269
wildemirt 20

wildemosterd 252, 253, 254
wildepoinsettia 268
wilderoos 56, 191
wilderosie 303
wilderosyntjie 18
wildeseldery 208
wildesenna 180
wildesering 18
wildesesam 289
wildeskorsenier 242
wildeslaai 232
wildesonneblom 242
wildespinasie 296
wildestokroos 276
wildetabak 52
wilgerhakea 49, 152
wilgerhout 157
willow hakea 49, 152
willow tea 157
willow-leaved showberry 47, 189
windmill pink 260
winter cherry 300, 304
winter senna 180
wintergras 70, 100
wintergrass 100
winterweed 323
wire grass 20, 83, 341
wire weed 341
witaandblom 284
witakasia 55, 130, 133
witbiesie 38
witblom-bloudissel 45
witchweed 285
witgemmerlelie 49, 307
witgousblom 215
withondebossie 261
witklawer 333
witloof 222
witmoerbei 52, 138
witnagblom 53, 284
witpluim-chloris 74
witpolgras 52, 89
witpopulier 54
witstinkklawer 271
woestyntamarisk 57, 162
Wolffia spp. 388
wolfsmelk 266
wolgras 88, 108
wollerige-eragrostis 83
wolwedoring 195
wonder tree 174
wondergrasperk 313, 328
wonderlawn 328
woolly clover 334
woolly grevillea 151
woolly nightshade 161
woolly plectranthus 54, 186
woolly senna 56, 179

wormseed goosefoot 263
wortelsaadgras 108
wurmbossie 38

X

Xanthium spinosum 58, 197, 243
Xanthium strumarium 58, 197, 244

Y

yellow bells 57, 172
yellow bunny-ears 53, 377
yellow cestrum 47, 158
yellow clover 333
yellow cockspur 220
yellow elder 172
yellow firethorn 55, 190
yellow flag 50
yellow floating heart 389
yellow ginger lily 49, 307
yellow goat's beard 242
yellow locust 133
yellow lupin 269
yellow nut-grass 60
yellow nutsedge 60
yellow oleander 57, 166
yellow pond lily 389
yellow sorrel 338
yellow star thistle 220
yellow sweet clover 271
yellow trefoil 333
yellow trumpet bush 172
yellow tumbleweed 236
yellow water lilies 52
yellow water lily 389
yellow weed 229
yellow-flowered 286
yellow-flowered Mexican poppy 45, 286
yhepe 205
ystervarkgras 38

Z

Zaleya pentandra 310
Zea mays 14
ZINGIBERACEAE 307
Zinnia peruviana 199, 245
Ziziphus mucronata 19
Ziziphus zeyheriana 19
ZYGOPHYLLACEAE 349

Biocontrol agents

C
Cactoblastis cactorum 25, 376, 378
Carposina autologa 27
Charidotis auroguttata 27
Chrysolina quadrigemina 26
Colletotrichum gloeosporioides 27
Cylindrobasidium leave 27, 33
Cyrtobagous salviniae 26, 393

D
Dactylopius austrinus 25, 26
Dactylopius ceylonicus 25
Dactylopius opuntiae 25

E
Eccritotarsus catarinensis 26
Erytenna consputa 27

G
Gargaphia decoris 27

L
Leptinotarsa defecta 27
Leptinotarsa texana 27

M
Melanterius acaciae 27
Melanterius maculates 26
Melanterius servulus 27
Melanterius ventralis 26

N
Neochetina bruchi 26
Neochetina eichhorniae 26
Neodiplogrammus quadrivittatus 26

P
Parectopa thalassias 27

R
Rhyssomatus marginatus 26

S
Sameodes albiguttalis 26

T
Trichapion lativentre 26
Trichoglaster acacialongifoliae 26

U
Uromycladium tepperianum 28, 127

Z
Zeuxidiplosis giardi 26